Chemical Demonstrations

Volume 3

Volume 3 Collaborators and Contributors

JERRY A. BELL, PH.D.
Professor of Chemistry, Simmons College; Director, Institute for Chemical Education, University of Wisconsin–Madison, from 1986

GLEN E. DIRREEN, PH.D.
Associate Director, Institute for Chemical Education, University of Wisconsin–Madison

FREDERICK H. JUERGENS, M.A.T.
Lecture Demonstrator, University of Wisconsin–Madison

RONALD I. PERKINS, M.S.T.
Senior Teacher, Greenwich High School, Greenwich, Connecticut; Visiting Lecturer, University of Wisconsin–Madison, Summer 1983; Resident Fellow and Assistant Director, Institute for Chemical Education, University of Wisconsin–Madison, June 1984–July 1985; Summer 1986, Summer 1987, Summer 1988

RODNEY SCHREINER, PH.D.
Project Associate; Resident Fellow, Institute for Chemical Education, University of Wisconsin–Madison

EARLE S. SCOTT, PH.D.
Professor of Chemistry, Ripon College; Visiting Professor, University of Wisconsin–Madison, June–December 1980, Summer 1981, Summer 1982, Summer 1984, Summer 1987, Summer 1988

DAVID B. SHAW, PH.D.
Instructor, Madison Area Technical College; Honorary Fellow, University of Wisconsin–Madison, 1985; Visiting Assistant Professor, University of Wisconsin–Madison, Summer 1986, Summer 1987, Summer 1988

WORTH E. VAUGHAN, PH.D.
Professor of Chemistry, University of Wisconsin–Madison

Chemical Demonstrations

A Handbook for
Teachers of Chemistry

Bassam Z. Shakhashiri

VOLUME **3**

THE UNIVERSITY OF WISCONSIN PRESS

The University of Wisconsin Press
1930 Monroe Street, 3rd floor
Madison, Wisconsin 53711-2059
uwpress.wisc.edu

3 Henrietta Street
London WC2E 8LU, England
eurospanbookstore.com

13   12   11   10   9

Printed in the United States of America

Library of Congress Cataloging-in-Publication Data
(Revised for volume 3)
Shakhashiri, Bassam Z.
    Chemical demonstrations.
    Includes bibliographies.
    1. Chemistry—Experiments. I. Title.
QD43.S5   1983   540'.7'8   81-70016
ISBN 0-299-11950-5 (v. 3)

ISBN-13: 978-0-299-11950-8 (V. 3: alk. paper)

To
Hubert N. Alyea,
an international grand master of lecture demonstrations
and a delightful communicator of chemistry.

B.Z.S.

Contents

Preface

This is the third in a series of volumes aimed at providing teachers of science at all educational levels with detailed instructions and background information for using chemical demonstrations in the classroom and in public lectures. Volumes 1 and 2 included demonstrations in the areas of thermochemistry, chemiluminescence, polymers, metal ion precipitates and complexes, physical behavior of gases, chemical behavior of gases, and oscillating chemical reactions. The demonstrations in this volume deal with acids and bases and with liquids, solutions, and colloids. Additional volumes, now in preparation, will include demonstrations on electrochemistry, clock reactions, spectroscopy and color, and other topics.

The introduction to each chapter in this volume includes material aimed at reinforcing and expanding the knowledge base of the user. We believe firmly that whenever demonstrations are presented the phenomena should be discussed and explained at a level suitable to the audience. A number of demonstrations included in this volume involve quite complex chemical concepts, and to some readers the mathematical treatment provided in the chapter introductions may seem intimidating. We offer it as an aid to teachers and their advanced students who choose to delve more deeply into questions raised by the demonstrations. Teachers in elementary and secondary schools, as well as college teachers, are urged to make use of the demonstrations in this volume, for many of them are designed to serve as an introduction to the chemistry of everyday life.

The reception accorded to both Volumes 1 and 2 has been gratifying. Teachers of chemistry and other sciences have commented most favorably about this series and its usefulness. We are pleased with the unequivocal statements about the quality of our work, and have undertaken to continue our commitment of time and effort to maintain the same standards of excellence. We reaffirm the importance of reliable and effective source materials and the exercise of good judgment in the use of demonstrations in communicating chemistry.

I am thankful for the blessing of having expert coauthors. Rod Schreiner, as principal coauthor, deserves major credit for his skills and judgment, which have made this volume possible. David Shaw, an excellent two-year college teacher, contributed to both chapters in this volume. Fred Juergens developed several new demonstrations and contributed to chapter 8. Worth Vaughan's insight and experience have made chapter 9 a reality. Ron Perkins, an outstanding high school teacher, was a major contributor to the completion of this volume; his enthusiasm is infectious, and his insistence on maintaining quality is commendable.

I am grateful for the suggestions and comments made by numerous colleagues from around the country. In particular, I thank Jerry A. Bell of Simmons College, Glen E. Dirreen and Lee R. Sharpe of the University of Wisconsin–Madison, and Robert Becker of Greenwich High School, Greenwich, Connecticut, for their contributions to chapter 8; Jeanne Dyer of Vivian Gaither High School in Tampa, Florida, Valerie Wilcox of the Boston Science Museum, C. Marvin Lang and Donald L. Showalter of the University of Wisconsin–Stevens Point, and Alfred A. Rottino of Half Hollow Hills High School, Dix Hills, New York, for their contributions to chapter 9; Doris K. Kolb of Bradley University, Richard W. Ramette of Carleton College, and Helen Stone of Ben L. Smith High School in Greensboro, North Carolina, for reviewing chapter 8; Henry A. Bent of North Carolina State University for reviewing chapter 9; and Jerry Bell and Ron Perkins for reviewing chapter 9. Their advice and contributions have been very helpful.

Others at the University of Wisconsin–Madison provided invaluable assistance: Vince Genna in testing many of the procedures, Edwin M. Turner in helping develop procedures for chapter 9, Patti Puccio in proofreading and in other matters which helped to complete this volume, copy editor Robin Whitaker in clarifying the text with her perceptive suggestions, and Gardner Wills of the University of Wisconsin Press in expediting production of the volume.

I wish to express special gratitude to Elizabeth A. Steinberg, Assistant Director of the University of Wisconsin Press, for all her efforts, her professionalism, and her kindness.

Washington, D.C.
November 1988

Bassam Z. Shakhashiri
Assistant Director for Science
   and Engineering Education
National Science Foundation

Introduction[†]

Bassam Z. Shakhashiri

Lecture demonstrations help to focus students' attention on chemical behavior and chemical properties, and to increase students' knowledge and awareness of chemistry. To approach them simply as a chance to show off dramatic chemical changes or to impress students with the "magic" of chemistry is to fail to appreciate the opportunity they provide to teach scientific concepts and descriptive properties of chemical systems. The lecture demonstration should be a process, not a single event.

In lecture demonstrations, the teacher's knowledge of the behavior and properties of the chemical system is the key to successful instruction, and the way in which the teacher manipulates chemical systems serves as a model not only of technique but also of attitude. The instructional purposes of the lecture dictate whether a phenomenon is demonstrated or whether a concept is developed and built by a series of experiments. Lecture experiments, which some teachers prefer to lecture demonstrations, generally involve more student participation and greater reliance on questions and suggestions, such as "What will happen if you add more of . . .?" Even in a lecture demonstration, however, where the teacher is in full control of directing the flow of events, the teacher can ask the same sort of "what if" questions and can proceed with further manipulation of the chemical system. In principle and in practice, every lecture demonstration is a situation in which teachers can convey their attitudes about the experimental basis of chemistry, and can thus motivate their students to conduct further experimentation, and lead them to understand the interplay between theory and experiment.

Lecture demonstrations should not, of course, be considered a substitute for laboratory experiments. In the laboratory, students can work with the chemicals and equipment at their own pace and make their own discoveries. In the lecture hall, students witness chemical changes and chemical systems as manipulated by the teacher. The teacher controls the pace and explains the purposes of each step. Both kinds of instruction are integral parts of the education we offer students.

In teaching and in learning chemistry, teachers and students engage in a complex series of intellectual activities. These activities can be arranged in a hierarchy which indicates their increasing complexity [1]:
  (1) observing phenomena and learning facts
  (2) understanding models and theories
  (3) developing reasoning skills
  (4) examining chemical epistemology
This hierarchy provides a framework for the purposes of including lecture demonstrations in teaching chemistry.

At the first level, we observe chemical phenomena and learn chemical facts. For example, we can observe that, at room temperature, sodium chloride is a white crys-

[†] Reprinted with minor modifications from Volume 1.

talline solid and that it dissolves in water to form a solution with characteristic properties of its own. One such property, electrical conductivity, can be readily observed when two wire electrodes connected to a light bulb and a source of current are dipped into the solution. There are additional phenomena and facts that can be introduced: the white solid has a very high melting point; the substance is insoluble in ether; its chemical formula is NaCl; etc.

At the second level, we explain observations and facts in terms of models and theories. For example, we teach that NaCl is an ionic solid compound and that its aqueous solution contains hydrated ions: sodium cations, $Na^+(aq)$, and chloride anions, $Cl^-(aq)$. The solid, which consists of Na^+ and Cl^- particles, is said to have ionic bonds, that is, there are electrostatic forces between the oppositely charged particles. The ions are arranged throughout the solid in a regular three-dimensional array called a face-centered cube. Here, the teacher can introduce a discussion of the ionic bond model, bond energy, and bond distances. Similarly, a discussion of water as a molecular covalent substance can be presented. The ionic and covalent bonding models can be compared and used to explain the observed properties of a variety of compounds.

At the third level, we develop skills which involve both mathematical tools and logic. For example, we use equilibrium calculations in devising the steps of an inorganic qualitative analysis scheme. We combine solubility product, weak acid dissociation, and complex ion formation constants for competing equilibria which are exploited in analyzing a mixture of ions. The logical sequence of steps is based on understanding the equilibrium aspects of solubility phenomena.

At the fourth level, we are concerned with chemical epistemology. We examine the basis of our chemical knowledge by asking questions such as, "How do we know that the cation of sodium is monovalent rather than divalent?" and "How do we know that the crystal structure of sodium chloride can be determined from x-ray data?" At this level we deal with the limits and validity of our fundamental chemical knowledge.

Across all four levels, the attitudes and motivations of both teacher and student are crucial. The attitude of the teacher is central to the success of interactions with students. Our motivation to teach is reflected in what we do and, as well, in what we do not do, both in and out of the classroom. Our modes of communicating with students affect their motivation to learn. All aspects of our behavior influence students' confidence and their trust in what we say. Our own attitudes toward chemicals and toward chemistry itself are reflected in such matters as how we handle chemicals, adhere to safety regulations, approach chemical problems, and explain and illustrate chemical principles. In my opinion, the single most important purpose that lectures serve is to give teachers the opportunity to convey an attitude toward chemistry—to communicate to students an appreciation of chemistry's diversity and usefulness, its cohesiveness and value as a central science, its intellectual excitement and challenge.

PRESENTING EFFECTIVE DEMONSTRATIONS

In planning a lecture demonstration, I always begin by analyzing the reasons for presenting it. Whether a demonstration is spectacular or quite ordinary I undertake to use the chemical system to achieve specific teaching goals. I determine what I am going to say about the demonstration and at what stage I should say it. Prior to the lecture, I practice doing the demonstration. By doing the demonstration in advance, I often see aspects of the chemical change which help me formulate both statements and questions that I then use in class.

Because one of the purposes of demonstrations is to increase the students' ability to make observations, I try to avoid saying, "Now I will demonstrate the insolubility of barium sulfate by mixing equal volumes of 0.1M barium chloride and 0.1M sodium sulfate solutions." Instead, I say, "Let us mix equal volumes of 0.1M barium chloride and 0.1M sodium sulfate solutions and observe what happens." Rather than announcing what should happen, I emphasize the importance of observing all changes. Often, I ask two or three students to state their observations to the entire class before I proceed with further manipulations. In addition, I help students to sort out observations so that relevant ones can be used in formulating conclusions about the chemical system. Some valid observations may not be relevant to the main purpose of the demonstration. For example, when the above-mentioned solutions are mixed, students may observe that the volumes are additive. However, this observation is not germane to the main purpose of the demonstration, which is to show the insolubility of barium sulfate. However, this observation is relevant if the purposes include teaching about the additive properties of liquids.

Every demonstration that I present in lectures is aimed at enhancing the understanding of chemical behavior. In all cases, the chemistry speaks for itself more eloquently than anything I can describe in words, write on a chalk board, or show on a slide.

Wesley Smith of Ricks College, who was a visiting faculty member at the University of Wisconsin–Madison from 1974 to 1977, has outlined six characteristics of effective demonstrations which best promote student understanding [2]:

1. *Demonstrations must be timely and appropriate.* Demonstrations should be done to meet a specific educational objective. For best results, plan demonstrations that are immediately germane to the material in the lesson. Demonstrations for their own sake have limited effectiveness.

2. *Demonstrations must be well prepared and rehearsed.* To ensure success, you need to be thoroughly prepared. *All* necessary material and equipment should be collected well in advance so that they are ready at class time. You should rehearse the entire demonstration from start to finish. Do not just go through the motions or make a dry run. Actually mix the solutions, throw the switches, turn on the heat, and see if the demonstration really works. Only then will you know that all the equipment is present and that all the solutions have been made up correctly. Always practice your presentation.†

3. *Demonstrations must be visible and large-scale.* A demonstration can help only those students who experience it. Hence, you need to set up the effect for the whole class to see. If necessary, rig a platform above desktop level to ensure visibility.

Perhaps the most important factor to consider is the size of what you are presenting. Only in the very tiniest of classes can the students see phenomena on the milligram and milliliter scale. Many situations require the use of oversized glassware and specialized equipment. Solutions and liquids should be shown in full-liter volumes, and solids should be displayed in molar or multi-molar amounts.

Contrasting backgrounds help emphasize chemical changes. A collection of large white and black cards to place behind beakers and other equipment is a valuable addition to your demonstration equipment. These are inexpensive, easy to use, and can provide an extra bit of polish to your demonstration.

4. *Demonstrations must be simple and uncluttered.* A common source of distraction is clutter on the lecture bench. Make sure that the demonstration area is neat and

† As Fred Juergens likes to say, "Prior practice prevents poor presentation."

free of extraneous glassware, scattered papers, and other disorder. All attention must be focused on the demonstration itself.

5. *Demonstrations must be direct and lively.* Action is an important part of a good demonstration. It is the very ingredient that makes demonstrations such efficient attention-grabbers. Students are eager to see something happen, but if nothing perceptible occurs within a few seconds you may lose their attention. The longer they have to wait for results, the less likely it is that the demonstration will have maximum educational value.

6. *Demonstrations must be dramatic and striking.* Usually, a demonstration can be improved by its mode of presentation. A lecture demonstration, according to Alfred T. Collette, is like a stage play. "A demonstration is 'produced' much as a play is produced. Attention must be given to many of the same factors as stage directors consider: visibility, audibility, single centers of attention, audience participation, contrasts, climaxes" [3]. The presentation of effective demonstrations is such an important part of good education that "no instructor is doing his best unless he can use this method of teaching to its fullest potential" [4].

USING THIS BOOK

The demonstrations in this volume are grouped in topical chapters dealing with the physical behavior of gases, the chemical behavior of gases, and chemical oscillating reactions. Each chapter has an introduction which covers the chemical background for the demonstrations that follow. We confine the discussion of relevant terminology and concepts to the introduction rather than repeating it in the discussion section of each demonstration. Accordingly, when teachers read the discussion section of any particular demonstration, they may find it necessary to refer to the chapter introduction for background information. For additional information teachers may wish to consult the sources listed at the end of each chapter's introduction.

Each demonstration has seven sections: a brief summary, a materials list, a step-by-step account of the procedure to be used, an explanation of the hazards involved, information on how to store or dispose of the chemicals used, a discussion of the phenomena displayed and principles illustrated by the demonstration, and a list of references. The brief summary provides a succinct description of the demonstration. The materials list for each procedure specifies the equipment and chemicals needed. Where solutions are to be used, we give directions for preparing stock amounts larger than those required for the procedure. The teacher should decide how much of each solution to prepare for practicing the demonstration and for doing the actual presentation. The availability and cost of chemicals may also affect decisions about the volumes to be prepared.

The procedure section often contains more than one method for presenting a demonstration. In all cases, the first procedure is the one the authors prefer to use. However, the alternative procedures are also effective and valid pedagogically.

The hazards and disposal sections include information compiled from sources believed to be reliable. We have enumerated many potentially adverse health effects and have called attention to the fact that many of the chemicals should be used only in well-ventilated areas. In all instances teachers should inquire about and follow local disposal practices and should act responsibly in handling potentially hazardous material. We recognize that several chemicals such as silver and mercury can be recovered and reused and have given references to recovery and purification procedures.

Chemical Demonstrations

Volume 3

Sources Containing Descriptions
of Lecture Demonstrations

We call attention to the following sources of information about lecture demonstrations. These lists, updated from Volume 2, are not intended to be comprehensive. Some of the books are out of print but may be available in libraries.

BOOKS

Alyea, H. N. *TOPS in General Chemistry,* 3d ed., Journal of Chemical Education: Easton, Pennsylvania (1967).

Alyea, H. N., and F. B. Dutton, Eds. *Tested Demonstrations in Chemistry,* 6th ed., Journal of Chemical Education: Easton, Pennsylvania (1965).

Arthur, Paul. *Lecture Demonstrations in General Chemistry,* McGraw-Hill: New York (1939).

Blecha, M. T. Ph.D. Dissertation, "The Development of Instructional Aids for Teaching Organic Chemistry," Kansas State University, Manhattan, Kansas (1981).

Chemical Demonstrations Proceedings, Western Illinois University and Quincy-Keokuk Section of the American Chemical Society, Macomb, Illinois, May 5–6, 1978.

Chemical Demonstrations Proceedings, Western Illinois University and Quincy-Keokuk Section of the American Chemical Society, Macomb, Illinois, May 4–5, 1979.

Chemical Demonstrations Proceedings, Western Illinois University and Quincy-Keokuk Section of the American Chemical Society, Macomb, Illinois, May 1–2, 1981.

Chemical Demonstrations Proceedings, Fifth Annual Symposium at 16th Great Lake Regional Meeting of the American Chemical Society, Normal, Illinois, June 8, 1982.

Chen, Philip S. *Entertaining and Educational Chemical Demonstrations,* Chemical Elements Publishing Co.: Camarillo, California (1974).

Faraday, M. *The Chemical History of a Candle: A Course of Lectures Delivered Before a Juvenile Audience at the Royal Institution,* The Viking Press: New York (1960).

Ford, L. A. *Chemical Magic,* T. S. Denison & Co.: Minneapolis, Minnesota (1959).

Fowles, G. *Lecture Experiments in Chemistry,* 5th ed., Basic Books, Inc.: New York (1959).

Frank, J. O., assisted by G. J. Barlow. *Mystery Experiments and Problems for Science Classes and Science Clubs,* 2d ed., J. O. Frank: Oshkosh, Wisconsin (1936).

Freier, G. D., and F. J. Anderson. *A Demonstration Handbook for Physics,* 2d ed., American Association of Physics Teachers: Stony Brook, New York (1981).

Gardner, R. *Magic Through Science,* Doubleday & Co., Inc.: Garden City, New York (1978).

Hartung, E. J. *The Screen Projection of Chemical Experiments,* Melbourne University Press: Carlton, Victoria (1953).

Herbert, D. *Mr. Wizard's Supermarket Science,* Random House: New York (1980).

Herbert, D., and H. Ruchlis. *Mr. Wizard's 400 Experiments in Science,* Revised Edition, Book-Lab: North Bergen, New Jersey (1983).

Joseph, A., P. F. Brandwein, E. Morholt, H. Pollack, and J. Castka. *A Sourcebook for the Physical Sciences,* Harcourt, Brace, and World, Inc.: New York (1961).

Lippy, J. D., Jr., and E. L. Palder. *Modern Chemical Magic,* The Stackpole Co.: Harrisburg, Pennsylvania (1959).

Meiners, H. F., Ed. *Physics Demonstration Experiments,* Vols. 1 and 2, The Ronald Press Company: New York (1970).

My Favorite Lecture Demonstrations, A Symposium at the Science Teachers Short Course, W. Hutton, Chairman; Iowa State University, Ames, Iowa, March 6–7, 1977.

Newth, G. S. *Chemical Lecture Experiments,* Longmans, Green and Co.: New York (1928).

Sharpe, S., Ed. *The Alchemist's Cookbook: 80 Demonstrations,* Shell Canada Centre for Science Teachers, McMaster University: Hamilton, Ontario, undated.

Siggins, B.A. M.S. Thesis, "A Survey of Lecture Demonstrations/Experiments in Organic Chemistry," University of Wisconsin–Madison, Wisconsin (1978).

Summerlin, L. R., and others. *Chemical Demonstrations: A Sourcebook for Teachers,* Vols. 1 and 2, American Chemical Society: Washington, DC (1985, 1987).

Walker, J. *The Flying Circus of Physics—With Answers,* Interscience Publishers, John Wiley and Sons: New York (1977).

Weisbruch, F. T., *Lecture Demonstration Experiments for High School Chemistry,* St. Louis Education Publishers: St. Louis, Missouri (1951).

Wilson, J. W., J. W. Wilson, Jr., and T. F. Gardner. *Chemical Magic,* J. W. Wilson: Los Alamitos, California (1977).

ARTICLES

Bailey, P. S., C. A. Bailey, J. Anderson, P. G. Koski, and C. Rechsteiner. Producing a chemistry magic show. *J. Chem. Educ.* 52:524–25 (1975).

Castka, J. F. Demonstrations for high school chemistry. *J. Chem. Educ.* 52:394–95 (1975).

"Chem 13 News" 81. The November issue contained a collection of chemical demonstrations. (1976).

Gilbert, G. L., Ed. Tested demonstrations. Regular column in *J. Chem. Educ.* since 1976.

Hanson, R. H. Chemistry is fun, not magic. *J. Chem. Educ.* 53:577–78 (1976).

Hughes, K. C. Some more intriguing demonstrations. *Chem. in Australia* 47:458–59 (1980).

Kolb, D. K., Ed. Overhead projector demonstrations. Regular column in *J. Chem. Educ.*

McNaught, I. J., and C. M. McNaught. Stimulating students with colourful chemistry. *School Sci. Review* 62:655–66 (1981).

Rada Kovitz, R. The SSP syndrome. *J. Chem. Educ.* 52:426 (1975).

Schibeci, R. A., J. Webb, and F. Farrel. Some intriguing demonstrations. *Chem. in Australia* 47:246–47 (1980).

Schwartz, A. T., and G. B. Kauffman. Experiments in alchemy, Part I: Ancient arts. *J. Chem. Educ.* 53:136–38 (1976).

Schwartz, A. T., and G. B. Kauffman. Experiments in alchemy, Part II: Medieval discoveries and "transmutations". *J. Chem. Educ.* 53:235–39 (1976).

Shakhashiri, B. Z., G. E. Dirreen, and W. R. Cary. Lecture Demonstrations, in *Sourcebook for Chemistry Teachers,* pp. 3–16, W. T. Lippincott, Ed., American Chemical Society, Division of Chemical Education: Washington, D.C. (1981).

Steiner, R., Ed. Chemistry for kids. Regular column in *J. Chem. Educ.*

Wilson, J. D., Ed. Favorite demonstrations. Regular column in *J. College Science Teaching.*

Sources of Information on Hazards and Disposal

In preparing the Hazards and Disposal sections of Volume 3, we have used the following references. The order of listing reflects our degree of utilization.

Bretherick, L., Ed. *Hazards in the Chemical Laboratory,* 3d ed.; The Royal Society of Chemistry: London (1981).

Windholz, M., Ed. *The Merck Index,* 10th ed., Merck & Co., Inc.: Rahway, New Jersey (1983).

Laboratory Waste Disposal Manual, Manufacturing Chemists Association (1975). This book is out of print, but some of the information was available in the 1981 Reagent Catalog of MCB Manufacturing Chemists, Inc., 2909 Highland Avenue, Cincinnati, Ohio 45212.

Registry of Toxic Effects of Chemical Substances, Dept. of Health, Education and Welfare (NIOSH): Washington, D.C., revised annually. Available from Superintendent of Documents, U.S. Government Printing Office, Washington, D.C. 20402.

Laboratory Waste Disposal and Safety Guide, University of Wisconsin–Madison Safety Department: Madison, Wisconsin (1984).

Prudent Practices for Handling Hazardous Chemicals in Laboratories, Committee on Hazardous Substances in the Laboratory, National Research Council (1981).

Fire Protection Guide on Hazardous Materials, 6th ed., National Fire Protection Association: 470 Atlantic Ave., Boston, Massachusetts 02210 (1975). New editions are published at intervals.

Safety in Academic Chemistry Laboratories, 3d ed., American Chemical Society Committee on Chemical Safety: Washington, D.C. (1979). A fourth edition is scheduled for publication. The bibliography lists many journal articles and books.

Health and Safety Guidelines for Chemistry Teachers. American Chemical Society Dept. of Educational Activities: Washington, D.C. (1979). The bibliography lists journal articles and books.

Steere, N. V. *Handbook of Laboratory Safety,* 2d ed., CRC Press: Cleveland, Ohio (1971).

Guide for Safety in the Chemical Laboratory, 2d ed., Van Nostrand Reinhold Co., Litton Educational Publishing, Inc.: New York (1972).

Steere, N. V., Ed. *Safety in Chemical Laboratory,* Journal of Chemical Education: Easton, Pennsylvania, Vol. 1 (1967), Vol. 2 (1971), Vol. 3 (1974).

Renfrew, M. M., Ed. *Safety in the Chemical Laboratory,* Journal of Chemical Education: Easton, Pennsylvania, Vol. 4 (1981).

Sax, N. I. *Dangerous Properties of Industrial Materials,* 3d ed., Van Nostrand Reinhold Co., Litton Educational Publishing, Inc.: New York (1968).

Chemical Demonstrations

Volume 3

8

Acids and Bases

Rodney Schreiner, Bassam Z. Shakhashiri, David B. Shaw,
Ronald I. Perkins, and Frederick H. Juergens

The demonstrations in this chapter are composed of representative examples that illustrate many of the properties of acids and bases. Some of these demonstrations are revised versions of ones that have been in use for many years and are favorites among teachers. Others are new and appear for the first time in this volume.

Many of the demonstrations employ acid-base indicator solutions. Therefore, the directions for the preparation of these solutions are gathered together at the end of the introduction (see pages 27–29). Most of the demonstrations in this chapter also involve solutions of the common laboratory acids and bases at standard concentrations. Because it is most convenient to have stocks of these solutions on hand, the directions for preparing them have been also assembled in a section at the end of the introduction (see pages 30–32).

Many demonstrations of acid-base properties are suitable for presentation with an overhead projector [1, 2]. Nearly half of the demonstrations in this chapter include directions for overhead presentation. The procedures that involve overhead projection can be easily recognized because the Materials section for each begins with "overhead projector."

One of the most visually impressive properties of acids and bases is their ability to change the colors of indicators; therefore, many demonstrations in this chapter involve the use of indicators. Demonstration 8.1 involves a broad range of common laboratory indicators to show the variety of color changes that can be produced with them, and the range of pH values over which these changes occur. The second demonstration shows how indicators can be combined to produce mixtures that undergo several color changes at various pH values and how a combination of indicators can produce nearly any color of the rainbow at a particular pH value. In the third demonstration, an invisible pattern created with colorless indicators on a sheet of paper is revealed when the paper is sprayed with a basic solution. Demonstration 8.4 shows that a variety of plant extracts change color in response to pH changes.

The next group of demonstrations deals with additional properties of acids and bases. The fifth demonstration illustrates many of the characteristic properties of acids and bases, and Demonstration 8.6 shows that many household products exhibit these properties. To dispel the notion that all acids are alike, Demonstration 8.7 shows that each of four common acids has some unique and striking properties all its own. The eighth demonstration shows the glass-etching property of hydrogen fluoride. The ninth demonstration illustrates the ability of nitric acid to dissolve several metals, and in one procedure, the product gases cause colorful solutions to flow from one flask to another and back. Demonstration 8.10 presents a multiple-flask variation of the ammonia fountain, which produces multiple colors through the use of universal indicator. The next two demonstrations also deal with acid-base reactions involving gases. Demonstrations 8.13 and 8.14 reveal that certain salts and certain oxides have properties of acids or bases. Several reactions involving the acidic or basic properties of a variety of gases are described in Demonstrations 8.15 and 8.16. Demonstrations 8.17 and 8.18 deal with properties of acids related to environmental concerns. The following demonstration, 8.19, illustrates that some substances possess properties of both acids and bases; these substances are amphoteric.

Most of the remaining demonstrations deal with quantitative aspects of acids and bases. Demonstration 8.20 illustrates that acids with the same molar concentration can have different acidities and different capacities to neutralize a base. In Demonstrations 8.21 through 8.23, the conductivity of acid solutions is shown and related to the acidity of the solutions. The extent of ionization of weak acids is investigated in Demonstrations 8.24 and 8.25. In Demonstration 8.26 the neutralizing capacity of an antacid tablet is determined, and a titration curve is generated instrumentally in Demonstration 8.27. The properties of buffer solutions are shown in Demonstration 8.28, and the buffering action of an antacid is presented in Demonstration 8.29. The significance of pH on the solubility of a protein is presented in Demonstration 8.30. The last two demonstrations illustrate reactions that can be regarded as Lewis acid-base reactions.

Although we assume that most readers are familiar with the concepts of acids and bases, we present in this introduction a brief discussion of their properties and of several theoretical explanations for these properties. Included in this discussion are the Arrhenius, Brønsted-Lowry, and Lewis concepts. Along with these theoretical concepts, the electrical properties of acids and bases and some quantitative aspects of acid-base equilibria are also presented. Additional information on acid-base theories can be found in two excellent articles by Professor Doris Kolb [3, 4].

CLASSIFICATION OF ACIDS AND BASES

From the nature of what we call acid comments, acid looks, acid tests, and acid indigestion, the meaning of acid is clear: something severe, biting, sour, and generally unpleasant. Some of the common uses of the word *acid* derive from its Latin origin: *acidus,* meaning "sour." Others, however, reflect the chemical notion of acids. Acid indigestion is a condition in which the stomach contents, containing hydrochloric acid, flow up into the esophagus, causing a burning sensation [5]. Acid tests are severe and crucial tests; the term for them derives from the use of nitric acid as a test for gold: gold does not dissolve in nitric acid, but brass does. Among all the technical concepts of chemistry, the concept of acid is perhaps the one most familiar to the general public.

The concepts of acid and base developed from classification schemes founded on the observed properties of substances. Acids are substances that share a common set of properties; bases share a set of properties different from that of acids. However, acids and bases are related by their ability to interact. These properties and relationships were recognized early in the development of chemistry, but theoretical explanations for their origin developed slowly. In the 19th century, first the theory of the German chemist Justus Liebig dominated, and then the theory of Svante Arrhenius, a Swedish chemist. Although these theories were essentially limited to aqueous acids and bases, they proved adequate in accounting for most qualitative and quantitative properties of aqueous solutions. Early in the 20th century, the acid-base concept was expanded through the theory developed by the English chemist Thomas Lowry and the Danish chemist

Johannes Brønsted and through that developed by the American chemist G. N. Lewis to include nonaqueous solutions and even solvent-free systems. These two theories expanded the classes of acids and bases to include, in addition to the acids and bases recognized by the Liebig and Arrhenius systems, substances which are not classified as acids or bases under either of these earlier systems. Thus, as the theories of acids and bases developed and the understanding of the underlying origins of acid and base behaviors grew, more and more substances were recognized as having acidic or basic properties.

The idea of classifying substances into the categories of acid and base is one of the oldest in chemistry. The terms *acid* and *base* occur in the writings of medieval alchemists [6]. Acids, such as acetic and tannic acids, are easily detected because of their sour taste. In fact, the word for acid in several languages derives from words meaning sour: English *acid* and French *acide* fcrom Latin *acidus,* as mentioned earlier, German *Säure* from *saure,* and Russian *kislota* from *kisly.* However, a classification based only on a sour taste would be rather trivial. Acids share other properties as well. They have the ability to change a variety of blue vegetable pigments (such as litmus) to red, to dissolve certain metals (such as iron and tin) while producing bubbles, and to cause effervescence when combined with certain minerals (such as limestone). This combination of diverse properties suggests that there is a common cause for these properties and, therefore, some significance to the classification of substances as acids.

Bases can be readily recognized because they have a slippery feel. They also reverse the color change brought about by acids on blue vegetable pigments: they turn them from red back to blue. This effect on the color of vegetable pigments suggests that bases are related in some way to acids. Other properties of bases confirm this relationship. When combined with an acid in the proper proportion, a base destroys the properties of the acid, that is, it neutralizes the acid. The mixture no longer tastes sour, affects the color of blue vegetable pigments, dissolves metals, or causes effervescence with limestone. Furthermore, the properties of the base are also destroyed (the mixture no longer feels slippery), which indicates that the acid in turn neutralizes the base. Bases are characterized mainly by their ability to neutralize acids.

When a base neutralizes an acid and the mixture is left uncovered until all of the water has evaporated and it is dry, there remains a crystalline substance called a *salt* because of its saline taste. Therefore, it can be said that a salt is the product of a reaction between an acid and a base.

Many of the substances familiar to alchemists and early chemists can be classified as acids or bases. Common acids included muriatic (hydrochloric) acid, oil of vitriol (sulfuric acid), aquafortis (nitric acid), vinegar (acetic acid), and tannin (tannic acid). Similarly, many common substances have basic properties, for example, soda (sodium carbonate), lime (calcium oxide), and potash (potassium carbonate). However, the lack of any apparent physical origin for the acidic or basic properties of these substances hindered the development of an explanation of these properties. Also impeding this development was the fact that some of the products of the reaction of an acid with a base (i.e., salts) themselves have acidic or basic properties. For example, the salt resulting from the combination of vinegar with soda has basic properties. However, this salt is not as strong a base as is soda itself, which is revealed by its less dramatic effects on the colors of vegetable pigments.

Further complicating the matter are substances which cannot be uniquely classified as acids or bases, because they have properties of both. In the presence of an acid, such a substance behaves as a base, and in the presence of a base, it acts as an acid. For example, the calx of lead (lead oxide) dissolves in an acid, neutralizing the acid, and it dissolves in a base, neutralizing the base. Such a substance is called *amphoteric,* which comes from the Greek word *amphoteros* meaning "both," because it has properties of both an acid and a base.

Some quantitative aspects of the neutralization process were also discovered quite early. A particular amount of an acid always requires the same amount of a base to neutralize it, and the neutralization process always produces the same amount of salt. For example, 10 g of tartaric acid always requires 7 g of sodium carbonate to neutralize it and always produces 13 g of sodium tartrate.† Furthermore, 10 g of tartaric acid requires 9 g of potassium carbonate and produces 15 g of potassium tartrate. Thus, it appears that in some way 7 g of sodium carbonate is equivalent to 9 g of potassium carbonate. This is an illustration of one of the earliest types of chemical equivalence to have been recognized. Another sort of quantitative relationship between acids is revealed by the ability of one acid to displace another from a salt. For example, treating sodium phosphate with sulfuric acid will produce phosphoric acid, and mixing phosphoric acid with sodium acetate will produce acetic acid. This makes it possible to arrange acids in order of apparent strength: a "stronger" acid can displace a "weaker" one from a salt. In this sense, sulfuric acid is stronger than phosphoric acid, which is stronger than acetic acid. However, these reactions in which one acid displaces another from its salt do not necessarily yield an amount of displaced acid equivalent to the amount of stronger acid used. The stronger acid may displace only a portion of the weaker acid. Furthermore, under certain conditions, a weaker acid can release a small amount of stronger acid from one of its salts. For example, when acetic acid and sodium phosphate are combined, the mixture contains not only acetic acid and sodium phosphate but also small amounts of phosphoric acid and sodium acetate. Therefore, it appears that these displacement reactions are reversible, and that rather than going to completion, they approach a balance or equilibrium instead. (Characteristics of this equilibrium are discussed later in the introduction.)

EARLY THEORIES OF ACIDS

One of the first attempts at a theoretical interpretation of acid behavior was made by the 18th-century French chemist Antoine Lavoisier, who believed that all acids contain oxygen. In fact, the name of this element reflects Lavoisier's belief: it is taken from the Greek *oxys* (sour) and *genes* (born). After it was learned that hydrochloric acid contains no oxygen, the English chemist Humphry Davy in 1815 declared that the acid-forming element is hydrogen. However, not all compounds of hydrogen are acids, and it was not until 1838 that the first adequate theoretical description of an acid was proposed by the German chemist Justus Liebig. Liebig's description of an acid states that an acid is a substance containing hydrogen which can be replaced by a metal. This proved to be a workable definition for over 50 years and is still compatible with modern definitions. In Liebig's time, however, there was no theoretical notion of bases. These were still treated empirically as substances which neutralized acids, even though there was no explanation for how they were capable of doing this.

---

†The equation for the reaction of sodium carbonate with tartaric acid is

$$Na_2CO_3 \text{ (106 g/mol)} + H_2C_4H_4O_6 \text{ (150 g/mol)} \longrightarrow Na_2C_4H_4O_6 \text{ (194 g/mol)} + H_2O + CO_2$$

ELECTRICAL CONDUCTIVITY OF AQUEOUS SOLUTIONS

During the 19th century another property of acid, base, and salt solutions was investigated, and the investigations led to a theory of aqueous solutions that encompassed both acid and base behavior. The property in question is the ability of these aqueous solutions to conduct an electric current. A substance whose solution is more electrically conductive than pure water is called an electrolyte. Therefore, acids, bases, and salts are electrolytes.

Because an electric current is a flow of charge, a solution that conducts an electric current must possess some means of moving charge through it. In the 1880s, the Swedish chemist Svante Arrhenius proposed his theory of ionization to explain this movement of charge. According to this ionization theory, solutions of electrolytes conduct electricity because they contain charged atoms or charged groups of atoms that can move about in the solution. These charged atoms and groups of atoms are called *ions*. As the ions move through the solution, they carry electrical charge with them. Because conductive solutions do not have a net charge, they must contain ions of both positive and negative charges, and furthermore, the total charge of the positive ions must be in exact balance with the total charge of the negative ions.

When two electrodes are placed in a solution and a sufficiently large voltage is applied across these electrodes (see Figure 1), an electrical current flows between the electrodes. The magnitude of the current (I) is measured with an ammeter and expressed in amperes. This magnitude depends on the magnitude of the potential, V, applied to the electrodes and on the electrical resistance of the solution, R:

$$I = V/R$$

The potential is expressed in volts and the resistance in ohms. The resistance of any material, liquid or solid, increases with its length, l, and decreases with its cross-sectional area, A:

$$R = rl/A$$

For a solution, the length is the distance between the electrodes, and the cross-sectional area is the area of the electrodes. In this equation, r is a proportionality constant called resistivity. The conductivity, k, is the inverse of resistivity, so

$$R = (1/k)l/A \qquad \text{or} \qquad k = l/RA$$

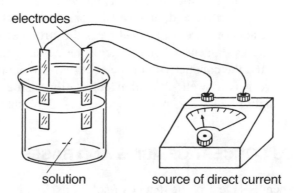

electrodes

solution   source of direct current

Figure 1.

Table 1. Molar Conductivities (ohm^{-1}cm^{2}mol^{-1}) of Electrolytes in Aqueous Solution at 18°C [5]

Electrolyte	Concentration (mol · liter^{-1})					
	0.0010	0.0050	0.010	0.050	0.10	0.50
Acetic acid	41	20.0	14.3	6.48	4.60	2.01
Ammonia	28.0	13.2	9.6	4.6	3.3	1.35
Citric acid	88.4	54	42.5	22	16.1	7.3
Hydrochloric acid	377	373	370	360	351	327
Hydrofluoric acid	—	90	60	35.9	31.3	27.0
Nitric acid	375	371	368	357	350	324
Potassium carbonate	266	243	231	201	188	156
Potassium hydroxide	234	230	228	219	213	197
Sodium acetate	75.2	72.4	70.2	64.2	61.1	49.4
Sodium chloride	106.5	103.8	102.0	95.7	92.0	80.9
Sodium hydroxide	208	203	200	190	183	172
Sulfuric acid	722	660	616	506	450	410

The conductivity of pure water at 18°C is 4×10^{-8} ohm^{-1}cm^2.

Length can be expressed in centimeters and area in square centimeters. Because resistance is expressed in units of ohms, the units of k are ohm^{-1}cm^{-1}. The conductivity of any solution depends on the concentration of the solution. To adjust for this concentration dependence, the conductivity of a solution is often reported as a molar conductivity, represented by the capital Greek lambda, Λ. The molar conductivity of a solution is calculated by dividing its conductivity by its molar concentration, c:

$$\Lambda_m = k/c$$

The molar conductivities of several electrolyte solutions at various concentrations are listed in Table 1. These molar conductivities show two types of behavior as the concentration varies. Some electrolytes have a molar conductivity that changes little (by less than a factor of 2) as the concentration of its solution changes from 0.50M to 0.0010M. In Table 1, these include hydrochloric acid, nitric acid, potassium carbonate, potassium hydroxide, sodium acetate, sodium chloride, sodium hydroxide, and sulfuric acid. These substances are called strong electrolytes. For other substances called weak electrolytes, the molar conductivity depends very much on the concentration of the solution. For these, the molar conductivity increases by a factor greater than 10 as the concentration decreases from 0.50M to 0.0010M. From the concentration dependence of the molar conductivities in Table 1, the weak electrolytes can be identified as acetic acid, ammonia, citric acid, and hydrofluoric acid. Each of these four substances is either a weak acid or a weak base, whereas the strong electrolytes are salts, strong acids, or strong bases. In general, among acids there is a correlation between the acidity of the solution and the conductivity of the solution. The more acidic a solution is, the more conductive it is as well. A similar correlation also exists among bases: solutions that are strongly basic are more conductive than those that are weakly basic.

THE ARRHENIUS CONCEPT OF ACIDS AND BASES

In light of Arrhenius's electrolytic theory of solutions, and because acid behavior has been associated with hydrogen, acids appear to be substances which produce "hydrogen ions" in water. These ions are generally represented by the symbol H$^+$, although, as discussed later, this is not their actual chemical form. Those substances whose solutions are highly acidic and highly conductive produce a high concentration of H^+ in solution, whereas those that are weakly acidic and poorly conductive give rise to only a low concentration of H^+. Because basic solutions also conduct electricity, they too contain a characteristic ion responsible for their properties. Furthermore, because bases neutralize acids, this ion reacts with hydrogen ions to form a substance which is neither acidic nor basic. Arrhenius proposed that the hydroxide ion, OH^-, is responsible for the basic properties of a solution. Then, in the reaction between an acid and a base which results in the neutralization of both, hydrogen ions combine with hydroxide ions to form neutral water molecules, as represented by $H^+ + OH^- \rightarrow H_2O$. According to the Arrhenius theory, acids are those substances which produce hydrogen ions when dissolved in water, and bases are those substances which form hydroxide ions in aqueous solution. This was the dominant theory of acids and bases from the 1880s to the 1920s. Its description of acids is in practice equivalent to Liebig's theory of acids, but the Arrhenius theory does add an interpretation of the origin of the properties of bases.

Although the hydrogen ion is frequently represented as H^+, bare protons do not exist in aqueous solutions. Instead, the "hydrogen ion" in aqueous solution is actually attached to a water molecule, and a more accurate formulation of the hydrogen ion in aqueous solutions is H_3O^+, called a *hydronium ion* [7]. However, the hydronium ion is surrounded by other water molecules that are loosely attached to it by hydrogen bonds. The congregation of water molecules around the hydronium ion can be represented as $(H_2O)_nH_3O^+$, where n is about 3 [8]. This group of water molecules is not a fixed entity, because the water molecules in the group are continually exchanging with those in the surrounding solution. For the sake of convenience in writing, the hydronium ion is called a hydrogen ion so that it can be abbreviated accordingly as H^+.

The conductivities of the strong electrolytes in water, as listed in Table 1, are not identical, because the ions contained in their solutions do not all possess the same current-carrying ability. The factors influencing this ability include the charge and size of the ion. When travelling a certain distance, ions with a greater charge move more charge than an equal number of ions with a lesser charge. Small ions move through solution more quickly than large ions. The conductivity of a solution is the sum of the conductivities of the individual ions in the solution. Comparing the conductivities of different solutions, some of whose ions are the same, has enabled the conductivities of individual ions to be determined. The differences between the molar conductivities of sodium chloride and hydrochloric acid solutions listed in Table 1 can be attributed to the differences between the molar conductivities of sodium ions and hydrogen ions. The most highly conductive ion in aqueous solutions is the hydrogen ion, which has a conductivity of 349.8 $ohm^{-1}cm^2mol^{-1}$ at 25°C [9]. In contrast, the conductivity of the sodium ion is 50.1 $ohm^{-1}cm^2mol^{-1}$, which is about on a par with that of most ions having a single charge. The conductivity of most ions is limited by the rate at which the individual ions travel through the solution. However, the electrical conductivity of hydrogen ions is relatively high, because hydrogen ions can be passed bucket-brigade fashion from one molecule of water to another [10]. This transfer process can be represented as follows:

$$
\begin{array}{cccccc}
H & H & H & H & H & H \\
| & | & | & | & | & | \\
H\!-\!O\!-\!H & O\!-\!H & O\!-\!H & H\!-\!O & H\!-\!O\!-\!H & O\!-\!H \\
+ & & & & + & \\
\end{array}
$$

As this shows, charge can move through water without being carried by an individual hydrogen ion: charge can move independently of mass. This transfer of hydrogen ions from one molecule of water to another is the fastest known reaction in aqueous solution. The reaction is so fast that in 1 liter of an aqueous solution containing 1 mole of hydrogen ions, 1.4×10^{11} *moles* of hydrogen ions are transferred per second [11]. Therefore, hydrogen ions have a much higher conductivity than ions such as sodium, which carry charge through a solution only by their own movement. The hydroxide ion, which can interact with water through proton transfers and participate in a charge conduction mechanism similar to that of the hydrogen ion, also has an unusually high conductivity of 197.6 ohm^{-1}cm^{2}mol^{-1}.

THE BRØNSTED-LOWRY ACID-BASE CONCEPT

The Arrhenius concept of acids and bases is very successful in accounting for most qualitative and quantitative properties of aqueous acids and bases. However, because the Arrhenius concept defines acids and bases in terms of aqueous solutions, it does not apply to solutions in which the solvent is not water. Yet, some nonaqueous solutions possess properties that appear to be acid-base properties. For example, hydrogen chloride is soluble in carbon tetrachloride. So are some dyes that behave as pH indicators when they are dissolved in water. When hydrogen chloride is dissolved in carbon tetrachloride, it will affect the color of these dyes in the same way it does in aqueous solution. Therefore, the solution of hydrogen chloride in carbon tetrachloride has a characteristic property of an acid: it changes the color of an indicator. In order to treat this solution like an acid, the theory of acids and bases must be expanded.

It may be tempting to extend the Arrhenius concept to other solvents by defining an acid as a substance that produces hydrogen ions in solution, whether or not the solvent is water. However, the solution of hydrogen chloride in carbon tetrachloride is completely nonconductive. Therefore, there are no ions in the solution, and in particular, there are no hydrogen ions. Hence, this extension of the Arrhenius acid-base concept does not apply to the solution of hydrogen chloride in carbon tetrachloride. In order to treat this solution like an acid, the acid-base concept must be expanded in some other way. In 1923, an expanded theory was proposed independently by the English chemist Thomas Lowry and the Danish chemist Johannes Brønsted [12–14]. Their definition is: an acid is a species that loses a hydrogen ion and a base is a species that gains a hydrogen ion during a chemical reaction. Acids and bases are defined in terms of the reactions they undergo. Any reaction in which the net effect can be regarded as the transfer of a hydrogen ion from one species to another is a Brønsted-Lowry acid-base reaction. In the Brønsted-Lowry system, the "hydrogen ion" is exactly that, a hydrogen atom with its electron removed, H$^+$. In fact, the hydrogen ion in the Brønsted-Lowry system is often referred to as a proton.

The Brønsted-Lowry acid-base concept encompasses the Arrhenius scheme. An Arrhenius acid is a substance that produces hydrogen ions in solution. These Arrhenius "hydrogen ions" are the product of a reaction of some acid-forming substance (e.g., hydrogen chloride) with water:

$$HCl(g) + H_2O(l) \longrightarrow H_3O^+(aq) + Cl^-(aq)$$

In this reaction, the hydrogen chloride has transferred a "proton" to the water. Therefore, in the Brønsted-Lowry scheme, HCl is an acid, as it is in the Arrhenius scheme. An Arrhenius base is a substance that forms OH^- in water (e.g., ammonia):

$$NH_3(g) + H_2O(l) \longrightarrow NH_4^+(aq) + OH^-(aq)$$

Here, the ammonia accepts a "proton" from water. Therefore, it is a base in the Brønsted-Lowry scheme, as it is in the Arrhenius. What is new in the Brønsted-Lowry scheme is that acids and bases are defined in terms of the reactions they undergo, not by the condition of their aqueous solutions. This frees the notion of acids and bases from the properties of an aqueous solution; acid-base reactions can occur in any solvent or, indeed, without a solvent, as in the gas or solid phases.

This advances the acid-base concept as well, because now a single substance can be an acid or a base, depending on the material with which it reacts. For example, in the reaction with HCl, water behaves as a base, whereas in the reaction with ammonia, it reacts as an acid. The Brønsted-Lowry concept encompasses amphoteric properties of substances, such as some salts. These salts, the products of the neutralization of an acid by a base, have residual properties of an acid or a base. For example, the solution of sodium acetate in water has basic properties. These properties are easily rationalized by the Brønsted-Lowry concept. The acetate ion, being the anion of acetic acid, is capable of bonding to a hydrogen ion ("proton") and forming a molecule of acetic acid. When sodium acetate is placed in water, some of the acetate ions react with water, forming acetic acid:

$$C_2H_3O_2^-(aq) + H_2O(l) \leftrightarrows HC_2H_3O_2(aq) + OH^-(aq)$$

In this reaction, the acetate ion has gained a proton; therefore it is a base, and sodium acetate has basic properties. Additionally, the water molecules in this reaction lose protons to the acetate ion; therefore, water acts as a Brønsted-Lowry acid in this reaction.

Other salts, such as ammonium chloride, have acidic properties. The ammonium ion (NH_4^+) can lose a hydrogen ion to form ammonia, and is therefore a Brønsted-Lowry acid. When ammonium chloride is dissolved in water, some of the ammonium ions react with water, forming ammonia:

$$NH_4^+(aq) + H_2O(l) \leftrightarrows NH_3(aq) + H_3O^+(aq)$$

In this reaction, ammonium ions lose protons to water molecules. Therefore, the ammonium ion is an acid and water is a base in this reaction. As the previous two chemical equations show, in the Brønsted-Lowry system, water can be either an acid or a base, depending on the substance with which it reacts. Thus, water is an amphoteric substance in the Brønsted-Lowry system.

Perhaps the most significant advantage of the Brønsted-Lowry concept is that it provides a more comprehensive view of acid-base reactions. Whenever a Brønsted-Lowry acid-base reaction occurs, the substance behaving as an acid becomes a base, and the one acting as a base turns into an acid. For example, in the reaction of acetate ions with water, the acetate ions gain a proton to become acetic acid molecules: the base becomes an acid. Furthermore, the water molecules lose a proton, forming hydroxide ions: the acid becomes a base. The product of the mixture of an acid and a base (water and acetate ions) is also a mixture of a base and an acid (hydroxide ions and acetic acid). Therefore, it is not surprising that the reaction is reversible. It is an acid-base reaction in either direction! The extent to which the reaction proceeds in either direction is related to the relative strengths of the acids and bases in the mixture. In general, the stronger acid and base will react to form the weaker acid and base until reaching a balance where the mixture contains more of the weaker acid and base than of the stronger pair. Because the reaction between HCl and NaOH goes virtually to completion, the base formed from a strong acid (Cl⁻ formed from HCl) must be a weak base, and the acid formed from a strong base (H_2O formed from OH^-) must be a weak acid.

$$HCl(aq) + NaOH(aq) \longrightarrow NaCl(aq) + H_2O(l)$$

However, the mixture of acetic acid and water contains mainly acetic acid molecules and relatively few acetate ions.

$$HC_2H_3O_2(aq) + H_2O(l) \rightleftharpoons C_2H_3O_2^-(aq) + H_3O^+(aq)$$

This indicates that acetic acid is a weaker acid than H_3O^+, and water is a weaker base than acetate ions.

THE LEWIS ACID-BASE CONCEPT

By defining acids and bases in terms of what happens in the reaction rather than in terms of the solvent, the Brønsted-Lowry system allows the extension of the acid-base concept to reactions that occur in nonaqueous solvents. Yet, like the Arrhenius system, it requires that an acid contain hydrogen. However, there are reactions in which substances that do not contain hydrogen exhibit behavior characteristic of acids. Such a substance is aluminum chloride. It can change the color of a common acid-base indicator, methyl violet, from its basic color to its acidic color. Methyl violet is violet in basic aqueous solutions and yellow in very acidic solutions. Methyl violet is soluble in carbon tetrachloride, and its solution is violet. When aluminum chloride is added to the solution, the mixture turns yellow. Aluminum chloride produces the same color change in methyl violet as that produced by hydrochloric acid. Therefore, aluminum chloride appears to behave as an acid. However, it cannot be considered an acid in the Brønsted-Lowry system, because aluminum chloride does not contain hydrogen and, therefore, cannot be a proton donor.

The American chemist G. N. Lewis proposed in 1923 that an acid be defined as a species containing an atom that can form a bond to another atom through a pair of electrons belonging to the other atom [15]. The species containing this other atom, which donates its pair of electrons to form the bond, is to be considered the base in the reaction. For example, when aluminum chloride reacts with methyl violet, a bond is formed between the aluminum atom and a nitrogen atom in methyl violet:

$$
\begin{array}{ccc}
\overset{\displaystyle CH_3}{\underset{\displaystyle CH_3}{R-\overset{|}{\underset{|}{N}}:}} \;+\; \overset{\displaystyle Cl}{\underset{\displaystyle Cl}{Al-Cl}} & \longrightarrow & \overset{\displaystyle CH_3\quad Cl}{\underset{\displaystyle CH_3\quad Cl}{R-\overset{|}{\underset{|}{N}}\!\!-\!\!\overset{|}{\underset{|}{Al}}-Cl}}
\end{array}
$$

The aluminum atom forms a bond to the nitrogen atom using a pair of electrons from the nitrogen atom. In common parlance, aluminum has accepted a pair of electrons and nitrogen has donated a pair of electrons in forming the bond. Thus, a base is an "electron-pair donor" and an acid is an "electron-pair acceptor" in the Lewis concept of acids and bases. The Lewis concept of acids and bases does not depend on the presence of a particular element (hydrogen) as do the Brønsted-Lowry and Arrhenius systems. Acidic or basic properties are instead ascribed to the behavior of valence electrons, which all substances possess.

Because the Lewis acid-base system incorporates a broader range of reactions as acid-base reactions, there is a greater range of reactivity patterns between Lewis acids and bases. The reactions in which metal ions react with complexing agents to form complex ions are Lewis acid-base reactions. Such a reaction is that in which water molecules in tetraaquo copper(II) ions are displaced by ammonia molecules to form the tetraammine copper(II) complex ion:

$$[Cu(H_2O)_4]^{2+}(aq) + 4 NH_3(aq) \longrightarrow [Cu(NH_3)_4]^{2+}(aq) + 4 H_2O(l)$$
$$\text{light blue} \hspace{7cm} \text{deep blue}$$

In complex-formation reactions such as these, the metal ion is a Lewis acid and the complexing agent is the Lewis base. Many more such Lewis acid-base reactions are presented in the demonstrations in Chapter 4 in Volume 1 of this series. Unlike the situation with Arrhenius acids and bases, not every Lewis acid will react with a particular Lewis base. Some Lewis acids react more readily with some bases than with others. The observed reactivities of Lewis acids and bases can be correlated by the system of hard and soft acids and bases, as proposed by Pearson [16, 17]. "Hard" acids tend to react with "hard" bases, and "soft" acids react with "soft" bases. Hard acids are ions that have a small size, a high positive charge, and no easily excited valence electrons (e.g., Al^{3+}). Soft acids are generally large, have little or no positive charge, and contain electrons (often d-subshell electrons) that are relatively easily excited to higher energy levels (e.g., Tl^+). On the other hand, soft bases are generally large, easily oxidized, of low electronegativity, and have low-energy empty orbitals (e.g., I^-), and hard bases are small, difficult to reduce, and are those whose lowest empty orbitals are of high energy (e.g., NH_3). These generalizations correlate a great many of the reactivity patterns of Lewis acids and bases.

EQUILIBRIUM CONSTANT EXPRESSION

As Table 1 shows, the molar conductivity of a solution of a weak electrolyte, such as acetic acid, increases as its concentration decreases. This can be attributed to an increase in the ratio of ion concentration to overall concentration as the overall concentration of the weak electrolyte decreases. This increase in ion concentration can be explained by characterizing the weak electrolyte as only partly ionized when it is dissolved in water. As the overall concentration decreases, the degree of ionization increases. This increased ionization is the result of an equilibrium between the ions and the nonionized form of the electrolyte. For example, when acetic acid is dissolved in water, some of its molecules dissociate into hydrogen ions and acetate ions, as represented in the equation

$$HC_2H_3O_2(aq) \rightleftharpoons H^+(aq) + C_2H_3O_2^-(aq)$$

Simultaneous with the dissociation of acetic acid molecules, the hydrogen ions and acetate ions recombine to form molecules.

There is a mathematical relationship between the concentration of the ions and the concentration of the undissociated molecules in the solution. For a weak electrolyte, this relationship is represented by the equilibrium constant expression. For a weak acid, HA,

$$HA \rightleftharpoons H^+ + A^-$$

the equilibrium constant expression is

$$K_a = \frac{a_{H^+} \, a_{A^-}}{a_{HA}}$$

where a_X represents the activity of each solute, X.

The subscript a in K_a stands for "acid" to distinguish this equilibrium constant from other types of equilibrium constants (e.g., solubility products). The activity of a solute is related to its concentration by

$$a_i = \gamma_i m_i$$

where m_i is the molality of i
and γ_i is a factor called the activity coefficient.

The activity coefficient is a function of concentration, and as m_i approaches zero, γ_i approaches unity. The value of the activity coefficient for a particular solute is also affected by the temperature of the solution and by the concentrations of other solutes in the solution. The activities of ionic solutes are particularly affected by the concentration of ions in the solution, that is, by the "ionic strength" of the solution. Therefore, it is not possible to measure the activity of a particular ion in solution, because that ion must be accompanied by counterions that affect its activity. What can be measured is a mean activity coefficient, $\gamma_\pm$, which combines the activity coefficients of the ions: $\gamma_\pm^2 = \gamma_+ \gamma_-$. Then, the equilibrium constant expression can be written as

$$K_a = \frac{\gamma_\pm^2 \, m_{H^+} \, m_{A^-}}{\gamma_u \, m_{HA}}$$

For uncharged solutes, such as HA, in a dilute solution, $\gamma_u \simeq 1$, and the equilibrium constant expression becomes

$$K_a = \gamma_\pm^2 \frac{m_{H^+} \, m_{A^-}}{m_{HA}}$$

Mean activity coefficients, $\gamma_\pm$, are tabulated for several aqueous ionic solutes at various concentrations in Table 2 [18]. The data in Table 2 indicate that as the concentration of the ions decreases, $\gamma_\pm$ approaches 1. The data also show that at a given concentration there is considerable variation in $\gamma_\pm$ from one solute to another. As the concentration of the solute in an aqueous solution becomes small, the value of the molality approaches the value of the molarity. Therefore, for dilute solutions, the equilibrium constant expression is also written as

$$K_a = \frac{[H^+][A^-]}{[HA]}$$

where the square brackets represent molar concentrations.

This ratio of concentrations is constant only when the concentrations are rather low. As a general rule, this expression is accurate only for ion concentrations below 0.001M.

The value of K_a is a constant for a particular weak acid at a given temperature. This means that the degree of ionization of a weak acid must increase as the concentration of the acid decreases. Suppose, for a particular acid, $K_a = 1 \times 10^{-5}$. If the concentration of HA is 0.1M, then the concentration of H^+ is 0.001M, and the ratio of their concentrations is $0.001/0.1 = 0.01$. If the solution is diluted until [HA] = 0.01M, then [H^+] = 0.0003M, and the concentration ratio is $0.0003/0.01 = 0.03$. When the

Table 2. Activity Coefficients at Various Concentrations for Several Aqueous Acids, Bases, and Salts[a]

Solute	Concentration							
	0.001	0.005	0.01	0.05	0.10	0.50	1.00	2.00
Hydrochloric acid	0.966	0.930	0.906	0.833	0.798	0.769	0.811	1.011
Sodium hydroxide	—	—	—	0.82	—	0.69	0.68	—
Sodium bromide	0.966	0.934	0.914	0.844	0.800	0.695	0.686	—
Potassium chloride	0.966	0.927	0.902	0.816	0.770	0.652	0.607	0.577
Potassium hydroxide	—	0.92	0.90	0.82	0.80	0.73	0.76	—
Calcium chloride	0.888	0.789	0.732	0.584	0.524	0.510	0.725	—
Sulfuric acid	—	0.643	0.545	0.341	0.266	0.155	0.131	0.125
Potassium sulfate	0.89	0.78	0.71	0.52	0.43	—	—	—
Copper (II) sulfate	0.74	0.53	0.41	0.21	0.16	0.068	0.047	—
Lanthanum (III) chloride	0.853	0.716	0.637	0.417	0.356	0.303	0.583	0.954
Indium (III) sulfate	—	0.16	0.11	0.035	0.025	0.014	—	—

[a]These activity coefficients are for 25°C, and the concentrations are expressed in molality (mol solute per kg solvent).

solution is diluted even more, so that [HA] = 0.001M, then [H^+] = 0.0001M, and the ratio of [H^+]/[HA] = 0.1. Thus, the degree of ionization increases as the concentration of a weak acid decreases.

Each weak acid has a characteristic value of K_a, which is, for practical purposes, independent of the concentration of the solution (but does depend on temperature). For any two weak acid solutions at the same concentration, the more acidic solution contains a higher concentration of hydrogen ions, the acid is more ionized, and the value of its K_a is greater. Therefore, the relative strengths of acids are expressed quantitatively by their K_a values.

Some acids are capable of releasing more than one hydrogen ion for each molecule of acid, and these acids are called *polyprotic acids*. Oxalic and phosphoric acids are such acids. Some molecules of oxalic acid release two hydrogen ions, as represented in the following equations:

Step 1: $H_2C_2O_4 \rightleftarrows H^+ + HC_2O_4^-$ K_{a1}

Step 2: $HC_2O_4^- \rightleftarrows H^+ + C_2O_4^{2-}$ K_{a2}

In the case of phosphoric acid, some of its molecules release three hydrogen ions, as represented in these equations:

Step 1: $H_3PO_4 \rightleftarrows H^+ + H_2PO_4^-$ K_{a1}

Step 2: $H_2PO_4^- \rightleftarrows H^+ + HPO_4^{2-}$ K_{a2}

Step 3: $HPO_4^{2-} \rightleftarrows H^+ + PO_4^{3-}$ K_{a3}

For each of these ionization steps, there is a K_a value. Because polyprotic acids can produce more than one hydrogen ion per molecule, comparing strengths of polyprotic acids is more complicated than simply comparing K_a values.

The K_a values for a wide range of acids have been determined and are tabulated in several reference books, and a number of them are given in Table 3 [5]. These tabulated values allow the concentration of hydrogen ions in a weak acid solution of known concentration to be calculated from the equilibrium constant expression.

Table 3. Equilibrium Constants for the Ionization of Weak Acids at 25°C [5]

Acid	Formula	K_{a1}	K_{a2}	K_{a3}
Acetic	$HC_2H_3O_2$	1.75×10^{-5}		
Benzoic	$HC_7H_5O_2$	6.25×10^{-5}		
Boric	H_3BO_3	5.8×10^{-10}	1.8×10^{-12}	1.6×10^{-14}
Carbonic	H_2CO_3	4.4×10^{-7}	4.7×10^{-11}	
Citric	$H_3C_6H_5O_7$	7.45×10^{-4}	1.73×10^{-5}	4.02×10^{-7}
Formic	$HCHO_2$	1.77×10^{-4}		
Hydrocyanic	HCN	6.2×10^{-10}		
Hydrofluoric	HF	6.6×10^{-4}		
Hydrosulfuric	H_2S	1.1×10^{-7}	1.2×10^{-13}	
Hypochlorous	$HClO$	2.9×10^{-8}		
Iodic	HIO_3	1.57×10^{-1}		
Lactic	$HC_3H_5O_3$	1.39×10^{-4}		
Nitrous	HNO_2	7.2×10^{-4}		
Oxalic	$H_2C_2O_4$	5.36×10^{-2}	5.34×10^{-5}	
Phenol	HC_6H_5O	1.0×10^{-10}		
Phosphoric	H_3PO_4	7.1×10^{-3}	6.3×10^{-8}	4.2×10^{-13}
Propionic	$HC_3H_5O_2$	1.33×10^{-5}		
Sulfuric	H_2SO_4		1.0×10^{-2}	
Sulfurous	H_2SO_3	1.3×10^{-2}	6.2×10^{-8}	
Tartaric	$H_2C_4H_4O_4$	6.0×10^{-4}	1.5×10^{-5}	

SELF-IONIZATION OF WATER

Electrical conductivity measurements can be made very accurately, and when measured, pure water is found to be a weak conductor, with a conductivity of 4×10^{-8} ohm^{-1}cm^{-1} at 18°C [5]. This indicates that even pure water contains ions. The presence of these ions can be explained by considering the nature of acid-base neutralization reactions. The essential process in an acid-base neutralization reaction is

$$H^+ + OH^- \longrightarrow H_2O$$

The reaction of hydrogen ions with hydroxide ions to form water molecules reaches an equilibrium point at which water molecules dissociate to form ions at the same rate that the ions recombine. This dissociation also occurs in pure water and is expressed by

$$H_2O \rightleftarrows H^+ + OH^-$$

The ions formed in this ionization of water account for the electrical conductivity of pure water. There is a mathematical relationship between the concentrations of the ions in water, namely:

$$K_w = \frac{a_{H^+} a_{OH^-}}{a_{H_2O}}$$

where K_w is the ionization constant of water.

For a pure liquid, the activity is 1. Therefore,

$$K_w = a_{H^+} a_{OH^-} = \gamma_\pm^2 m_{H^+} m_{OH^-}$$

Because the conductivity of pure water is low, the concentrations of ions must also be low. Therefore, $\gamma_\pm \simeq 1$, the molalities are nearly equal to the molar concentrations, and the K_w expression can be written as

$$K_w = [H^+][OH^-]$$

The measured conductivity of pure water along with the calculated conductivities of the individual H^+ and OH^- ions permits the calculation of the value of K_w. At 18°C,

$$4 \times 10^{-8} \text{ ohm}^{-1}\text{cm}^{-1} = [H^+](349.8 \text{ ohm}^{-1}\text{cm}^2\text{mol}^{-1})$$
$$+ [OH^-](197.6 \text{ ohm}^{-1}\text{cm}^2\text{mol}^{-1})$$

In pure water $[H^+] = [OH^-]$, so

$$[H^+] = (4 \times 10^{-8} \text{ ohm}^{-1}\text{cm}^{-1})/(349.8 \text{ ohm}^{-1}\text{cm}^2\text{mol}^{-1}$$
$$+ 197.6 \text{ ohm}^{-1}\text{cm}^2\text{mol}^{-1})$$

$$= 7 \times 10^{-11} \text{ mol}\cdot\text{cm}^{-3} = 7 \times 10^{-8} \text{ mol L}^{-1}$$

Therefore, the electrical conductivity of pure water at 18°C indicates that the hydrogen-ion concentration is 7×10^{-8}M, and the hydroxide-ion concentration is also 7×10^{-8} M. As the temperature of water increases, its conductivity and the concentration of ions increases; as its temperature decreases, its conductivity and the concentration of its ions decreases. At 100°C, $[H^+]$ is 7×10^{-7}M, and at 0°C it is 3×10^{-8}M. Because in pure water $[H^+] = [OH^-]$ and $[H^+] = 1 \times 10^{-7}$M at 25°C, the value of K_w at this temperature is $(1 \times 10^{-7})(1 \times 10^{-7}) = 1 \times 10^{-14}$.

Pure water is neutral, it is neither acidic nor basic, and it contains equal numbers of hydrogen ions and hydroxide ions. An acidic solution contains more hydrogen ions than hydroxide ions, and a basic solution contains more hydroxide ions than hydrogen ions. Therefore, in an acidic solution at 25°C, the concentration of hydrogen ions is greater than 1×10^{-7}M, and in a basic solution, the concentration of hydrogen ions is less than 1×10^{-7}M. There are hydrogen ions even in basic solutions, although they are outnumbered by hydroxide ions.

The concentration of hydrogen ions in a solution is a quantitative measure of the acidity of the solution. Because the concentration of hydrogen ions in aqueous solutions can vary over a very large range, it is more common to express the acidity of different solutions with the pH scale. The pH of a solution is the negative logarithm of the activity of the hydrogen ion, that is, $pH = -\log(a_{H^+})$. For example, in a solution which has a hydrogen-ion concentration of 0.00020M, the $pH = -\log(0.00020) = 3.70$. Because an acidic solution at 25°C has a hydrogen-ion concentration greater than 1×10^{-7}M, its pH is less than 7. A basic solution has a hydrogen-ion concentration of less than 1×10^{-7}M and, therefore, a pH greater than 7. It is common parlance to speak of the pH of a solution rather than the acidity or basicity of the solution. Figure 2 shows the range of the pH scale and the approximate pH values for several common substances.

When an acid is added to water, the concentration of hydrogen ions in the water increases. According to the K_w expression, which indicates that the product of the hydrogen-ion concentration and the hydroxide-ion concentration is constant ($K_w = [H^+][OH^-]$), when the hydrogen-ion concentration increases, the concentration of hydroxide ions decreases simultaneously. For example, when 0.01 mole of hydrogen chloride is dissolved in enough water to make 1.0 liter of solution, the concentration of hydrogen ions in the water changes from 1×10^{-7}M to 0.01M, and the pH of the water

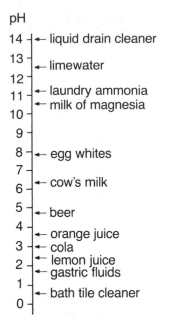

Figure 2. pH values for various common liquids.

changes from 7 to 2. Simultaneously, as revealed by the K_w value of 1×10^{-14} at 25°C, the concentration of hydroxide ions changes from 1×10^{-7}M to 1×10^{-12}M:

$$[OH^-] = K_w/[H^+] = (1 \times 10^{-14})/(0.01) = 1 \times 10^{-12}$$

Thus, by the addition of 0.01 mole of hydrogen chloride to a liter of water, the concentration of hydrogen ions in the water changes by a factor of 5 powers of 10, or 100,000.

ACID-BASE INDICATORS

An indicator is a material that gives a visible sign—usually through a color change—of the state or condition of another material. Substances that change color in response to an acid or a base are called acid-base indicators. Many vegetable pigments function as acid-base indicators, a well-known example being litmus, which is a mixture extracted from a lichen that grows in the Netherlands (the -*mus* in *litmus* has the same origin as the word *moss*). Vegetable pigments have been joined by a large group of synthetic dyes that also change color when an acid or a base is added to them. The behavior of these indicators can be interpreted, if they are considered to be weak acids themselves. The indicator molecules may be represented as HIn. These ionize when dissolved in water, and this ionization process is represented by the equation

$$HIn \rightleftarrows H^+ + In^-$$

A specific example of an indicator ionization represented by this equation is that of phenol red. In the equation at the top of page 19, the molecule of phenol red, on the left is represented in the above equation as HIn, and the anion of phenol red, on the right is represented above by In^-.

What differentiates an indicator from other weak acids is that the color of the undissociated molecules, HIn, is different from the color of the ions, In^-. In general, indicators are large organic molecules, as illustrated in the structure of phenol red. When a hydrogen ion is added or removed from these molecules or ions, their electronic structures change. This change in electronic structure leads to a change in the energy of the light absorbed by these molecules or ions and, therefore, to a change in the color of their solutions. Because indicators themselves are acids or bases, when they are used to assess the pH of a solution, they react with hydrogen ions or hydroxide ions in the solution. This affects the pH of the tested solution. Fortunately, indicators are intensely colored substances, and only a trace is needed to produce a visible color. Because only a tiny amount is needed, only a tiny fraction of the acid or base in the solution reacts with the indicator. Therefore, the effect of the indicator on the acidity of the tested solution can be negligible.

For a weak-acid indicator, as for any weak acid, the concentrations of the ions and molecules of the indicator are related by a K_a expression:

$$K_a = \frac{[H^+][In^-]}{[HIn]}$$

When acid is added to an indicator solution, the concentration of H^+ increases. Therefore, the ratio $[In^-]/[HIn]$ must decrease, in order to maintain the equality between the K_a value and the ratio $[H^+][In^-]/[HIn]$. This is accomplished when In^- ions react with the added H^+ ions to form HIn molecules. Because In^- and HIn have different colors, the color of the solution changes as the ratio $[In^-]/[HIn]$ changes. When a base is added to an indicator solution, the hydroxide ions of the base react with molecules of the indicator and form indicator ions:

$$OH^- + HIn \rightleftharpoons In^- + H_2O$$

The equilibrium constant of this reaction is related to the K_a of the indicator and to the K_w of water. This reaction is the sum of the ionization of the indicator and the reverse of the ionization of water:

$$
\begin{array}{lll}
HIn & \rightleftharpoons H^+ + In^- & K_a \\
OH^- + H^+ & \rightleftharpoons H_2O & K = 1/K_w \\
\hline
OH^- + HIn & \rightleftharpoons In^- + H_2O & K_{eq} = K_a/K_w
\end{array}
$$

Because K_w is generally much smaller than K_a for most indicators, the value of K_{eq} is very large, and the HIn reacts almost completely with excess OH^-. This affects the ratio $[In^-]/[HIn]$ and the color of the solution. Once the color of the solution has changed to that of In^-, the color can be restored to that of HIn by adding an acid to the solution; the hydrogen ions react with In^- to form molecules of the indicator:

$$H^+ + In^- \rightleftharpoons HIn$$

Whenever the pH of a solution containing an indicator changes, the ratio of $[In^-]/[HIn]$ changes. If the solution were to contain much more In^- than HIn, the ratio $[In^-]/[HIn]$ would be much greater than 1 and the color of the solution would be that produced by In^-. If the pH were to decrease, $[In^-]$ would decrease and $[HIn]$ would increase, resulting in a decrease in $[In^-]/[HIn]$. If the ratio $[In^-]/[HIn]$ were still much greater than 1, then the color of the solution would still be that of In^-, even though the pH has changed. However, if the ratio $[In^-]/[HIn]$ were close to 1, then there would be similar amounts of In^- and HIn in the solution, and the color of the solution would be a result of both In^- and HIn. If the ratio $[In^-]/[HIn]$ were much less than 1, then the color of the solution would be that of HIn. Therefore, the color of the solution changes when the value of the ratio $[In^-]/[HIn]$ passes through 1. As the K_a expression of the indicator shows, $[In^-]/[HIn] = K_a/[H^+]$, and the ratio $[In^-]/[HIn]$ is 1 when the $[H^+]$ of the solution is equal to the value of K_a of the indicator. Because different indicators have different K_a values, they change color at different $[H^+]$, and they can be used to assess the hydrogen-ion concentration in a solution. By analogy with the definition pH $= -\log[H^+]$, a quantity called the pK_a of an indicator is defined as $pK_a = -\log(K_a)$. This is done to facilitate the selection of indicators that are useful in a particular pH range. Because an indicator undergoes its color change when $[H^+]$ is comparable to the K_a of the indicator, the color change also occurs when the pH is comparable to the pK_a. Table 4 lists a number of indicators, along with their structural formulas, their pK_a values, and their colors.

In general, a single indicator will reveal only whether the hydrogen-ion concentration is greater than, less than, or about the same as its K_a value. If several indicators having a range of K_a values and various colors are combined, the mixture will undergo several color changes at various pH values. For example, the chart displays four indicators, their pK_a values, and their colors over a pH range of $2-10$. It also shows the color of the combination of these indicators over this pH range. The combination has a different color at each of the pH values.

	Color at pH:				
Indicator (pK_a)	2	4	6	8	10
methyl orange (3.8)	red	orange	yellow	yellow	yellow
methyl red (5.4)	red	red	yellow	yellow	yellow
bromothymol blue (6.8)	yellow	yellow	yellow	blue	blue
phenolphthalein (9.2)	colorless	colorless	colorless	colorless	pink
combination	red-orange	orange	yellow	green	purple

A combination of indicators such as this is called a universal indicator, because it reveals the approximate hydrogen-ion concentration over a wide range of possible values.

BUFFER SOLUTIONS

When only 0.36 g (0.010 mole) of hydrogen chloride gas or 0.56 g (0.010 mole) of solid potassium hydroxide is added to a liter of water, the hydrogen-ion concentration changes by a factor of 100,000, and the pH changes by 5 units. When a similar amount of hydrogen chloride or potassium hydroxide is added to an aqueous solution

Table 4. Structures,[a] Colors, and pK$_a$ Values of Selected Acid-Base Indicators

Indicator (pK$_a$)	Low-pH structure (color)	High-pH structure (color)
Alizarin (6.4)	(yellow)	(red)
Alizarin yellow GG (11.0)	(red)	(orange)
Bromocresol green (4.7)	(yellow)	(blue)
Bromocresol purple (6.0)	(yellow)	(purple)

[a] The low-pH structures are those of the predominant form of the indicator at pH values 1 unit below the pK$_a$ value of the indicator. Similarly, the high-pH structures are those of the predominant form at pH values 1 unit above the pK$_a$.

Continued on following page

Table 4, *continued*

Indicator (pK$_a$)	Low-pH structure (color)	High-pH structure (color)
Bromophenol blue (3.8)	(yellow)	(blue)
Bromothymol blue (6.8)	(yellow)	(blue)
Cresol red (7.9)	(yellow)	(red)
o-Cresolphthalein (9.1)	(colorless)	(red)
Crystal violet (1.0)	(yellow)	(blue)

Table 4, *continued*

Indicator (pK$_a$)	Low-pH structure (color)	High-pH structure (color)
Erythrosin B (2.9)	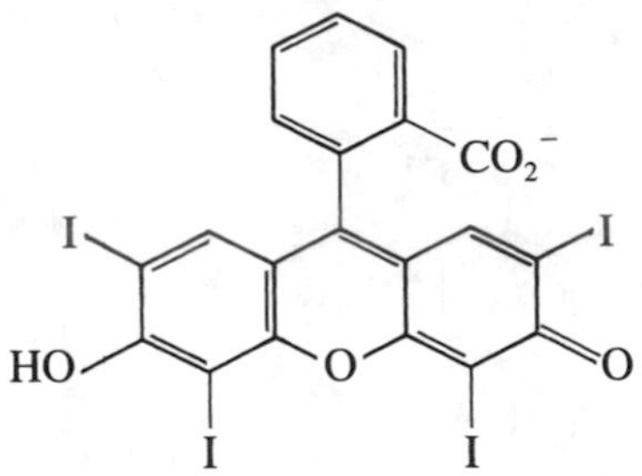 (orange)	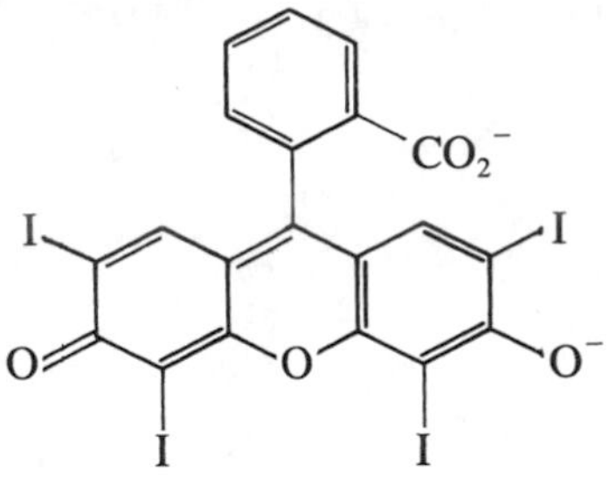 (red)
Litmus (6.4)	complex mixture of substances (red)	(blue)
Methyl orange (3.8)	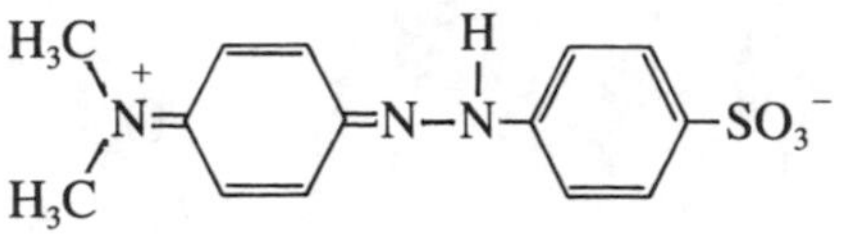 (red)	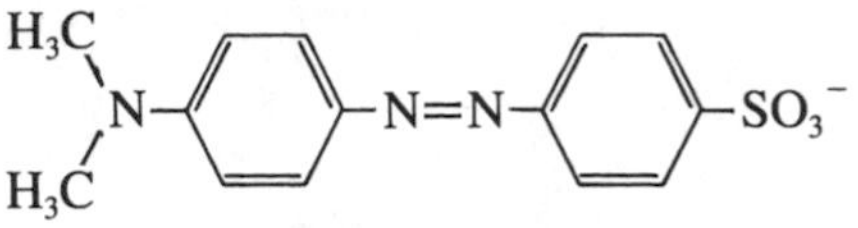 (yellow)
Methyl red (5.4)	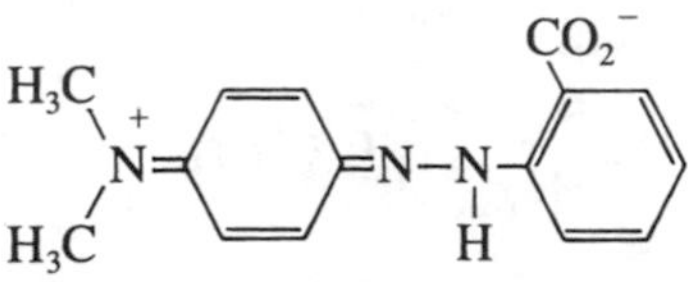 (red)	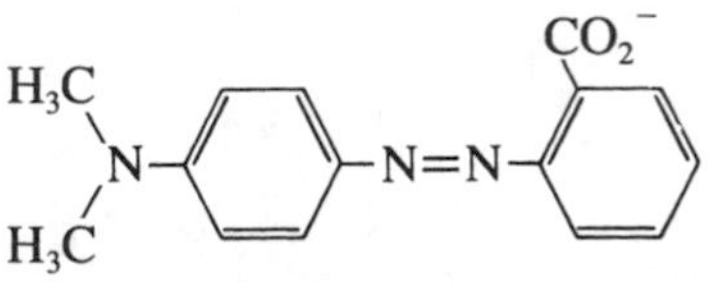 (yellow)
Methyl violet (1.7)	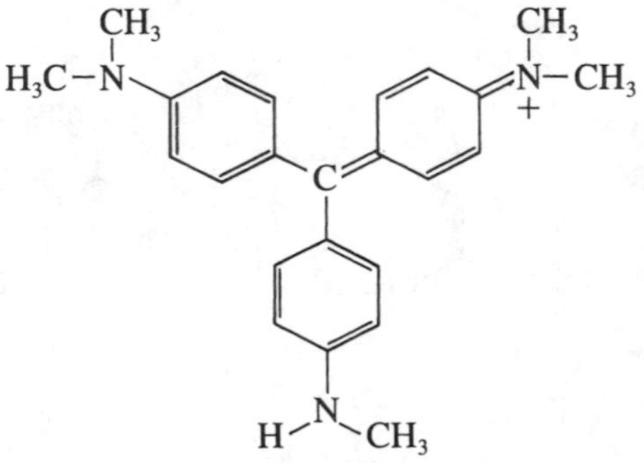 (yellow)	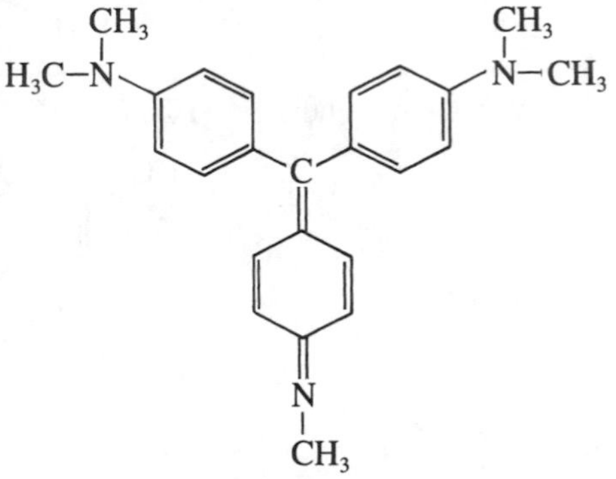 (violet)
m-Nitrophenol (8.3)	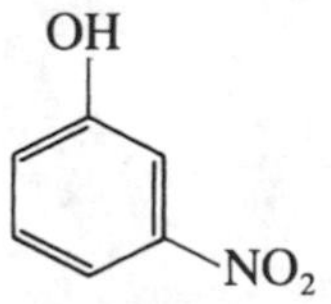 (colorless)	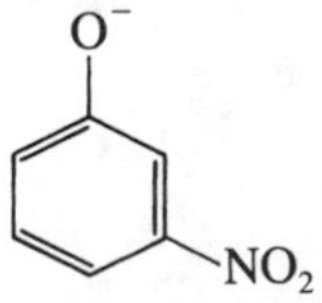 (yellow)

Continued on following page

Table 4, *continued*

Indicator (pK$_a$)	Low-pH structure (color)	High-pH structure (color)
Phenol red (7.3)	(yellow)	(red)
Phenolphthalein (9.2)	(colorless)	(red-purple)
Thymol blue (8.8)	(yellow)	(blue)
Thymolphthalein (10.1)	(colorless)	(blue)

made by dissolving both 0.05 mole of acetic acid and 0.05 mole of sodium acetate in a liter of solution, the hydrogen-ion concentration changes by less than a factor of 2, and the pH by less than 0.2 of a unit. The hydrogen-ion concentration of the acetic acid and sodium acetate solution changes by several powers of 10 less than that of the pure water when acid or base is added to it. A solution whose hydrogen-ion concentration changes relatively little when an acid or base is added is called a *buffer solution*. Buffer solutions are capable of resisting changes in pH when an acid or base is added, because they contain a species which reacts with hydrogen ions and another species which reacts with hydroxide ions. In the case of the acetic acid–sodium acetate mixture, the species that reacts with hydroxide ions is molecular acetic acid:

$$HC_2H_3O_2 + OH^- \leftrightarrows C_2H_3O_2^- + H_2O$$

The species that reacts with added hydrogen ions is the acetate ion (from sodium acetate):

$$C_2H_3O_2^- + H^+ \leftrightarrows HC_2H_3O_2$$

The mixture of a weak acid and the salt of the weak acid contains species that react with H^+ and with OH^-, and therefore the mixture is capable of consuming either. Because the reaction of H^+ with acetate ions and the reaction of OH^- with acetic acid are equilibrium reactions, these reactions do not go to completion (i.e., not all of the added acid or base is consumed by the buffer), and the pH of the buffer solution is not completely unaffected by the addition of acid or base. Furthermore, the buffer does not have an unlimited capacity to consume added acid or base. If enough acid or base is added, the *buffering capacity* will be exceeded and the pH will change drastically.

The pH of a buffer solution is related to the K_a of its weak acid and to the relative amounts of acid and salt in the solution. These relationships can be quantified with the K_a expression of the weak acid:

$$K_a = \frac{[H^+][A^-]}{[HA]}$$

In the solution, the concentration of the hydrogen ion is

$$[H^+] = \frac{K_a[HA]}{[A^-]}$$

In order for the buffer solution to be effective in resisting changes in the hydrogen-ion concentration, it must contain significant amounts of both HA and A^-. Because it is the HA in the buffer solution that consumes added base, the buffer solution is effective at maintaining its pH only when the total amount of added base is less than the amount of HA in the buffer solution. If more base is added, the buffering capacity will be exceeded, and the pH of the buffer solution will change drastically. Similarly, the total amount of added acid must be less than the amount of A^- in the buffer solution if the buffer solution is to retain its effectiveness. Therefore, the amount of HA and A^- in the solution determines the capacity of the buffer solution for consuming added acid or base. In order to make the buffer similarly effective at resisting changes in pH when either acid or base is added to it, the ratio $[HA]/[A^-]$ should be in the range of 10–0.1. This means that a buffer solution is effective at maintaining the hydrogen-ion concentration only when $[H^+]$ is near the value of K_a of the weak acid.

The brief outline of acids and bases presented in this introduction is not intended to be all inclusive. A number of other acid-base concepts have been proposed in addition to the Arrhenius, Brønsted-Lowry, and Lewis concepts—for example, solvent-systems theory and the Usanovich theory—and these are discussed in a book by Finston and Rychtman [*19*]. In addition, there are more extensive discussions of the Arrhenius theory [*20*], Brønsted-Lowry theory [*21*], and Lewis theory [*22–24*], as well as of the hard and soft acid-base system [*25*].

REFERENCES

1. E. Hartung, *The Screen Projection of Chemical Experiments*, Cambridge University Press: London (1953).
2. D. Kolb, *J. Chem. Educ.* 64:348 (1987).
3. D. Kolb, *J. Chem. Educ.* 55:459 (1978).
4. D. Kolb, *J. Chem. Educ.* 56:49 (1979).
5. D. F. Tapley, R. J. Weiss, and T. Q. Morris, Eds., *The Columbia University College of Physicians and Surgeons Complete Home Medical Guide*, Crown Publishers: New York (1985).
6. A. J. Ihde, *The Development of Modern Chemistry*, Dover Publications: New York (1984).
7. E. Fitzgerald and A. Lapworth, *J. Chem. Soc.* 93:2163 (1908).
8. P. F. Knewstubb and A. W. Tickner, *J. Chem. Phys.* 38:464 (1964).
9. J. A. Dean, Ed., *Lange's Handbook of Chemistry*, 13th ed., McGraw-Hill Book Co.: New York (1985).
10. S. Glasstone, K. J. Laidler, and H. Eyring, *The Theory of Rate Processes*, McGraw-Hill Book Co.: New York (1941).
11. M. Eigen, K. Kustin, and G. Maas, *Z. phys. Chem.* (Leipzig) 30:130 (1961).
12. T. M. Lowry, *Chem. Ind.* (London) 42:43 (1923).
13. T. M. Lowry, *Chem. Ind.* (London) 42:1048 (1923).
14. J. N. Brønsted, *Rec. Trav. Chim. Pays-Bas* 42:718 (1923).
15. G. N. Lewis, *Valence and the Structure of Atoms and Molecules*, Chemical Catalog Co.: New York (1923).
16. R. G. Pearson, *J. Chem. Educ.* 45:581 (1968).
17. R. G. Pearson, *J. Chem. Educ.* 64:561 (1987).
18. P. W. Atkins, *Physical Chemistry*, W. H. Freeman and Co.: San Francisco (1978).
19. H. L. Finston and A. C. Rychtman, *A New View of Current Acid-Base Theories*, John Wiley and Sons: New York (1982).
20. P. Walden, *Salts, Acids, and Bases*, McGraw-Hill Book Co.: New York (1929).
21. R. P. Bell, *The Proton in Chemistry*, 2d ed., Cornell University Press: Ithaca, New York (1961).
22. W. F. Luder and S. Zuffanti, *The Electronic Theory of Acids and Bases*, 2d ed., Dover Publications: New York (1961).
23. R. S. Drago and N. A. Matwiyoff, *Acids and Bases*, D. C. Heath: Lexington, Massachusetts (1968).
24. W. B. Jensen, *The Lewis Acid-Base Concepts*, John Wiley and Sons: New York (1980).
25. R. G. Pearson, Ed., *Hard and Soft Acids and Bases*, Dowden, Hutchison and Ross, Inc.: Stroudsburg, Pennsylvania (1973).

Preparation of Indicator Solutions

These instructions are for the preparation of 100 mL of stock solution of each indicator. The solid indicators used in the preparations are available from laboratory supply companies. Several of the indicators are available as both pure indicator and in the salt form; take care to use the form specified. Failure to do so may result in the incomplete dissolving of the solid. Each of these solutions is stable for at least 6 months, and therefore, can be prepared well in advance of use.

HAZARDS

Ethanol is flammable and should be kept away from open flames. Its flash point is 12°C, and in air, its vapor is explosive at concentrations of 3–19%.

Wear plastic or rubber gloves while preparing any of these solutions, both for safety's sake and to prevent staining of the skin. A compilation of the results of most recent investigations into the toxicological and carcinogenic properties of many substances such as these indicators is contained in the *Registry of Toxic Effects of Chemical Substances,* published periodically by the U.S. Department of Health and Human Services.

PREPARATIONS

alizarin
> Dissolve 0.1 g of alizarin (1,2-dihydroxyanthraquinone) in 100 mL of distilled water.

alizarin yellow GG
> Dissolve 0.01 g of alizarin yellow GG (3-carboxy-4-hydroxy-3′-nitroazobenzene) in 100 mL of distilled water.

bromocresol green
> Dissolve 0.02 g of the sodium salt of bromocresol green (3′,3″,5′,5″-tetrabromo-*m*-cresolsulfonephthalein, sodium salt) in 100 mL of distilled water.

bromocresol purple
> Dissolve 0.04 g of the sodium salt of bromocresol purple (5′,5″-dibromo-*o*-cresolsulfonephthalein, sodium salt) in 100 mL of distilled water.

bromophenol blue
> Dissolve 0.04 g of the sodium salt of bromophenol blue (3′,3″,5′,5″-tetrabromo-phenolsulfonephthalein, sodium salt) in 100 mL of distilled water.

bromothymol blue
>   Dissolve 0.04 g of the sodium salt of bromothymol blue (3′,3″-dibromothymol-sulfonephthalein, sodium salt) in 100 mL of distilled water.

cresol red
>   Dissolve 0.04 g of cresol red (o-cresolsulfonephthalein) in 25 mL of 95% ethanol, and dilute the resulting solution to 100 mL with distilled water.

o-cresolphthalein
>   Dissolve 0.04 g of o-cresolphthalein in 100 mL of 95% ethanol.

crystal violet
>   Dissolve 0.02 g of crystal violet (hexamethyl-p-rosaniline chloride) in 100 mL of distilled water.

erythrosin B
>   Dissolve 0.1 g of the disodium salt of erythrosin B (2′,4′,5′,7′-tetraiodofluorescein, disodium salt) in 100 mL of distilled water.

litmus
>   Grind the litmus to a powder, boil 2 g of the powder in 100 mL of distilled water for 5 minutes, and reconstitute the resulting, cooled solution to 100 mL with distilled water.

methyl orange
>   Dissolve 0.01 g of methyl orange (sodium 4-[4-(dimethylamino)phenylazo]-benzenesufonate) in 100 mL of distilled water.

methyl red
>   Dissolve 0.02 g of methyl red (2-[4-(dimethylamino)-phenylazo]benzoic acid) in 60 mL of 95% ethanol and add 40 mL of distilled water.

methyl violet
>   Dissolve 0.03 g of methyl violet hydrochloride (pentamethyl-p-rosaniline hydrochloride) in 100 mL of distilled water.
>   **Caution!** Methyl violet is toxic when ingested; its oral LD50† in mice is 105 mg/kg. It also causes persistent staining of the skin.

m-nitrophenol
>   Dissolve 0.3 g of m-nitrophenol in 100 mL of distilled water.
>   **Caution!** m-Nitrophenol is a poison when ingested and is toxic when absorbed through the skin. Its oral LD50† in mice is 1.07 g/kg.

phenol red
>   Dissolve 0.04 g of the sodium salt of phenol red (phenolsulfonephthalein, sodium salt) in 100 mL of distilled water.

phenolphthalein
>   Dissolve 0.05 g of phenolphthalein (3,3-bis(p-hydroxyphenyl)phthalide) in 50 mL of 95% ethanol, and dilute the resulting solution to 100 mL with distilled water.

---

†LD50 stands for lethal dose in 50% of the cases. It is calculated by determining the size of the dose that causes death in half of the organisms to which it is administered, and it is stated in terms of the mass of that dose per kilogram of the organism's body mass.

thymol blue

> Dissolve 0.04 g of the sodium salt of thymol blue (sodium thymolsulfonephthalate) in 100 mL of distilled water.

thymolphthalein

> Dissolve 0.04 g of thymolphthalein in 50 mL of 95% ethanol and dilute the resulting solution to 100 mL with distilled water.

universal indicator, pH range 1–13

> Dissolve 0.02 g of methyl orange, 0.04 g of methyl red, 0.08 g of bromothymol blue, 0.10 g of thymol blue, and 0.02 g of phenolphthalein in 100 mL of 95% ethanol.

Yamada's universal indicator

> Dissolve 0.0025 g of thymol blue, 0.006 g of methyl red, 0.030 g of bromothymol blue, and 0.05 g of phenolphthalein in 50 mL of 95% ethanol. Add 0.01M sodium hydroxide until the mixture is green, and dilute the resulting solution to 100 mL with distilled water.

Preparation of
Acid and Base Stock Solutions

The following instructions are for the preparation of 1 liter of each of the stock solutions.

HAZARDS

Concentrated solutions ($> 2M$) of strong acids and bases are caustic and can irritate the skin. Therefore they should be handled with great care; wear plastic or rubber gloves while preparing these solutions. Spills should be covered with solid sodium bicarbonate ($NaHCO_3$) and then wiped up with plenty of water. When diluting concentrated acids, add the acid to the water to avoid spattering. Spattering occurs when water is added to a concentrated acid, because the dilution process releases enough heat to vaporize some of the water as it is added.

PREPARATIONS

acetic acid, $HC_2H_3O_2$

2M $HC_2H_3O_2$ Pour 115 mL of glacial (17.5M) $HC_2H_3O_2$ into 600 mL of distilled water, and dilute the resulting solution to 1.0 liter.

1M $HC_2H_3O_2$ Pour 57 mL of glacial (17.5M) $HC_2H_3O_2$ into 600 mL of distilled water, and dilute the resulting solution to 1.0 liter.

0.1M $HC_2H_3O_2$ Pour 5.7 mL of glacial (17.5M) $HC_2H_3O_2$ into 600 mL of distilled water, and dilute the resulting solution to 1.0 liter, *or* pour 100 mL of 1M $HC_2H_3O_2$ into 600 mL of distilled water and dilute the resulting solution to 1.0 liter.

Caution! Glacial acetic acid can irritate the skin, and its vapors are irritating to the eyes and respiratory system. It should be handled only in a well-ventilated area.

aqueous ammonia, NH_3

5M NH_3 Pour 330 mL of concentrated (15M) NH_3 into 400 mL of distilled water, and dilute the resulting solution to 1.0 liter.

1M NH_3 Pour 66 mL of concentrated (15M) NH_3 into 600 mL of distilled water and dilute the resulting solution to 1.0 liter.

Caution! Concentrated aqueous ammonia can irritate the skin, and its vapors are harmful to the eyes and mucous membranes. It should be handled only in a well-ventilated area.

hydrochloric acid, HCl

6M HCl   Slowly pour 500 mL of concentrated (12M) HCl into 400 mL of distilled water and dilute the resulting, cooled solution to 1.0 liter.

2M HCl   Slowly pour 170 mL of concentrated (12M) HCl into 600 mL of distilled water and dilute the resulting solution to 1.0 liter.

0.5M HCl   Slowly pour 42 mL of concentrated (12M) HCl into 600 mL of distilled water and dilute the resulting solution to 1.0 liter, *or* pour 250 mL of 2M HCl into 600 mL of distilled water and dilute the resulting solution to 1.0 liter.

0.1M HCl   Slowly pour 8.5 mL of concentrated (12M) HCl into 600 mL of distilled water and dilute the resulting solution to 1.0 liter, *or* pour 200 mL of 0.5M HCl into 600 mL of distilled water and dilute the resulting solution to 1.0 liter.

Caution! Hydrochloric acid can irritate the skin. Hydrochloric acid vapors are extremely irritating to the eyes and respiratory system. Therefore, it should be handled only in a well-ventilated area.

nitric acid, HNO_3

1M HNO_3 Pour 63 mL of concentrated (16M) HNO_3 into 600 mL of distilled water and dilute the resulting solution to 1.0 liter.

0.1M HNO_3 Pour 6.3 mL of concentrated (16M) HNO_3 into 600 mL of distilled water and dilute the resulting solution to 1.0 liter, *or* pour 100 mL of 1M HNO_3 into 600 mL of distilled water and dilute the resulting solution to 1.0 liter.

Caution! Concentrated nitric acid is both a strong acid and a powerful oxidizing agent. Contact with combustible materials can cause fires. Contact with the skin can result in severe burns. The vapor irritates the respiratory system, eyes, and other mucous membranes, and therefore, concentrated nitric acid should be handled only in a well-ventilated area.

phosphoric acid, H_3PO_4

0.1M H_3PO_4 Pour 6.8 mL of 85% (14.6M) H_3PO_4 into 600 mL of distilled water and dilute the resulting solution to 1.0 liter.

Caution! Phosphoric acid can cause severe burns to the eyes and mucous membranes and is irritating to the skin.

sodium hydroxide, NaOH

6M NaOH   Set a 2-liter beaker containing 600 mL of distilled water in a pan of ice water. While stirring the water, add six 40-gram portions (a total of 240 grams) of NaOH, waiting for each portion to dissolve before adding the next. Cool the resulting solution and dilute it to 1.0 liter.

2M NaOH   Set a 2-liter beaker containing 600 mL of distilled water in a pan of ice water. While stirring the water, add two 40-gram portions (a total of 80 grams) of NaOH, waiting for the first portion to dissolve before adding the second. Cool the resulting solution and dilute it to 1.0 liter.

1M NaOH   Set a 2-liter beaker containing 600 mL of distilled water in a pan of ice water. While stirring the water, add 40 g of NaOH. Cool the resulting solution and dilute it to 1.0 liter.

0.1M NaOH   Dissolve 4.0 g of NaOH in 600 mL of distilled water and dilute the resulting solution to 1.0 liter.

Caution! Solid sodium hydroxide and concentrated solutions can cause severe burns to the eyes, skin, and mucous membranes. Dust from solid sodium hydroxide is very irritating to the eyes and respiratory system.

sulfuric acid, H_2SO_4

   2M H_2SO_4 Set a 2-liter beaker containing 600 mL of distilled water in a pan of ice water. While stirring the water, slowly pour 110 mL of concentrated (18M) H_2SO_4 into the beaker. Cool the resulting solution and dilute it to 1.0 liter.

   0.1M H_2SO_4 Slowly pour 5.5 mL of concentrated (18M) H_2SO_4 into 600 mL of distilled water and dilute the resulting solution to 1.0 liter, *or* pour 50 mL of 2M H_2SO_4 into 600 mL of distilled water and dilute the resulting cooled solution to 1.0 liter.

Caution! Because sulfuric acid is both a strong acid and a powerful dehydrating agent, it must be handled with great care. The dilution of concentrated sulfuric acid is a highly exothermic process and releases sufficient heat to cause burns. Therefore, when preparing dilute solutions from the concentrated acid, always add the acid to the water, slowly and with stirring. The receiving beaker should be immersed in crushed ice if the concentration of the resulting solution will be more than 6M.

8.1

Colorful Acid-Base Indicators

Several indicator solutions are mixed with colorless solutions having a range of pH values, and the resulting colored solutions are displayed directly (Procedure A) or with an overhead projector (Procedure B). When it is mixed with six different colorless solutions, a single indicator produces six colors of the rainbow, which are observed directly (Procedure C) or with an overhead projector (Procedure D).

MATERIALS FOR PROCEDURE A

For preparation of indicator solutions, see pages 27–29.

50 mL each of pH-standard solutions with pH values of 1, 2, 3, 4, 5, 6, 7, 8, 9, 10, 11, 12, and 13 (For preparation, see Procedure A.)

10 mL of *one* of the following indicator solutions, which change color below pH 7: methyl violet, crystal violet, cresol red, thymol blue, erythrosin B, bromophenol blue, methyl orange, bromocresol green, methyl red, or bromocresol purple

10 mL of *one* of the following indicator solutions, which change color near pH 7: alizarin, bromothymol blue, litmus, phenol red, *m*-nitrophenol, or cresol red

10 mL of *one* of the following indicator solutions, which change color above pH 7: *o*-cresolphthalein, phenolphthalein, thymolphthalein, or alizarin yellow GG

lighted background (see Procedure A for description), or white background (e.g., 80-cm × 20-cm poster board)

13 100-mL beakers

13 labels for beakers

13 glass stirring rods

3 droppers

MATERIALS FOR PROCEDURE B

For preparation of indicator solutions, see pages 27–29.

overhead projector, with transparency and marker

15 mL each of pH standard solutions with pH values of 1, 2, 3, 4, 5, 6, 7, 8, 9, 10, 11, 12, and 13 (For preparation, see Procedure A.)

1 mL of *one* of the following indicator solutions, which change color below pH 7: methyl violet, crystal violet, cresol red, thymol blue, erythrosin B, bromo-

phenol blue, methyl orange, bromocresol green, methyl red, or bromocresol purple

1 mL of *one* of the following indicator solutions, which change color near pH 7: alizarin, bromothymol blue, litmus, phenol red, *m*-nitrophenol, or cresol red

1 mL of *one* of the following indicator solutions, which change color above pH 7: *o*-cresolphthalein, phenolphthalein, thymolphthalein, or alizarin yellow GG

13 60-mm petri dishes

13 glass stirring rods

25-mL graduated cylinder

3 droppers

MATERIALS FOR PROCEDURE C

For preparation of indicator solutions, see pages 27–29.
For preparation of stock solutions of acids and bases, see pages 30–32.

160 mL distilled water

71 mL 6M hydrochloric acid, HCl

70 mL 6M sodium hydroxide, NaOH

60 mL 1M ammonium chloride, NH_4Cl (To prepare 1 liter of solution, dissolve 53 g of NH_4Cl in 600 mL of distilled water and dilute the resulting solution to 1.0 liter.)

6 mL bromothymol blue indicator solution

6 100-mL beakers

6 labels for beakers

6 stirring rods

dropper

MATERIALS FOR PROCEDURE D

For preparation of indicator solutions, see pages 27–29.
For preparation of stock solutions of acids and bases, see pages 30–32.

overhead projector, with transparency and marker

160 mL distilled water

71 mL 6M hydrochloric acid, HCl

70 mL 6M sodium hydroxide, NaOH

60 mL 1M ammonium chloride, NH_4Cl (To prepare 1 liter of solution, dissolve 53 g of NH_4Cl in 600 mL of distilled water and dilute the resulting solution to 1.0 liter.)

6 mL bromothymol blue indicator solution

6 100-mL beakers

6 labels for beakers

6 stirring rods

dropper

PROCEDURE A

Preparation

Consult the table of indicators (Table 4 on pages 21–24) in the introduction to this chapter to find the pK_a values of the indicators. Select one indicator that has a pK_a below 7 (e.g., methyl orange), another indicator with a pK_a around 7 (e.g., bromothymol blue), and a third with one above 7 (e.g., phenolphthalein). Be sure that the pK_a values for the indicators you select differ from each other by at least 2 units.

If pH-standard solutions are not available, they can be prepared with commercial buffer tablets, or 500 mL of stock solution of each can be prepared as described in the following table and list [1].

Preparation of pH-Standard Solutions[a]

pH	Components
1	125 mL of 0.20M KCl and 335 mL of 0.20M HCl
2	125 mL of 0.20M KCl and 33 mL of 0.20M HCl
3	250 mL of 0.10M $KHC_8H_4O_4$ and 56 mL of 0.20M HCl
4	250 mL of 0.10M $KHC_8H_4O_4$ and 0.25 mL of 0.20M HCl
5	250 mL of 0.10M $KHC_8H_4O_4$ and 56 mL of 0.20M NaOH
6	250 mL of 0.10M KH_2PO_4 and 14 mL of 0.20M NaOH
7	250 mL of 0.10M KH_2PO_4 and 73 mL of 0.20M NaOH
8	250 mL of 0.10M KH_2PO_4 and 115 mL of 0.20M NaOH
9	250 mL of 0.025M $Na_2B_4O_7$ and 12 mL of 0.20M HCl
10	250 mL of 0.025M $Na_2B_4O_7$ and 27 mL of 0.20M NaOH
11	250 mL of 0.050M $NaHCO_3$ and 57 mL of 0.20M NaOH
12	125 mL of 0.20M KCl and 30 mL of 0.20M NaOH
13	125 mL of 0.20M KCl and 330 mL of 0.20M NaOH

[a]Combine the indicated components and dilute the resulting solutions to 500 mL with distilled water.

One liter of each of the component solutions can be prepared as follows:

0.20M potassium chloride, KCl: Dissolve 14.9 g of KCl in 500 mL of distilled water and dilute the resulting solution to 1.0 liter.

0.20M hydrochloric acid, HCl: Add 16.7 mL of concentrated (12M) HCl to 500 mL of distilled water and dilute the resulting solution to 1.0 liter.

0.10M potassium hydrogen phthalate, $KHC_8H_4O_4$: Dissolve 20.4 g of $KHC_8H_4O_4$ in 500 mL of distilled water and dilute the resulting solution to 1.0 liter.

0.20M sodium hydroxide, NaOH: Dissolve 8.0 g of NaOH in 500 mL of distilled water and dilute the resulting solution to 1.0 liter.

0.10M potassium dihydrogen phosphate, KH_2PO_4: Dissolve 13.6 g of KH_2PO_4 in 500 mL of distilled water and dilute the resulting solution to 1.0 liter.

0.025M borax, $Na_2B_4O_7\cdot10H_2O$: Dissolve 9.5 g of $Na_2B_4O_7\cdot10H_2O$ in 500 mL of distilled water and dilute the resulting solution to 1.0 liter.

0.050M sodium bicarbonate, $NaHCO_3$: Dissolve 4.2 g of $NaHCO_3$ in 500 mL of distilled water and dilute the resulting solution to 1.0 liter.

A lighted background or white backdrop can enhance the visibility of the colors of the indicators. A lighted background can be constructed as illustrated in Figure 1. This unit illuminates the solutions by means of 2-foot fluorescent lamps located behind and below translucent, white plastic panels. Switches allow the lamps below and those behind the solutions to be operated independently. Support for the plastic panels is provided by an aluminum box. The back and base of the unit should be completely enclosed and properly grounded to avoid electrical shocks and damage from spilled chemicals. The dimensions in the figure are only suggestions; the size can be adjusted to suit local applications.

Label each of 13 100-mL beakers with one each of the following labels: "pH 1," "pH 2," "pH 3," "pH 4," "pH 5," "pH 6," "pH 7," "pH 8," "pH 9," "pH 10," "pH 11," "pH 12," and "pH 13." Arrange the beakers in a row on the lighted background or before the white background in order of increasing pH. Pour 50 mL of the appropriate pH standard into each beaker and place a stirring rod in each.

Presentation

Place 6 drops of the indicator that changes color *below* pH 7 in the "pH 1" beaker and stir the mixture. Repeat this with the same indicator in each sequential beaker until the indicator produces a different color in the mixture. Place 6 drops of the indicator that changes color *near* pH 7 in the next beaker and stir the mixture. Repeat this with the next several beakers until this indicator changes color. Place 6 drops of the indicator

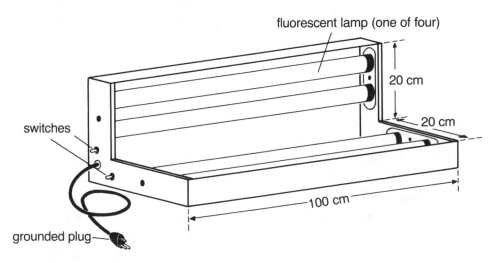

Lighted background. Top and front to be covered with 20-cm × 100-cm sheets of ¼-inch translucent white acrylic.

that changes color *above* pH 7 in the next beaker and stir the mixture. Repeat this with the remaining beakers.

PROCEDURE B

Preparation

Consult the table of indicators (Table 4 on pages 21–24) in the introduction to this chapter to find the pK_a values of the indicators. Select one indicator that has a pK_a below 7 (e.g., methyl orange), another indicator with a pK_a around 7 (e.g., bromothymol blue), and a third with one above 7 (e.g., phenolphthalein). Be sure that the pK_a values for the indicators you select differ from each other by at least 2 units.

If pH-standard solutions are not available, they can be prepared with commercial buffer tablets, or 500 mL of stock solution of each can be prepared as described in Procedure A.

Write the numbers 1 through 13 on the overhead-projector transparency so that each number is adjacent to a 60-mm petri dish when the dishes are arranged on the transparency. Place the transparency on the overhead projector and set the petri dishes on the transparency. Pour 15 mL of the appropriate standard solution, pH 1 through 13, into each dish.

Presentation

Place 4 drops of the indicator that changes color *below* pH 7 in the dish containing the pH 1 standard and swirl the dish to mix the solution. Repeat this with the same indicator in the dishes containing sequentially higher pH values until the indicator produces a different color in the mixture. Place 4 drops of the indicator that changes color *near* pH 7 in the next dish and swirl the mixture. Repeat this with the next several dishes until this indicator changes color. Place 4 drops of the indicator that changes color *above* pH 7 in the next dish and swirl the mixture. Repeat this with the remaining dishes.

PROCEDURE C [2]

Preparation

Label six 100-mL beakers with "6M HCl," "1M HCl," "0.1M HCl," "1M NH₄Cl," "1M NaOH," and "6M NaOH."

Pour 50 mL of distilled water into the beaker labelled "1M HCl" and 60 mL into the beaker labelled "0.1M HCl." Pour 60 mL of 6M HCl into the beaker labelled "6M HCl," 10 mL into the beaker labelled "1M HCl," and 1 mL into the beaker labelled "0.1M HCl," and stir each mixture. Each beaker now contains hydrochloric acid of the appropriate concentration.

Pour 60 mL of 6M NaOH into the beaker labelled "6M NaOH" and 10 mL into the beaker labelled "1M NaOH." Pour 50 mL of distilled water into the latter beaker and stir the mixture.

Pour 60 mL of 1M NH₄Cl into the remaining beaker.

Presentation

Arrange the beakers in a row in the order the labels are listed above. Put 10 drops of bromothymol blue indicator solution in each beaker and stir the mixtures. Each mixture will be a different color:

Solution	Color
6M HCl	red
1M HCl	orange
0.1M HCl	yellow
1M NH_4Cl	green
1M NaOH	blue
6M NaOH	purple

PROCEDURE D [2]

Preparation

Label six 100-mL beakers with "6M HCl," "1M HCl," "0.1M HCl," "1M NH_4Cl," "2M NaOH," and "6M NaOH."

Pour 50 mL of distilled water into the beaker labelled "1M HCl" and 60 mL of distilled water into the beaker labelled "0.1M HCl." Pour 60 mL of 6M HCl into the beaker labeled "6M HCl," 10 mL into the beaker labelled "1M HCl," and 1 mL into the beaker labelled "0.1M HCl" and stir each mixture. Each beaker now contains hydrochloric acid of the appropriate concentration.

Pour 60 mL of 6M NaOH into the beaker labelled "6M NaOH" and 10 mL into the beaker labelled "1M NaOH." Pour 50 mL of distilled water into the latter beaker and stir the mixture.

Pour 60 mL of 1M NH_4Cl into the remaining beaker.

Write the labels "6M HCl," "1M HCl," "0.1M HCl," "1M NH_4Cl," "1M NaOH," and "6M NaOH" on a transparency so that each label is adjacent to a beaker when the beakers are arranged in a circle on the transparency.

Presentation

Arrange the beakers in a circle on the transparency on the overhead projector, with each beaker adjacent to its appropriate label. Put 10 drops of bromothymol blue indicator solution in each beaker and stir the mixtures. Each mixture will be a different color as listed in the chart in Procedure C.

HAZARDS

Hydrochloric acid can irritate the skin. Hydrochloric acid vapors are extremely irritating to the eyes and respiratory system.

Solid sodium hydroxide and concentrated solutions can cause severe burns to the eyes, skin, and mucous membranes. Dust from solid sodium hydroxide is very irritating to the eyes and respiratory system.

If methyl violet and/or *m*-nitrophenol are used, see the cautions included with their Preparation instructions in the introduction to this chapter.

DISPOSAL

The waste solutions should be combined to neutralize them and then flushed down the drain with water.

DISCUSSION

A chemical indicator is a substance which gives a visible sign, usually a color change, of the presence or absence of a threshold concentration of another chemical species. Certain indicators respond to the concentration of hydrogen ion (or hydroxide ion) in a solution (i.e., to the pH of the solution) and are therefore called pH indicators. One pH indicator, methyl red, imparts a red color to an acidic solution; if a base is added to the solution, the solution remains red until the acid has been neutralized, at which point the color suddenly turns yellow. Procedures A and B of this demonstration show, not only that different indicators undergo different color changes, but also that they do so at different pH values. (Procedures A and B do not include a display of each indicator at all pH values used, because to do this would require a large number of beakers.) A certain threshold value of hydrogen-ion concentration must be reached before the color change occurs, and this threshold value differs from one indicator to another. Procedures C and D show that a single indicator can undergo more than one color change as the pH changes.

Each pH indicator has a characteristic pH range through which it undergoes its color change. The range in which the color change occurs depends on several factors, including the acidity of the indicator itself and the relative intensity of its two colors, as discussed in the introduction to the chapter.

As illustrated in Procedures C and D, some indicators undergo more than one color change. For example, bromothymol blue has a different color in each of the solutions used: red in 6M HCl, orange in 1M HCl, yellow in 0.1M HCl, green in 1M NH_4Cl, blue in 1M NaOH, and purple in 6M NaOH. Not all of these colors represent different forms of the indicator. Some, namely orange, green, and purple, are combinations of the colors of two forms: green results from the combination of yellow and blue, orange from the combination of red and yellow, and purple from blue and red. Bromothymol blue is a sulfonephthalein indicator, which takes four different forms as the pH of its solution varies [3]. These forms are represented by the structures shown on page 40.

red

yellow

blue

red

In 6M HCl, with a pH less than 0 and a very high concentration of hydrogen ions, bromothymol blue takes the red form on the top left, where a hydrogen ion is attached to one of the hydroxyl groups. In 1M HCl the concentration of hydrogen ions is lower than in 6M HCl, and bromothymol blue takes the second form, the yellow form, in which both hydroxyl groups are free of hydrogen ions. At the very low hydrogen ion concentrations in 1M NaOH, one of the hydrogens of the hydroxyl groups ionizes, leaving bromothymol blue in the blue form. In 6M NaOH, the hydrogen ion concentration is even lower, and the second hydroxyl group begins to lose its hydrogen. Therefore, some bromothymol blue takes the red form on the bottom right, and the solution appears purple.

REFERENCES

1. R. C. Weast, Ed., *CRC Handbook of Chemistry and Physics,* 66th ed., CRC Press: Boca Raton, Florida (1985).
2. D. Kolb, *J. Chem. Educ.* 64:348 (1987).
3. O. Tomicek, *Chemical Indicators,* Butterworths Scientific Publishers: London (1951).

8.2

Rainbow Colors
with Mixed Acid-Base Indicators

When a colorless liquid is poured into six beakers, the liquid in each beaker turns a different color of the rainbow. The colors disappear and reappear when other colorless liquids are added to the beakers (Procedure A) [1, 2]. A mixture of indicators is prepared to have a different color at different pH values, and the colors of the mixture are compared with those of each of its components at these pH values, displayed directly (Procedure B) or by overhead projector (Procedure C).

MATERIALS FOR PROCEDURE A

30 mL viscous acid solution (To prepare 30 mL of viscous acid solution, combine 10 mL of concentrated sulfuric acid and 20 mL of glycerin. Stir the mixture cautiously; it will become hot. Allow the mixture to cool.)

30 mL 9M sodium hydroxide, NaOH (To prepare this solution, gradually add 11 g of NaOH to 20 mL of distilled water, stirring during the addition until all of the NaOH has dissolved. The mixture will become hot; allow it to cool.)

3 liters 0.012M sodium hydroxide, NaOH (To prepare 3 liters of solution, dissolve 1.5 g of NaOH in 3 liters of distilled water in a 3-liter Erlenmeyer flask.)

500 mL 0.050M sulfuric acid, H_2SO_4 (To prepare 1 liter of solution, slowly pour 2.7 mL of concentrated [18M] H_2SO_4 into 600 mL of distilled water, and dilute the resulting solution to 1.0 liter.)

100 mL distilled water

2 drops "violet" indicator solution (To prepare 30 mL of solution, dissolve 0.9 g of phenolphthalein and 0.4 g of thymolphthalein in 30 mL of 95% ethanol.)

1 drop "blue" indicator solution (To prepare 30 mL of solution, dissolve 1.5 g of thymolphthalein in 30 mL of 95% ethanol.)

1 drop "green" indicator solution (To prepare 30 mL of solution, dissolve 0.6 g of thymolphthalein and 6.0 g of m-nitrophenol in 30 mL of 95% ethanol.)

1 drop "yellow" indicator solution (To prepare 30 mL of solution, dissolve 6.0 g of m-nitrophenol in 30 mL of 95% ethanol.)

1 drop "orange" indicator solution (To prepare 30 mL of solution, dissolve 0.45 g of phenolphthalein and 6.0 g of m-nitrophenol in 30 mL of 95% ethanol.)

1 drop "red" indicator solution (To prepare 30 mL of solution, dissolve 1.5 g of phenolphthalein and 3.0 g of m-nitrophenol in 30 mL of 95% ethanol.)

2 30-mL dropping bottles

50-mL buret, with stand

5.0-mL volumetric pipette

250-mL beaker

50-mL graduated cylinder

6 500-mL beakers, graduated

2 3-liter Erlenmeyer flasks

6 glass stirring rods

MATERIALS FOR PROCEDURE B

For preparation of indicator solutions, see pages 27–29.

100 mL each of pH-standard solutions with pH of 2, 4, 6, 8, and 10 (See Procedure A of Demonstration 8.1.)

1 mL methyl orange indicator solution

1 mL methyl red indicator solution

1 mL bromothymol blue indicator solution

1 mL thymol blue indicator solution

1 mL phenolphthalein indicator solution

25 50-mL beakers

25 labels for 50-mL beakers

6–10 100-mL beakers, with a stirring rod for each

5 droppers

MATERIALS FOR PROCEDURE C

For preparation of indicator solutions, see pages 27–29.

overhead projector, with transparency and marker

50 mL each of pH-standard solutions with pH of 2, 4, 6, 8, and 10 (See Procedure A of Demonstration 8.1.)

1 mL methyl orange indicator solution

1 mL methyl red indicator solution

1 mL bromothymol blue indicator solution

1 mL thymol blue indicator solution

1 mL phenolphthalein indicator solution

25 50-mL beakers

25 labels for 50-mL beakers

6–10 100-mL beakers, with stirring rod for each

5 droppers

PROCEDURE A [1, 2]

Preparation

Pour 30 mL of viscous acid solution into one of the 30-mL dropping bottles and 30 mL of 9M NaOH into the other.

Fill the buret with 0.012M NaOH solution. Pipette 5.0 mL of the 0.050M H_2SO_4 into the 250-mL beaker and add 100 mL distilled water and 1 drop of the violet indicator solution. Titrate the H_2SO_4 with the 0.012M NaOH to the first appearance of a persistent color in the beaker. This will take about 40 mL of NaOH. To compute the volume of H_2SO_4 needed to neutralize 350 mL of NaOH, divide 1750 by the actual volume of NaOH used.†

Pour the computed volume of H_2SO_4 (about 44 mL) into each of the six 500-mL beakers. Place the remaining acid in one of the 3-liter flasks and locate it out of view in an accessible location. Arrange the beakers in a row. To the first beaker add 1 drop of the violet indicator solution, to the second 1 drop of blue, to the third 1 drop of green, to the fourth yellow, the fifth orange, and the last red. Stir the mixtures; they will all be colorless. Turn the beakers so their graduations face you. Pour the remaining 0.012M NaOH into the other 3-liter Erlenmeyer flask and set the flask next to the row of beakers. Place nearby the six stirring rods and the dropper bottles of viscous acid solution and of 9M NaOH.

Presentation

Slowly pour 0.012M NaOH solution from the flask into each beaker until the liquid reaches the 200 mL mark. Stir the contents of each beaker. The mixtures will be colorless. While stirring each solution, slowly pour more 0.012M NaOH into each beaker until the liquid in each becomes colored, but do not add an excess. (The volume of the liquid in the beakers will be about 400 mL.) Conceal the now nearly empty 3-liter flask near the hidden flask containing H_2SO_4.

Without stirring the contents of the beakers, add 2 drops of the viscous acid solution to each. Stir each of the solutions vigorously until it becomes colorless. Again without stirring, add 3 drops of 9M NaOH to each beaker. Stir each of the solutions vigorously until the color reappears. Set the flask of 0.050M H_2SO_4 next to the row of beakers. Pour the contents of the six beakers, two at a time, into the flask of H_2SO_4. All of the colors will disappear.

PROCEDURE B

Preparation

Label the 25 50-mL beakers, marking 5 of them with "pH 2" labels, 5 with "pH 4," 5 with "pH 6," and so on with "pH 8," and "pH 10." Pour 10 mL of the appropriate pH-standard solution into each beaker. Arrange the beakers in five groups, so all beakers containing the same pH standard are grouped together.

Place a stirring rod in each of the 100-mL beakers.

---

† If y mL of NaOH is required to neutralize 5.0 mL of H_2SO_4, the ratio of their volumes is $5/y$. If x mL of H_2SO_4 is required to neutralize 350 mL of NaOH, then $x/350 = 5/y$, and $x = 350 \cdot 5/y = 1750/y$.

Presentation

Place 2 drops of methyl orange indicator in one beaker of each group and swirl each beaker to mix its contents. Record the color of the solution at each pH. Repeat this with the other indicators: methyl red, bromothymol blue, thymol blue, and phenolphthalein. When completed, the results of this process can be displayed in a chart of color as it is affected by different pH values.

Indicator	Color at pH:				
	2	4	6	8	10
methyl orange	red	orange	yellow	yellow	yellow
methyl red	red	red	yellow	yellow	yellow
bromothymol blue	yellow	yellow	yellow	blue	blue
thymol blue	orange	yellow	yellow	yellow	blue
phenolphthalein	colorless	colorless	colorless	colorless	pink
combination	red-orange	orange	yellow	green	purple

Pour all five of the pH 2 solutions into a 100-mL beaker. Record the color of this combined mixture. Repeat this with the other pH standards. Each of the combined solutions will be a different color (also shown in the color chart).

Place 2 drops of each indicator together in one 100-mL beaker. This mixture forms a "universal" indicator. Pour 50 mL of one of the pH-standard solutions into this beaker and stir the mixture. Referring to the chart, use the color of the mixture to ascertain the pH of the standard. This can be repeated with the other standard solutions.

PROCEDURE C

Preparation

Label the 25 50-mL beakers, marking 5 of them with "pH 2" labels, 5 with "pH 4," 5 with "pH 6," and so on with "pH 8" and "pH 10." Write one of each of the pH labels on the overhead-projector transparency so that a beaker can be positioned adjacent to each label. Pour 10 mL of the appropriate pH-standard solution into each beaker. Arrange the beakers in five groups, so all beakers containing the same pH standard are grouped together.

Place a stirring rod in each of the 100-mL beakers.

Presentation

Place the transparency on the overhead projector, and set one beaker from each pH group next to its appropriate label. Place 2 drops of methyl orange indicator in one beaker of each group and swirl each beaker to mix its contents. Record the color of the solution at each pH. Repeat this with the other indicators: methyl red, bromothymol blue, thymol blue, and phenolphthalein. When completed, the results of this process with these five indicators can be displayed in a chart of color as it is affected by different pH values. Such a chart is presented in Procedure B.

Place all five pH 2 beakers on the overhead projector along with a 100-mL beaker. Pour all five of the pH 2 solutions into a 100-mL beaker. Record the color of this combined mixture. Repeat this with the other pH standards. Each of the combined solutions will be a different color.

Set a 100-mL beaker on the overhead projector. Place 2 drops of each indicator together in the 100-mL beaker. This mixture forms a "universal" indicator. Pour 50 mL of one of the pH-standard solutions into this beaker and stir the mixture. Referring to the chart, use the color of the mixture to ascertain the pH of the standard. This can be repeated with the other standard solutions.

HAZARDS

Because sulfuric acid is both a strong acid and a powerful dehydrating agent, it must be handled with great care. The dilution of concentrated sulfuric acid is a highly exothermic process and releases sufficient heat to cause burns. Therefore, when preparing dilute solutions from the concentrated acid, always add the acid to the water, slowly and with stirring. The receiving beaker should be immersed in crushed ice if the concentration of the resulting solution will be more than 6M. Spills should be neutralized by covering with solid sodium bicarbonate ($NaHCO_3$), then be wiped up with plenty of water.

Solid sodium hydroxide and concentrated solutions can cause severe burns to the eyes, skin, and respiratory system. Dust from solid sodium hydroxide is very irritating to the eyes and respiratory system.

m-Nitrophenol is a poison when ingested and is toxic when absorbed through the skin. Its oral LD50† in mice is 1.07 g/kg.

Ethanol is flammable and should be kept away from open flames.

DISPOSAL

The combined solution from Procedure A should be flushed down the drain with water. Any "viscous acid" that will not be stored should be disposed of by pouring it into 500 mL of water, stirring the mixture, and adding solid sodium bicarbonate ($NaHCO_3$) until no more fizzing occurs and the acid is neutralized; then flush the solution down the drain with water. The remaining 9M NaOH solution should be poured into 500 mL of water, the mixture stirred and flushed down the drain with water.

The solutions from Procedures B and C should be combined to neutralize them, and the resultant mixture flushed down the drain with water.

DISCUSSION

This demonstration shows that a mixture of indicators can produce a color change different from that obtained with the individual indicators. In Procedure A, mixtures of indicators are used to cause the colors of the rainbow to appear and disappear. This procedure is known by some as the "magic pitcher" demonstration. In Procedures B and C, the colors obtained by mixing indicators are compared with the colors of the individual indicators, demonstrating the preparation of a "universal" indicator.

The three indicators used in Procedure A of this demonstration are all colorless in

† See the note on page 28.

acidic solution, but each produces a different color in basic solution. The indicators, their colors, and their pK$_a$ values are given in the following chart [3]:

Indicator	Basic color	pK$_a$
phenolphthalein	red-purple	9.4
m-nitrophenol	yellow	8.3
thymolphthalein	blue	10.0

Because these are close to the three primary colors, virtually any color of the rainbow can be produced by some combination of these indicators in basic solution, as described in Procedure A. The indicator solutions used in this demonstration are very concentrated, more concentrated than standard indicator solutions. These concentrated solutions are used to produce colors more intense than those produced by standard indicator solutions.

At the start of Procedure A, each beaker contains one of the six rainbow color indicator solutions along with some sulfuric acid. Because the contents of the beakers are acidic, the indicators are colorless. Then the beakers are half filled with sodium hydroxide solution. The amount of acid in the beaker is more than the amount of base added, and the indicators remain colorless. With the second addition of NaOH, an excess of base results, the solutions become alkaline, and the indicators become colored. Then several drops of viscous acid solution, a very dense solution of sulfuric acid in glycerin, are added to each beaker. Because the acid is so dense, it sinks to the bottom without mixing with the solutions. When the mixtures are stirred, the acid becomes evenly dispersed in the solution, the base is neutralized, and the indicators return to colorless. When drops of 9M NaOH, which is also dense, are added to each beaker, these too sink to the bottom without mixing. When the mixtures are stirred, the NaOH disperses, the solution becomes basic, and the colors of the indicators return. Then the contents of the beakers are combined with sulfuric acid, which makes the mixture acidic, and the colors disappear.

Although Procedure A can be used as a "magic" show, it has great pedagogic potential. It can be quite instructive to ask students to explain what is happening at each step. It shows that there are several indicators that are colorless in acid solution, and illustrates how several indicators can be combined to produce different colors, as is done in universal indicators.

Procedures B and C provide a more direct demonstration of how several indicators can be combined to produce a "universal" indicator. They demonstrate the colors of five individual indicators at a range of pH values from 2 through 10. Then, the same five indicators are grouped by these pH values, and the indicators of each group are combined, illustrating that the color is different at each of the pH values. An indicator whose color differs from one pH value to another over the entire pH range is referred to as a universal indicator.

REFERENCES

1. J. W. Wilson, J. W. Wilson, Jr., and T. F. Gardner, *Chemical Magic,* Chemical Magic: Los Alamitos, California (1959).
2. B. Hutton, *J. Chem. Educ.* 61:172 (1984).
3. J. A. Dean, Ed., *Lange's Handbook of Chemistry,* 13th ed., McGraw-Hill Book Co.: New York (1985).

8.3

Invisible Painting

A pattern is "painted" on a large sheet of paper with colorless indicator solution and the pattern is revealed when the paper is sprayed with a developing solution of sodium hydroxide.

MATERIALS

0.05 g phenolphthalein or *o*-cresolphthalein

60 mL 95% ethanol, C_2H_5OH

0.05 g thymolphthalein

2 g *m*-nitrophenol

0.2 g sodium hydroxide, NaOH

500 mL water

spray bottle (e.g., from window cleaner)

white poster board or paper, ca. 60 cm square or larger

3 paint brushes (e.g., 1 inch wide, flat)

pencil (optional)

PROCEDURE

Preparation

For red "paint," dissolve 0.05 g of phenolphthalein or *o*-cresolphthalein in 20 mL of 95% ethanol. For blue, dissolve 0.05 g of thymolphthalein in 20 mL of 95% ethanol. For yellow, dissolve 2 g of *m*-nitrophenol in 20 mL of 95% ethanol.

Prepare the developing solution by dissolving 0.2 g of NaOH in 500 mL of water; this solution is 0.01M NaOH. Fill the spray bottle with water and pump the sprayer to rinse it with water and clear it of any residue from prior use. Then fill the spray bottle with 0.01M NaOH. Pump the sprayer several times to fill it with solution.

"Paint" a picture or message on the poster board or large sheet of paper using a separate brush for each color. Lightly drawing the picture on the board with pencil before painting will make painting easier, because at this stage the painting is invisible. Colors other than red, blue, and yellow can be obtained by painting over a previously painted region with a different color. For example, green can be produced by painting yellow over a region that was previously painted blue. As with any endeavor, practice and experimentation will improve the results and reveal new techniques.

Presentation

Display the prepared board and spray it lightly with the developing solution. **(Caution: do not allow the spray to fall on yourself or any observers—it is irritating to the skin and harmful to the eyes.)** The painted designs will appear. Avoid spraying too heavily, or the colors will run. Over a period of about a minute, the blue will fade. Spraying the painting again will regenerate the blue.

HAZARDS

m-Nitrophenol is a poison when ingested and is toxic when absorbed through the skin. The oral LD50† in mice is 1.07 g/kg.

Dust from solid sodium hydroxide is very irritating to the skin, eyes, and respiratory system. Avoid breathing mist from the developing solution. Although it is rather dilute, prolonged exposure to it can irritate mucous membranes. If the developing solution should enter the eyes, flush them with water for 10 minutes and then get medical attention.

Ethanol is flammable and should be kept away from open flames.

DISPOSAL

The surplus "paints" can either be stored in sealed bottles for use in repeated performances of this demonstration or be flushed down the drain with water. The board or paper should be discarded in a solid-waste receptacle. The spray bottle should be emptied, and the solution either stored for future use or discarded by flushing down the drain with water. The sprayer should be rinsed by pumping about 25 mL of water through it.

DISCUSSION

This demonstration uses acid-base indicators that are colorless when acidic or neutral but colored when sufficiently basic. A painting or message is formed on a blank sheet of poster board or paper with concentrated solutions of the colorless indicators. When a basic solution is sprayed on the white board, the painting or message appears, because the basic solution causes the indicators to take on their basic colors.

The indicators used are listed in the following chart, along with their colors and pK_a values. At pH values below their pK_a values, all of these indicators are colorless. At pH values above the pK_a, the indicators become colored.

Indicator	Basic color	pK_a
phenolphthalein	red-purple	9.2
o-cresolphthalein	red	9.1
thymolphthalein	blue	10.1
m-nitrophenol	yellow	8.3

† See the note on page 28.

The image produced by the colored form of these indicators is not very long-lasting. In fact, the blue produced by thymolphthalein fades quite quickly, and in less than a minute may disappear altogether. Eventually, all of the colors will fade. The fading is caused by carbon dioxide in the atmosphere. Carbon dioxide, which dissolves in water to form carbonic acid, gradually neutralizes the colored indicators returning them to their colorless states. This occurs more rapidly with thymolphthalein than with the other indicators, because thymolphthalein has a higher pK_a value. Therefore, it takes less acid (carbon dioxide) to return it to its colorless state. The acidic properties of carbon dioxide are also shown in Demonstration 8.14 and in Demonstration 6.2 in Volume 2.

8.4

Acid-Base Indicators
Extracted from Plants

A purple liquid is extracted from red cabbage, and this liquid changes color when added to various substances found around the house (Procedure A). A colored liquid is extracted from plant material, and this extract has different colors at different pH values, as observed directly (Procedure B) or by overhead projection (Procedure C). "Indicator paper" is prepared by soaking a sheet of paper in the plant extract; the paper changes color when it is dampened with solutions of various pH values (Procedure D). A multicolored design is created when a sheet of paper treated with plant extract is "painted" with colorless liquids (Procedure E).

MATERIALS FOR PROCEDURE A

Provide household materials in their original containers.

ca. 100 g red cabbage (half of a small head is sufficient)

ca. 2 liters distilled water

125 mL vinegar

125 mL laundry ammonia

5 mL (1 teaspoon) baking soda

125 mL colorless carbonated beverage (e.g., lemon-lime flavor)

5 mL (1 teaspoon) laundry detergent

125 mL milk

knife to cut cabbage

electric blender or food processor

kitchen sieve or colander

1-liter beaker

7 (or more) 250-mL beakers

spoon or stirring rod

MATERIALS FOR PROCEDURE B

For preparation of stock solutions of acids and bases, see pages 30–32.

100 g plant material, such as:

   vegetables (e.g., red cabbage, beets, red onions, radishes, rhubarb) fresh or frozen fruit (e.g., blueberries, cherries, red grapes)

flowers (e.g., day lilies, roses)

plants (e.g., carrot greens, tomato plants, black tea)

100 mL 95% ethanol or distilled water

600 mL distilled water

200 mL 2M hydrochloric acid, HCl

200 mL 2M sodium hydroxide, NaOH

electric blender or food processor

kitchen sieve or colander

2 1-liter beakers

50-mL graduated cylinder

glass stirring rod

MATERIALS FOR PROCEDURE C

For preparation of stock solutions of acids and bases, see pages 30–32.

overhead projector

100 g plant material, such as:

vegetables (e.g., red cabbage, beets, red onions, radishes, rhubarb)

fresh or frozen fruit (e.g., blueberries, cherries, red grapes)

flowers (e.g., day lilies, roses)

plants (e.g., carrot greens, tomato plants, black tea)

100 mL 95% ethanol or distilled water

10 mL distilled water

20 mL 2M hydrochloric acid, HCl

20 mL 2M sodium hydroxide, NaOH

electric blender or food processor

kitchen sieve or colander

1-liter beaker

50-mL graduated cylinder

50-mL beaker

MATERIALS FOR PROCEDURE D

For preparation of stock solutions of acids and bases, see pages 30–32.

100 g plant material (See Materials for Procedure B.)

100 mL 95% ethanol or distilled water

10 mL each of buffer solutions with pH of 1, 3, 5, 7, 9, 11, and 13 (For preparation, see Procedure A of Demonstration 8.1.)

10 mL 2M hydrochloric acid, HCl

10 mL 2M sodium hydroxide, NaOH

electric blender or food processor

kitchen sieve or colander

1-liter beaker

sheet of white cotton cloth or filter paper, 50 cm × 50 cm

2 ring stands

100 cm adhesive tape (optional)

9 20-mL test tubes (optional)

meter stick (optional)

MATERIALS FOR PROCEDURE E

For preparation of stock solutions of acids and bases, see pages 30–32.

100 g red cabbage

100 mL distilled water

10 mL each of buffer solutions with pH of 1, 5, 7, 11, and 13 (For preparation, see
      Procedure A of Demonstration 8.1.)

10 mL 2M sodium hydroxide, NaOH

knife to cut cabbage

electric blender or food processor

kitchen sieve or colander

1-liter beaker

sheet of white cotton cloth or filter paper, 50 cm × 50 cm

2 ring stands

6 50-mL beakers

6 small paint brushes

PROCEDURE A

Preparation and Presentation

Prepare red cabbage extract as follows. Cut the red cabbage into 1-inch cubes.
Place the cabbage cubes in the blender and add enough water to cover the cabbage.
Blend the mixture until the cabbage has been chopped into uniformly tiny pieces. Using
the sieve, strain the liquid from the mixture into the 1-liter beaker, pressing the cabbage
to speed the straining. The strained liquid is the red cabbage extract.

Pour 125 mL of vinegar into one of the 250-mL beakers. Add about 5 mL of red
cabbage extract and stir the mixture. Record the color of the mixture.

Pour 125 mL of laundry ammonia into another 250-mL beaker. Add about 5 mL of
red cabbage extract and stir the mixture. Record the color of the mixture.

Place about 5 mL (1 teaspoon) of baking soda in another beaker. Add 125-mL of distilled water and stir the mixture until the soda has dissolved. Add about 5 mL of red cabbage extract and stir the mixture. Record the color of the mixture.

Pour 125 mL of colorless carbonated beverage into another beaker. Add about 5 mL of red cabbage extract and stir the mixture. Record the color of the mixture.

Dissolve about 5 mL of laundry detergent (liquid or solid) in 125 mL of distilled water. Add about 5 mL of red cabbage extract and stir the mixture. Record the color of this mixture.

Pour 125 mL of milk into another 250-mL beaker. Add about 5 mL of red cabbage extract and stir the mixture. Record the color of the mixture.

Test other household materials in a similar fashion. Solids and viscous liquids should be tested by dissolving them first in distilled water as described for baking soda. Substances that can be tested include sugar, lemon juice, table salt, hair shampoo, hair rinse, milk of magnesia, antacid tablets, and aspirin. Use a different beaker for each substance so the colors can be compared.

PROCEDURE B [1, 2]

Preparation and Presentation

Place 100 g of plant material in the blender and add 100 mL of either ethanol or distilled water. **Caution: Do not use ethanol in a food processor—see Hazards section.** (For most plant materials, ethanol will yield a clear extract, whereas water produces a cloudy one. Refer to the table in the Discussion section.) Blend the mixture until the solid particles have been chopped into fine pieces. Pour the mixture from the blender into the sieve and strain at least 75 mL of liquid (plant indicator solution) into the 1-liter beaker.

Pour about 600 mL of distilled water and about 10 mL of the plant indicator solution into another 1-liter beaker, and stir the mixture with the stirring rod. To see the acid color of the indicator, slowly pour 25 mL of 2M HCl from the graduated cylinder into the solution while stirring it. Then, to see the base color, slowly stir in 50 mL of 2M NaOH. Continue to shift the color back and forth by alternately adding acid and base.

PROCEDURE C [1, 2]

Preparation and Presentation

Place 100 g of plant material in the blender and add 100 mL of either ethanol or distilled water. **Caution: Do not use ethanol in a food processor—see Hazards section.** (For most plant materials, ethanol will yield a clear extract, whereas water produces a cloudy one. Refer to the table in the Discussion section. For overhead projection, a clear extract is required.) Blend the mixture until the solid particles have been chopped into fine pieces. Pour the mixture from the blender into the sieve and strain at least 75 mL of liquid (plant indicator solution) into a 1-liter beaker.

Place a 50-mL beaker on the overhead projector and add 5 mL of the plant extract. Pour 10 mL of distilled water into the beaker and swirl the beaker to mix the solution. While swirling the beaker, add 2M HCl drop by drop until the color changes. Continue swirling the beaker while adding 2M NaOH drop by drop. Continue to shift the color back and forth by alternately adding acid and base.

PROCEDURE D

Preparation

At least one day before the demonstration is to be presented, prepare an extract as described in the first paragraph of Procedure B. Soak a cloth or large piece of filter paper in the extracted liquid and hang the cloth or paper until it is dry.

Suspend the dried cloth or paper like a curtain between two ring stands.

Presentation

Dribble 10 mL of each of the buffer solutions, in separate streaks, as well as 10 mL of 2M NaOH and 10 mL of 2M HCl, down the vertically mounted, extract-stained cloth or paper. (One way to accomplish this is by attaching nine upright test tubes with adhesive tape at intervals along with a 30-cm portion of a meter stick. Starting at one end of the row of test tubes, pour 10 mL of 2M HCl into the first tube, 10 mL of pH 1 buffer into the second, 10 mL of pH 3 buffer in the third, and so forth through 10 mL of pH 13 buffer solution in the eighth tube. Pour 10 mL of 2M NaOH into the ninth tube. Then tip the meter stick so that the solutions streak down the extract-stained material.)

PROCEDURE E

Preparation

At least one day before the demonstration is to be presented, prepare a red cabbage extract as described in the first paragraph of Procedure A. Then prepare the cloth or paper as described in the Preparation section of Procedure D.

Pour about 10 mL of each buffer solution of pH 1, 5, 7, 11, and 13 and 10 mL of 2M NaOH into separate 50-mL beakers.

Presentation

Dip a paint brush in one of the colorless buffer solutions or in the 2M NaOH, and "paint" with the brush on the vertically mounted, extract-stained cloth or paper. The color produced on the cloth when touched by the brush depends on the solution on the brush, as indicated below.

Solution	Color
pH 1	red
pH 5	violet
pH 7	blue
pH 11	blue-green
pH 13	yellow-green
2M NaOH	yellow

A multicolored picture can be created by painting with the brushes, dipping each in a different one of the colorless buffer solutions or 2M NaOH.

HAZARDS

Care should be taken to avoid spilling ethanol near the blender. Sparks in the electric motor of the blender can ignite ethanol vapor. Ethanol should not be used in a food processor because vapors are free to enter the motor housing from the bowl of the processor; sparks from the motor can ignite ethanol vapors.

Concentrated solutions of sodium hydroxide can cause severe burns to the eyes, skin, and mucous membranes.

Hydrochloric acid can irritate the skin. Its vapors are extremely irritating to the eyes and respiratory system.

DISPOSAL

The waste solutions should be flushed down the drain with water.

DISCUSSION

An extract of red cabbage is prepared in Procedure A and this extract changes color when it is mixed with different household substances. In Procedures B and C, a plant extract is prepared and its color is changed reversibly by the alternating addition of acid and base. In Procedure D, the plant extract is used to prepare test paper, and solutions of various pH values are dribbled onto the paper to show the effect of each pH on the color of the paper. Procedure E involves "painting" with colorless buffer solutions on a sheet of paper that has been treated with red cabbage extract.

For well over 300 years investigators have been studying the effects of the addition of acids and bases to colored plant extracts. The ability of certain substances to change the colors of plant extracts was one of the earliest defining characteristics of acids. In 1664 Robert Boyle published in *The Experimental History of Colours* that the extracts of certain plants such as red roses and brazil wood changed color reversibly when made alternately basic and acidic. These extracts could be used as acid-base indicators. In 1670–1671 DuClos was perhaps the first to use "turnesole," an early impure form of litmus. Over a hundred years later James Watt complained that red rose paper didn't keep its color for more than a few months and suggested the use of red cabbage indicator paper. In 1801 Vauquelin spoke of using litmus and violets, and Welter advocated using litmus and radishes; Lampadius reported success using litmus, curcuma, red cabbage, alkanet, rhubarb, violets, columbine, and red roses.

The table lists the variations in the colors of several plant extracts with changes in their pH values. Most plant extracts will undergo at least a faint color change if the pH is changed sufficiently. Not all such color changes are reversible, so not all extracts are, strictly speaking, pH indicators. However, the extracts listed in the table do undergo reversible changes, although the initial color change may not be completely reversible. This is most likely due to the irreversible destruction of some pigment in the extract when the pH is changed. Subsequent changes in pH will result in reversible color

Colors[a] of Selected Plant Extracts at Various pH Values

Plant extract	Solvent	Initial appearance of extract[b]	Color at pH:												
			1	2	3	4	5	6	7	8	9	10	11	12	13
Beets	ethanol	dk rd	vt	rd-vt						rd-vt					br	yl-gn
Beets	water	dk rd	rd-vt				rd		rd					vt		yl-gn
Blackberry juice	water	dk rd			rd					vt	rd	br	vt		bl	gn
Blueberry juice	water	dk rd			rd					vt	br	vt			bl	bl-gn
Carrot greens	ethanol	*dk gn*	yl-br						gn-yl							
Carrot greens	water	*yl-br*	lt yl-br												dk yl-br	gn
Cherries	ethanol	*rd*		pk											br	
Cherries	water	*rd-br*	rd-br													gn-br
Daisy top	ethanol	*yl-gn*	pa yl												dk yl	
Daisy top	water	*br*	pa br												dk br	
Day lily	ethanol	*gn-br*	pk		pa br		pa yl					gn				yl-gn
Day lily	water	*br*	pk		pa br	rd				dk br	yl-gn	pa yl			dk yl	
Grape juice	water	dk rd	rd		dk vt			pa bl		vt	rd-bl		vt		bl	gn-bl
Red cabbage	ethanol	rd-br	rd						vt	gn	br	bl-gn				yl
Red cabbage	water	*vt*	rd		vt	rd-vt		vt	bl	gn-bl		gn				yl-gn
Red onions	ethanol	dk vt	rd			pa pk				pa yl	yl	gn				yl
Red onions	water	*rd-vt*	pk			pa pk				pa gn	yl	gn				yl
Radishes	ethanol	*pk*	pk		pa pk					cl			vt		gn	yl
Radishes	water	*pk*	pk		pa pk					cl			vt		gn	yl
Rhubarb	ethanol	*rd*	rd		pk						br		bl			dk br
Rhubarb	water	or	pk				cl			pa vt	br	vt	gr		gn	yl
Rose petals	ethanol	*yl-gn*	pk							pa yl	br	yl			gn	br
Rose petals	water	*br*	pa yl-br							br	gr		gr			br
Black tea	ethanol	*br*	pa yl									yl-br				br
Black tea	water	*br*	pa br												dk br	
Tomato leaves	ethanol	*gn*	pa br							dk yl-gr	dk yl-gr	gn				yl gn
Tomato leaves	water	*gn-br*	pa yl-br	pa yl-gr											dk yl-br	yl-br

[a]The following abbreviations have been used for the colors in this chart:

bl = blue	dk = dark	pa = pale	vt = violet
br = brown	gn = green	pk = pink	yl = yellow
cl = colorless	or = orange	rd = red	

[b]The abbreviations in regular type in this column indicate that the extract initially has a clear appearance; the abbreviations in *italic* type, that it has a cloudy appearance.

changes, as long as pH extremes (below 1 and above 11) are avoided. The initial pH of the extracts can be discerned by comparing the initial color of the extract with its color as it is affected by the different pH values (see the table).

Most plant extracts contain a mixture of pigments [3]. Because of this, there is usually no sharp color change with pH, but instead a gradual fading from one color into another over a range of several pH units. However, some extracts do change sharply, some at more than one pH value (e.g., red cabbage extract). As illustrated in Procedure A, red cabbage can be used to determine the approximate pH of a number of common household substances (or their solutions).

Most of the pH-sensitive red, blue, and violet pigments in plants are water-soluble anthocyanins which are easily extracted from the plant. A typical anthocyanin is red in acid solution, purple in neutral, and blue in basic. The blue cornflower, burgundy dahlia, and red rose all contain the same anthocyanin, but they differ in the acidity of their sap. Many white flowers contain anthoxanthin, which turns yellow when treated with base. A green color can result in a plant extract made basic from the combined effect of blue anthocyanin and yellow anthoxanthin pigments. More than one anthocyanin can be present in a flower, and changes in the pH of its sap can produce subtle changes in shade of the flower [4].

red color   yellow color
of rose   of rose

The pH-sensitive pigments in most plants are soluble in ethanol as well as in water. Many of the other components of the sap are not readily soluble in ethanol, and therefore, ethanol is a more effective extracting solvent for the pigments than is water. The extracts produced with ethanol are generally clearer and have more vivid colors than the water extracts.

REFERENCES

1. S. Sharpe, *The Alchemist's Cookbook: 80 Demonstrations,* Shell Canada Centre for Science Teachers, McMaster University: Hamilton, Ontario (undated), p. 20.
2. M. Forster, *J. Chem. Educ.* 55:107 (1978).
3. E. Bishop, *Indicators,* Pergamon Press: Oxford (1972), Ch. 1 of Vol. 51 in International Series of Monographs in Analytical Chemistry.
4. G. S. Losey, "Biological Coloration," in *Encyclopaedia Britannica,* 15th ed. (1983), Vol. 4, p. 917.

8.5

Classical Properties of Acids and Bases

Acids and bases change the color of litmus; acids react with metals, with calcium oxide, and with copper carbonate; and after an acid is mixed with a base, it no longer reacts with metals, calcium oxide, or a carbonate. These properties are displayed directly (Procedure A) and by overhead projection (Procedure B). An acid also precipitates sulfur from a polysulfide solution (Procedure C). A base turns vegetable oil to soap (Procedure D).

MATERIALS FOR PROCEDURE A

For preparation of stock solutions of acids and bases, see pages 30–32.

450 mL distilled water

450 mL 2M hydrochloric acid, HCl

450 mL 2M sulfuric acid, H_2SO_4

450 mL 2M acetic acid, $HC_2H_3O_2$

8 strips blue litmus paper, 2 cm × 25 cm (To prepare these strips, dissolve 0.5 g of solid litmus in about 20 mL of distilled water, saturate eight 2-cm × 25-cm strips of filter paper with the litmus solution, and air dry the strips.)

120 cm³ magnesium turnings

40 g copper(II) carbonate, $CuCO_3$, powder

600 mL 6M sodium hydroxide, NaOH

12 250-mL beakers, with stirring rods

12 labels for beakers

MATERIALS FOR PROCEDURE B

For preparation of stock solutions of acids and bases, see pages 30–32.

overhead projector, with transparency and marker

90 mL distilled water

90 mL 2M hydrochloric acid, HCl

90 mL 2M sulfuric acid, H_2SO_4

90 mL 2M acetic acid, $HC_2H_3O_2$

20 cm³ magnesium turnings

4 g copper(II) carbonate, $CuCO_3$, powder

2 mL litmus solution (To prepare 20 mL of solution, dissolve 0.5 g of solid litmus in 20 mL of distilled water.)

100 mL 6M sodium hydroxide, NaOH

12 50-mL beakers, with stirring rods

MATERIALS FOR PROCEDURE C

For preparation of stock solutions of acids and bases, see pages 30–32.

15 g sulfur powder

250 mL 1M sodium hydroxide, NaOH

10 drops liquid detergent

100 mL 0.1M acetic acid, $HC_2H_3O_2$

300 mL 0.1M hydrochloric acid, HCl

100 mL 0.1M sulfuric acid, H_2SO_4

100 mL 0.1M nitric acid, HNO_3

100 mL 0.1M phosphoric acid, H_3PO_4

100 mL distilled water

100 mL 0.1M sodium chloride, NaCl (To prepare 1 liter of solution, dissolve 5.4 g of NaCl in 600 mL of distilled water and dilute the resulting solution to 1.0 liter.)

100 mL 0.1M sodium hydroxide solution, NaOH

2 400-mL beakers

hot plate or Bunsen burner

glass stirring rod

8 250-mL beakers, with labels

MATERIALS FOR PROCEDURE D

For preparation of stock solutions of acids and bases, see pages 30–32.

50 g vegetable oil (e.g., olive oil, corn oil, cottonseed oil)

50 mL ethanol, C_2H_5OH

65 mL 6M sodium hydroxide, NaOH

250 mL saturated sodium chloride, NaCl (To prepare a stock solution, stir about 400 g of NaCl with 1 liter of boiling distilled water for about 10 minutes, cool the mixture to room temperature, and decant the liquid from any remaining undissolved solid.)

hot plate

600-mL beaker

glass stirring rod

1-liter beaker filled with ice water (optional)

Büchner filter funnel apparatus and filter paper

6 paper towels

plastic wrap (optional)

PROCEDURE A

Preparation

Arrange the 12 beakers in three sets of 4. Label one beaker in each set "H_2O," another "HCl," another "H_2SO_4," and the last "$HC_2H_3O_2$." Fill each beaker labelled "H_2O" about half full with distilled water. Pour a similar amount of 2M HCl, H_2SO_4, and $HC_2H_3O_2$ into each of their respective beakers. Place a glass stirring rod in each beaker.

Presentation

Dip a strip of blue litmus paper in each of the four substances in one set of beakers. The paper dipped in the beaker of water, because it is wet, will become a deeper blue, but the strips dipped in the acids will become pink. Save all four strips of litmus paper. Drop a quarter of the magnesium turnings into each of the four beakers in the same set and note what happens to the magnesium.

Place 5 g of $CuCO_3$ in each of the four beakers of the second set and stir the mixtures. Note what happens in each beaker.

Pour a quarter of the 6M NaOH into each of the four beakers in the last set and stir the mixtures. Test the liquids in each of these beakers with fresh strips of blue litmus paper *and* with the litmus paper used earlier to test the first set of beakers. Then, place about 5 g $CuCO_3$ in each beaker, stir the mixtures, and note how the results with this set differ from those obtained with the second set of beakers.

PROCEDURE B

Preparation

Arrange the 12 beakers in three sets of 4. Label one beaker in each set "H_2O," another "HCl," another "H_2SO_4," and the last "$HC_2H_3O_2$." Fill each beaker labelled "H_2O" about half full with distilled water. Pour a similar amount of 2M HCl, H_2SO_4, and $HC_2H_3O_2$ into each of their respective beakers. Place a glass stirring rod in each beaker.

Write the labels "H_2O," "HCl," "H_2SO_4," and "$HC_2H_3O_2$" on the overhead transparency, so that a beaker can be positioned adjacent to each label on the transparency.

Presentation

Place the transparency on the overhead projector, and position the beakers from one set next to the appropriate labels. Drop an eighth of the magnesium turnings into each of the four beakers in the set and note what happens to the magnesium.

Place another set of beakers on the overhead projector. Place 1 g of $CuCO_3$ in each of the four beakers of the second set and stir the mixtures. Note what happens in each beaker.

Place the last set of beakers on the overhead projector. Add 10 drops of litmus solution to each of four beakers. The mixture in the beaker of water will be blue, but the mixtures in the other beakers will be pink. Pour a quarter of the 6M NaOH into each of the four beakers in the last set and stir the mixtures. Then, place a quarter of the remaining magnesium in each beaker, and note how the results with this set differ from those obtained with the first set of beakers.

PROCEDURE C

Preparation

Add 15 g of powdered sulfur to 250 mL of 1M NaOH in a 400-mL beaker. Boil the mixture for 5–10 minutes. If the sulfur forms clumps while the solution is boiling, add a few drops of liquid detergent to the mixture and stir the mixture. After the sulfur has dissolved, the solution will be yellow to dark red. Remove the beaker from the heat source. If small particles of sulfur remain undissolved, decant the clear solution into another 400-mL beaker, leaving the sulfur particles behind. Allow this solution to cool.

Label the eight 250-mL beakers with one each of these labels: "$HC_2H_3O_2$," "HCl," "H_2SO_4," "HNO_3," "H_3PO_4," "H_2O," "NaCl," "NaOH." Pour about 100 mL of each of the eight sample liquids—0.1M $HC_2H_3O_2$, 0.1M HCl, 0.1M H_2SO_4, 0.1M HNO_3, 0.1M H_3PO_4, distilled water, 0.1M NaCl, and 0.1M NaOH—into the appropriate beakers.

Presentation

Add about 20–30 mL of the solution of sulfur in alkali to each of the 250-mL beakers. Note that the yellow to red color of the sulfur solution disappears and that a white precipitate appears in each of the beakers containing an acid ($HC_2H_3O_2$, HCl, H_2SO_4, HNO_3, and H_3PO_4), but that the neutral or basic solutions (water, NaCl, and NaOH) remain colored and clear.

Add 0.1M HCl to each of the beakers in which no precipitate appeared until a precipitate does appear.

PROCEDURE D

Preparation

Turn on the hot plate, adjust it to medium heat, and allow it to heat up before the demonstration.

Presentation

The procedure described in this paragraph requires about half an hour to complete. Pour 50 g of vegetable oil into the 600-mL beaker. Add 50 mL of ethanol and 65 mL of 6M sodium hydroxide solution to the beaker of vegetable oil. Place the beaker on the hot plate. Boil the mixture, stirring as necessary to minimize foaming. If the mixture seems about to boil over, momentarily remove the beaker from the heat. Continue stirring the boiling mixture until most of the liquid has evaporated, leaving a pasty mass.

Remove the beaker from the heat and allow it to cool. Placing the beaker in the beaker of ice water will hasten the cooling. Add 250 mL of saturated sodium chloride solution to the cooled mass, and mix the mass thoroughly into the liquid. Filter this mixture in the Büchner funnel and rinse the solid with about 50 mL of cold water.

Remove the caked solid from the funnel and press it between paper towels into the shape of a bar of soap. If the towels tear, use plastic wrap to surround the soap as you squeeze it into a bar shape. Wash your hands to test the efficacy of the soap at cleaning and sudsing. Rinse your hands thoroughly with water to remove any excess sodium hydroxide remaining in the soap.

HAZARDS

Concentrated solutions of sodium hydroxide can cause severe burns to the eyes, skin, and mucous membranes.

Copper compounds are harmful if taken internally. Dust from copper compounds can irritate mucous membranes.

Ethanol is flammable and should be kept away from open flames. Mixtures containing it should be heated only on a hot plate, not with a burner.

DISPOSAL

The waste solutions from Procedures A and B should be neutralized by adding sodium bicarbonate ($NaHCO_3$) until fizzing stops; all the waste solutions should be flushed down the drain with water.

DISCUSSION

Acids are a group of substances which have a sour taste, change the color of some plant pigments, and dissolve certain metals and minerals. Bases are substances that have a slippery feel, reverse the color changes of vegetable pigments altered by acids, and when combined with an acid in the proper proportions, destroy the properties of the acid. Procedures A and B of this demonstration illustrate these properties of acids and bases, using hydrochloric, sulfuric, and acetic acids, and using sodium hydroxide as a base. The other procedures demonstrate other properties of acids and bases: that bases dissolve sulfur and that acids liberate sulfur from the basic solution (Procedure C), and that a base combined with a vegetable oil forms soap (Procedure D).

The classification of some substances as acids and others as bases (alkalies) is one of the oldest in chemistry, and the terms *acid, alkali,* and *salt* can be found in writings of medieval alchemists. Acids were probably the more easily recognized of the two substances because of their sour taste. Other properties associated with acids include the ability to dissolve many metals, to change the color of certain vegetable pigments, and to produce bubbles when combined with some minerals (carbonates). On the other hand, the most notable property of bases is their ability to neutralize acids.

Procedures A and B demonstrate the acid properties of hydrochloric, sulfuric, and acetic acids and the base properties of sodium hydroxide. The acids change the color of litmus paper from blue to pink. Litmus paper is colored with litmus, a pigment obtained from lichens, particularly *Variolaria* and *Lecanora*. (Litmus is by no means the only vegetable pigment affected by acids; as illustrated in Demonstration 8.4, there are many vegetable extracts whose colors are changed by acids.) These procedures also show that the acids dissolve magnesium metal and, in the process, produce bubbles. Magnesium is used because it reacts quickly with hydrochloric and sulfuric acids and more slowly with acetic acid. The fizzing that occurs while a metal dissolves in acid is caused by the formation of hydrogen gas. Bubbles are also produced when copper(II) carbonate, which constitutes the mineral malachite, is mixed with an acid. In this case, the bubbles are of carbon dioxide, and the copper(II) carbonate dissolves in the acid, forming a clear green or blue solution. Copper(II) carbonate is used because it produces a colored solution, indicating that, in addition to liberating a gas, the acid dissolves the carbonate. When the acids are mixed with sodium hydroxide, they lose their ability to dissolve copper(II) carbonate and to change litmus paper from blue to pink, in fact, the mixture now changes pink litmus paper back to blue, showing a characteristic property of a base.

Procedure C illustrates a special additional property of strong bases, that they dissolve sulfur, and a property of acids, that they precipitate sulfur from such a basic solution. When sulfur is heated in a concentrated solution of sodium hydroxide, it dissolves to form polysulfide ions. The process is rather slow, because the hydroxide ions must react with S_8 molecules on the surface of solid sulfur. Starting with powdered sulfur shortens the time required for the sulfur to dissolve. The hydroxide ions in the solution attack one of the bonds in the S_8 molecule, forming a soluble ion containing a chain of eight sulfur atoms [1]:

$$HO^- + \begin{array}{c} S-S \\ S \qquad S \\ S \qquad S \\ S-S \end{array} \longrightarrow HO-S-S-S-S-S-S-S-S^-$$

Once this ion is in solution, it is rapidly decomposed by excess hydroxide ions and forms ions containing only one sulfur:

$$HO-S-S-S-S-S-S-S-S^- + 7\,HO^- \longrightarrow 8\,HO-S^-$$

These HOS^- ions then attack more S_8 molecules and take them into solution, as HOS_9^- ions. These are, in turn, decomposed by other HOS^- ions to HOS_2^-. This process continues, gradually increasing the length of the sulfur chains in the ions. As the length of the chains increases, the color of the solution darkens from yellow to orange to red-brown. When the solution is red-brown the chains in the ions contain as many as nine sulfur atoms. When an acid is added to a solution containing these ions, the hydroxide ions are neutralized, and the sulfur chains return to their S_8 elemental molecules:

$$H^+ + HO-S-S-S-S-S-S-S-S^- \longrightarrow H_2O + S_8$$

The elemental sulfur precipitates from the solution as a pale yellow suspension.

Procedure D illustrates a long-exploited property of bases—that they turn vegetable oils (and animal fats) to soap. Vegetable oils and animal fats are mixtures of triglycerides, mixed esters of the triol glycerol (glycerin). In the generalized structure of a vegetable oil molecule below, R, R′, and R″ represent hydrocarbon chains. The hydrocarbon chains found in natural oils and fats contain odd numbers of carbon atoms from 7 to 21.

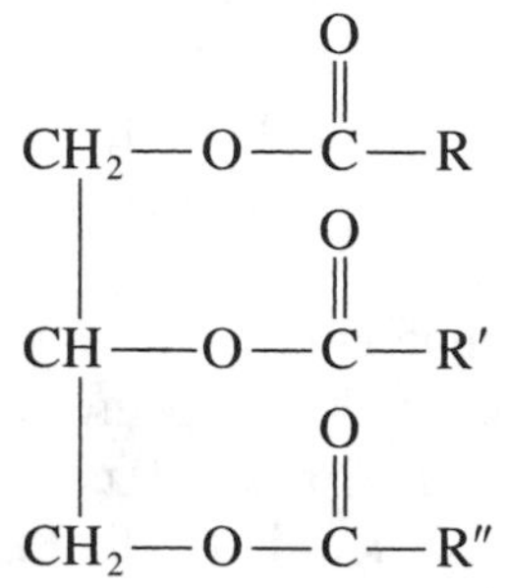

$$
\begin{array}{c}
\quad\quad\quad O \\
\quad\quad\quad \| \\
CH_2-O-C-R \\
\quad\quad\quad O \\
\quad\quad\quad \| \\
CH-O-C-R' \\
\quad\quad\quad O \\
\quad\quad\quad \| \\
CH_2-O-C-R''
\end{array}
$$

The base reacts with the triglyceride, breaking the ester linkages and producing salts of long-chain carboxylic acids:

$$
\begin{array}{c}
\quad\quad O \\
\quad\quad \| \\
CH_2-O-C-R \\
\quad\quad O \\
\quad\quad \| \\
CH-O-C-R' + 3\ NaOH \longrightarrow \\
\quad\quad O \\
\quad\quad \| \\
CH_2-O-C-R''
\end{array}
\quad
\begin{array}{c}
\quad\quad\quad\quad\quad\quad O \\
\quad\quad\quad\quad\quad\quad \| \\
CH_2-OH \quad NaO-C-R \\
\quad\quad\quad\quad\quad\quad O \\
\quad\quad\quad\quad\quad\quad \| \\
CH-OH + NaO-C-R' \\
\quad\quad\quad\quad\quad\quad O \\
\quad\quad\quad\quad\quad\quad \| \\
CH_2-OH \quad NaO-C-R''
\end{array}
$$

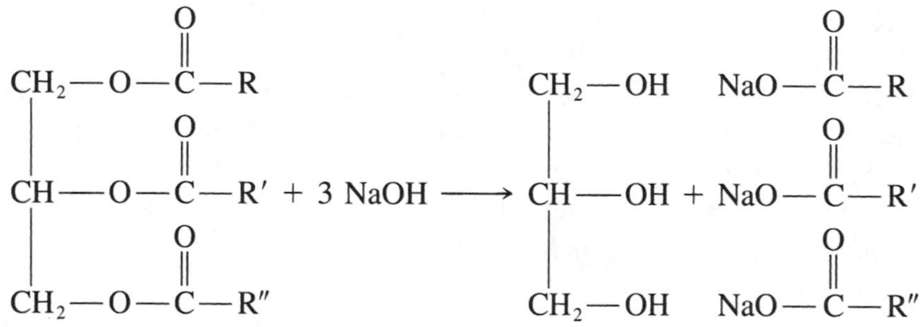

The salts of long-chain carboxylic acids produced in the reaction are soaps.

REFERENCE

1. W. A. Pryor, *Mechanism of Sulfur Reactions*, McGraw-Hill Book Co.: New York (1962).

8.6

Food Is Usually Acidic, Cleaners Are Usually Basic

The pH of several household chemicals is measured using either an indicator solution or a pH meter, revealing that most food products are acidic, and most cleaning agents are basic [1].

MATERIALS FOR PROCEDURE A

For preparation of indicator solutions, see pages 27–29.

either

> 10 mL universal indicator, 1–11 pH range
>
> 50 mL each of pH-standard solutions with pH of 1, 3, 5, 7, 9, and 11 (commercially available or prepared as described in Demonstration 8.1)
>
> rack of 6 test tubes, 25 mm × 200 mm, with stoppers
>
> 6 labels for test tubes
>
> dropper
>
> white background (e.g., 20-cm × 30-cm poster board)

or

> pH meter, with large display, standardized
>
> wash bottle filled with distilled water
>
> 250-mL beaker (optional)

at least 5 different household products as suggested in the following lists:

> (If universal indicator is used, these products should be nearly colorless; if a pH meter is used, these products can be colored.)
>
> 50 mL each of water-based liquids (e.g., fruit juice, vinegar, carbonated beverage, coffee, milk, mouthwash, ammonia, bleach, tile cleaner)
>
> ca. 1 teaspoon (5 mL) each of water-soluble solids or viscous liquids (e.g., laundry detergent, dishwashing detergent, hand soap, shampoo, scouring powder, drain cleaner, washing soda, baking soda, baking powder, toothpaste, antacid, aspirin)

100 mL distilled water for each solid or viscous liquid product to be tested

glass pitcher or Erlenmeyer flask, for distilled water

1 150-mL beaker with stirring rod for each solid or viscous liquid product to be tested

1 test tube, 25 mm × 200 mm, with stopper for each product to be tested

rack for test tubes

MATERIALS FOR PROCEDURE B

For preparation of indicator solutions, see pages 27–29.

overhead projector, with transparency and marker

10 mL universal indicator, 1–11 pH range

25 mL each of pH-standard solutions with pH of 1, 3, 5, 7, 9, and 11 (commercially available or prepared as described in Demonstration 8.1)

at least 5 different nearly colorless household products as suggested in the following lists:

> 25 mL each of water-based liquids (e.g., lemon juice, vinegar, carbonated beverage, ammonia, bleach, tile cleaner)

> ca. ½ teaspoon (2 mL) each of water-soluble solids or viscous liquids (e.g., laundry detergent, dishwashing detergent, hand soap, shampoo, scouring powder, drain cleaner, washing soda, baking soda, baking powder, toothpaste, antacid, aspirin)

25 mL distilled water for each solid or viscous liquid product to be tested

dropper

11 (or more) 50-mL beakers (one for each buffer plus one for each household product)

glass pitcher or Erlenmeyer flask, for distilled water

PROCEDURE A

Preparation

If universal indicator is used, prepare color standards for the indicator. Label six of the test tubes with the pH values of the pH-standard solutions: 1, 3, 5, 7, 9, and 11. Place 10 drops of universal indicator in each tube. Pour 50 mL of the appropriate pH-standard solution into each tube. Stopper each tube, invert it several times to mix the contents, and place the tubes in the rack in order of increasing pH. Set the rack before a white background.

If a pH meter is used, standardize it according to the manufacturer's instructions.

Gather the products to be tested, leaving them in their original packages. Fill the pitcher with 100 mL of distilled water for each of the solid products or liquid cleaning agents to be tested. Place a stirring rod in each of the 150-mL beakers.

Presentation

For each test of the water-based liquids, pour about 50 mL of the liquid into a 25-mm × 200-mm test tube. If universal indicator is to be used, add 10 drops of indicator to the test tube. Stopper the tube and invert it several times to mix the contents. Compare the color of the mixture with the color standards. Estimate the pH of the liquid to within 1 pH unit. If a pH meter is available, immerse its electrode in the liquid. Read the pH of the liquid from the meter. Remove the electrode from the liquid and, if there is no sink near enough, hold it over the 250-mL beaker while rinsing it with distilled water.

For each test of solids or of liquid cleaning products, place about 5 mL (1 teaspoon) of the product in a 150-mL beaker and add 100 mL of distilled water. Stir the mixture to dissolve some of the product. Pour 25 mL of the solution into a 25-mm × 200-mm test tube and add 10 drops of universal indicator. Stopper the tube and invert it several times to mix the contents. Compare the color of the mixture with the color standards in the test tube rack. Estimate the pH of the mixture to within 2 pH units. If a pH meter is available, immerse its electrode in the liquid in the test tube. Read the pH of the liquid from the meter. Remove the electrode from the liquid and rinse it with distilled water.

PROCEDURE B

Preparation

Prepare the color standards as follows: On a transparency write the pH values of the standard solutions: 1, 3, 5, 7, 9, and 11. Place 10 drops of universal indicator in each of the 50-mL beakers, and set each beaker adjacent to a pH value on the transparency atop the overhead projector. Pour 25 mL of the appropriate pH standard solution into each beaker. Swirl the beakers to mix the solutions.

Gather the products to be tested, leaving them in their original packages. Fill the pitcher with about 25 mL of distilled water for each of the solid products or liquid cleaning agents to be tested.

Presentation

For each test of the water-based liquids, pour about 25 mL of the liquid into a 50-mL beaker. Add 10 drops of indicator to the beaker. Swirl the beaker to mix the solution. Compare the color of the mixture to the color standards. Estimate the pH of the liquid to within 1 pH unit.

For each test of solids or of liquid cleaning products, place about 2 mL (½ teaspoon) of the product in a 50-mL beaker and add 25 mL of distilled water. Swirl the beakers to mix the liquids or dissolve some of the solid. Add 10 drops of universal indicator and swirl the beaker to mix the solutions. Place the beaker on the overhead projector. Compare the color of the mixture with the color standards and estimate the pH of the mixture to within 2 pH units.

HAZARDS

Some of the household products, particularly those that are basic, are harmful if swallowed or splashed into the eyes. Follow the precautions and first aid procedures given on the containers. Do not mix any of the cleaning agents together.

DISPOSAL

Dilute all solutions with large amounts of water when flushing them down the drain. Make certain that all the acidic materials are flushed down separately from the basic materials (cleaning agents) to avoid the formation of harmful and toxic gases.

DISCUSSION

This demonstration shows the broad range of pH values to be found in household products. The pH of products found around the house ranges from below 2 in a solution of bisulfate drain cleaner to over 12 in a lye solution. Patterns in the pH of certain kinds of products will develop if a broad range of substances is tested. For example, most foods have a pH in the neutral to slightly acidic range. Those foods that are sour (e.g., vinegar and lemon juice) generally are the most acidic. Cleaning products tend to be fairly basic, because oils and fats dissolve more readily in base than in acid. When testing products with universal indicator, only colorless products should be used, because a colored product will make the color of the indicator difficult to determine. Some products, such as bleach, can react with the components of the universal indicator and produce a colorless solution, showing that indicators are unsuitable for determining the pH of these products.

If a variety of shampoos and hair rinses is investigated, a range of pH values will be found. Most shampoos are slightly basic. Hair is sensitive to the pH of the shampoo because it is composed of protein molecules, amino acid chains held together by amide bonds, that are joined to each other by disulfide bonds between polymer chains. In a pH range of 8–9, some disulfide bonds that hold the amino acid chains together will break. This causes the cuticle, or outer surface of the hair strands to become roughened, causing the hair to look slightly dull. Repeated use of alkaline shampoos can lead to complete breakage of the disulfide bonds at the tips of the strands, resulting in "split ends." At pH values above 11, the amide bonds in the protein begin to break and hair literally dissolves. This demonstrates how strongly alkaline drain cleaners open clogged drains—by dissolving the hair that is causing the clog. Commercial depilatories also contain strongly alkaline ingredients that dissolve hair. Human hair strands are strongest at a pH of 4–5. Thus, to clean hair (remove the oils that are holding dirt), an alkaline shampoo is used. In order to restore the hair to the pH range of maximum strength, an acidic rinse is used. "Acid balanced" shampoos normally contain an acid such as citric acid (found, for example, in lemon juice) to counteract the alkaline nature of detergents [2–4].

Cleaning agents that contain ammonia carry a warning on the label that the cleaner should not be mixed with chlorine-type bleach. The reason for this is that ammonia and hypochlorite (the active ingredient in chlorine bleaches) react to produce a number of hazardous compounds [5]. One product is the flammable gas hydrazine:

$$2\ NH_3(aq) + ClO^-(aq) \longrightarrow Cl^-(aq) + H_2O(l) + N_2H_4(g)$$

The energy released in the combustion of hydrazine is so great that hydrazine is a common propellant in manmade satellites:

$$N_2H_4(l) + O_2(g) \longrightarrow N_2(g) + 2\ H_2O(l) \qquad \Delta H = -622\ kJ/mol\ N_2H_4$$

Another hazardous product is yellowish liquid nitrogen trichloride (NCl_3):

$$NH_3(aq) + 3\ ClO^-(aq) \longrightarrow NCl_3(l) + 3\ OH^-(aq)$$

Nitrogen trichloride is shock sensitive and decomposes to form nitrogen and chlorine gas:

$$2\ NCl_3(l) \longrightarrow N_2(g) + 3\ Cl_2(g)$$

Fortunately, nitrogen trichloride is not nearly as shock sensitive when it is damp as when it is dry.

The large-display pH meter called for in Procedure A can be fabricated in a number of ways. One method is to connect a projection meter (e.g., Central Scientific No. 82551-000), designed for use with an overhead projector, to the pH meter. Another method is to use one of several hardware-software packages for connecting a microcomputer to a pH meter and displaying the pH value on the video monitor. A third approach is to construct a large-scale digital display [6].

REFERENCES

1. L. H. Barrow, *J. Chem. Educ.* 62:339 (1985).
2. *Chem Matters* 1(2):8 (1983).
3. J. J. Griffin, R. F. Corcoran, and K. K. Akana, *J. Chem. Educ.* 54:553 (1977).
4. C. A. Rinzler, *Science* 82(3):54 (1982).
5. F. A. Cotton and G. Wilkinson, *Advanced Inorganic Chemistry*, 3d ed., John Wiley and Sons: New York (1972), pp. 351, 364.
6. G. H. Myers and R. J. Dugan, *J. Chem. Educ.* 54:495 (1977).

8.7

Differing Properties
of Four Common Acids

The four mineral acids—hydrochloric acid, sulfuric acid, phosphoric acid, and nitric acid—are heated with copper metal. Nitric acid liberates a red-brown, water-soluble gas and forms a blue solution. Sulfuric acid produces white, water-soluble fumes and a blue solution. Hydrochloric acid and phosphoric acid form colorless solutions (Procedure A). Concentrated sulfuric acid absorbs water from the air (Procedure B). Nitric acid turns egg white an orange color (Procedure C). Nitric acid turns paper yellow, and sulfuric acid chars it (Procedure D). When diluted with water, sulfuric acid liberates enough heat to vaporize the water, whereas the other three acids become only warm when diluted. The total volume of the mixtures decreases by 5% with sulfuric and phosphoric acids, 2% with nitric acid, and negligibly with hydrochloric acid (Procedure E).

MATERIALS FOR PROCEDURE A

ca. 3 liters tap water

15 mL concentrated (16M) nitric acid, HNO_3

4 3-g pieces of copper (Copper U.S. cents dated prior to 1983 can be used. Do *not* use a post-1982 copper-clad zinc cent, because it will cause the reaction to become too vigorous and difficult to control.)

1 liter crushed ice

80 mL concentrated (15M) aqueous ammonia, NH_3 (optional)

15 mL concentrated (12M) hydrochloric acid, HCl

5 mL 85% phosphoric acid, H_3PO_4

5 mL concentrated (18M) sulfuric acid, H_2SO_4

4 1-liter beakers

glass-working torch (for Pyrex), or Bunsen burner with wing top for soft glass

4 50-cm pieces of glass tubing, with outside diameter of 6 mm

4 1-holed rubber stoppers to fit Erlenmeyer flasks

hot plate

gloves, plastic or rubber

4 1-liter Erlenmeyer flasks

2 50-mL graduated cylinders

stand for beaker, the same height as the hot plate

tray, ca. 50 cm × 50 cm × 3 cm

white backdrop (e.g., 30-cm × 60-cm poster board)

MATERIALS FOR PROCEDURE B

10 mL concentrated (12M) hydrochloric acid, HCl

10 mL concentrated (18M) sulfuric acid, H_2SO_4

10 mL 85% phosphoric acid, H_3PO_4

10 mL concentrated (16M) nitric acid, HNO_3

4 250-mL beakers

4 labels for beakers

gloves, plastic or rubber

4 10-mL graduated cylinders

balance that reads to nearest centigram

MATERIALS FOR PROCEDURE C

4 hard-boiled eggs

10 mL concentrated (12M) hydrochloric acid, HCl

10 mL 85% phosphoric acid, H_3PO_4

10 mL concentrated (18M) sulfuric acid, H_2SO_4

10 mL concentrated (16M) nitric acid, HNO_3

4 petri dishes

gloves, plastic or rubber

4 10-mL graduated cylinders

MATERIALS FOR PROCEDURE D

1 mL concentrated (12M) hydrochloric acid, HCl

1 mL concentrated (16M) nitric acid, HNO_3

1 mL 85% phosphoric acid, H_3PO_4

1 mL concentrated (18M) sulfuric acid, H_2SO_4

4 test tubes, 10 mm × 75 mm

4 labels for test tubes

gloves, plastic or rubber

4 glass stirring rods

2 white paper towels

250-mL beaker

dropper

MATERIALS FOR PROCEDURE E

30 mL concentrated (12M) hydrochloric acid, HCl

30 mL concentrated (18M) sulfuric acid, H_2SO_4

30 mL 85% phosphoric acid, H_3PO_4

30 mL concentrated (16M) nitric acid, HNO_3

ca. 100 mL distilled water in a wash bottle

4 75-cm glass tubes, with outside diameter of 14 mm and sealed on one end

4 labels for glass tubes

15 cm brightly colored tape

4 ring stands with clamps

black backdrop (e.g., 30-cm × 100-cm poster board)

4 solid rubber stoppers to fit glass tubes

20 cm masking tape

4 strips of gauze, 5 cm × 60 cm

gloves, plastic or rubber

meter stick

PROCEDURE A

Preparation

Fill four 1-liter beakers about two-thirds full with tap water.

Using the glass-working torch, bend the four pieces of glass tubing into the shape illustrated in the figure. Insert these four tubes through the four rubber stoppers.

Turn on the hot plate and adjust it to its medium setting.

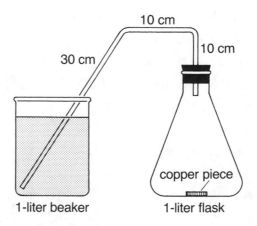

Presentation

Wearing gloves, carefully pour 15 mL of concentrated HNO_3 into a dry 1-liter Erlenmeyer flask. Drop a 3-g piece of copper into the flask and quickly stopper the flask with one of the stopper assemblies. Position the flask so the free end of the glass tubing is immersed in the water in one of the 1-liter beakers.

Brown fumes will fill the flask and bubbles of air will escape through the tubing into the beaker of water. When the brown gas fills the glass tubing, rub handfuls of ice on the sides of the Erlenmeyer flask, cooling the gas and lowering its pressure. Consequently, water from the beaker will slowly rise in the tubing. When the water reaches the flask, it will flow quickly into the flask, producing a blue solution. After the water has stopped flowing into the flask, the blue color of the solution can be intensified, if desired, by pouring 20 mL of concentrated aqueous NH_3 from a graduated cylinder into the flask.

Wearing gloves, carefully pour 15 mL of concentrated HCl into the second dry 1-liter Erlenmeyer flask. Drop a 3-g piece of copper into the flask and quickly stopper the flask with another of the stopper assemblies. Heat the flask on the hot plate until the hydrochloric acid has boiled for about 30 seconds. Then, remove the flask from the hot plate and position it so the free end of the glass tubing is immersed in the water in the second 1-liter beaker. Shortly, water will slowly flow up the glass tubing and then stream into the flask. The solution in the flask will be colorless. If 20 mL of concentrated aqueous NH_3 are added to the flask, white fumes of ammonium chloride will form, but the solution will remain colorless.

Wearing gloves, carefully pour 5 mL of 85% H_3PO_4 into the third dry 1-liter Erlenmeyer flask. Drop a 3-g piece of copper into the flask and quickly stopper the flask with a stopper assembly. Place another 1-liter beaker of water on a stand next to the hot plate, so the base of the beaker is level with the surface of the hot plate. Set the Erlenmeyer flask on the hot plate so the free end of the glass tubing is immersed in the water in the beaker. Agitate the flask to ensure that the piece of copper is covered with acid. Heat the flask until it is filled with white fumes and these fumes begin to bubble through the water in the beaker. Then, carefully lift the flask and the beaker, and place them in a shallow tray on the bench top. Water will flow slowly up the glass tubing and into the flask. As soon as the water level in the flask reaches about 3 cm, lift the flask so that no more water enters and hold the flask in front of a white backdrop. The solution will be colorless. After 20 mL of concentrated aqueous NH_3 are added to the flask, the solution remains colorless.

Wearing gloves, carefully pour 5 mL of concentrated H_2SO_4 into the last dry 1-liter Erlenmeyer flask. Drop a 3-g piece of copper into the flask and quickly stopper the flask with the remaining stopper assembly. Place the remaining 1-liter beaker of water on a stand next to the hot plate, so the base of the beaker is level with the surface of the hot plate. Set the Erlenmeyer flask on the hot plate so the free end of the glass tubing is immersed in the water in the beaker. Agitate the flask to ensure that the piece of copper is covered with sulfuric acid. Heat the flask until it is filled with dense white fumes and these fumes begin to bubble through the water in the beaker. Then, carefully lift the flask and the beaker, and place them in a shallow tray on the bench top. Water will flow slowly up the glass tubing and into the flask. As soon as the water drops onto the hot sulfuric acid, it will vaporize, causing rapid bubbling in the beaker of water. Then the water will again slowly flow up the tubing, and the first drops of water falling on the acid will vaporize and cause bubbling in the water. After a few more of these cycles, water will stream into the flask without being converted to steam. As soon as the water level in the flask reaches about 3 cm, lift the flask so that no more water enters and hold the flask in front of a white backdrop. The solution will be pale blue. The blue color can be intensified if 20 mL of concentrated aqueous NH_3 are added to the flask.

PROCEDURE B

Preparation

Label four 250-mL beakers as follows: "CAUTION—CONCENTRATED HCl," "CAUTION—CONCENTRATED H_2SO_4," "CAUTION—CONCENTRATED H_3PO_4," and "CAUTION—CONCENTRATED HNO_3."

Presentation

Wearing gloves, add about 10 mL of the appropriate acid to each beaker. Weigh each beaker and record the mass.

Set the open beakers aside where they will not be disturbed for at least 12 hours. The beakers should be placed in a well-ventilated area away from corrodible materials such as iron; the beakers of hydrochloric acid and nitric acid will emit corrosive vapors. Reweigh the beakers within 24 hours. The mass of the beaker of H_2SO_4 will be significantly greater. The masses of the beakers of HCl, HNO_3, and H_3PO_4 may increase or decrease, depending on the relative humidity and temperature of the air in which the beakers were stored.

PROCEDURE C

Preparation

Remove the shells from four hard-boiled eggs and place one egg in the center of each of four petri dishes.

Presentation

Wearing gloves, slowly pour about 10 mL of 12M HCl from a graduated cylinder over one egg. Then pour about 10 mL of 85% H_3PO_4 over the second egg. Slowly drip about 10 mL of 18M H_2SO_4 over the third egg. Finally, slowly drip about 10 mL of 16M HNO_3 over the last egg. This last egg turns yellow and within a few minutes becomes bright orange.

PROCEDURE D [1]

Preparation

Label four test tubes with one each of the following: "HCl," "HNO_3," "H_3PO_4," and "H_2SO_4." Pour about 1 mL of the appropriate acid into each test tube.

Presentation

Wearing gloves, dip a stirring rod in the test tube of HCl. Carefully remove the rod and wipe a few of the clinging drops of acid onto a white paper towel. Place the stirring rod in a 250-mL beaker. Using clean stirring rods, repeat this process with HNO_3, H_3PO_4, and H_2SO_4. The portions of the paper under the HCl and the H_3PO_4 show no apparent change, whereas the paper under the HNO_3 turns yellow, and the paper under the H_2SO_4 slowly becomes charred and is eaten away.

With a dropper, withdraw about half of the H_2SO_4 from the test tube and "write" a message on a fresh paper towel with the tip of the dropper. Hold up the towel as the message appears.

PROCEDURE E [2]

Preparation

Label the four 75-cm tubes near the open ends with one each of the following: "HCl," "H_2SO_4," "H_3PO_4," and "HNO_3." Wrap a piece of brightly colored tape 50 cm from the closed end of each tube. Cautiously pour 30 mL of the appropriate acid into each tube, and fasten each tube to a ring stand in front of a black backdrop. Using the wash bottle, slowly run distilled water down the inside of each tube, taking care not to mix the water with the acid. Add water until the liquid reaches the level of the colored tape. Then, stopper each tube. Using masking tape, fasten the end of a 5-cm $\times$ 60-cm gauze strip about 10 cm below each tape mark. Then, spiral a single layer of gauze around each tube, ending at or near the bottom. Fasten this end with masking tape.

Presentation

Using the wash bottle, wet the gauze on each tube. Wearing gloves, hold a finger over the stopper of one tube, unfasten the tube from the ring stand, quickly invert it a few times, then reclamp the tube to the ring stand. Repeat this process with the other three tubes. A cloud of condensed water vapor will spread from the water-soaked gauze wrapped around the H_2SO_4 tube. The other three tubes become slightly warm.

Measure the depth of the solution in each tube. The depths of the liquids in the H_2SO_4 and H_3PO_4 tubes will be about 3 cm less than before mixing, and the depth of the solution in the HNO_3 tube will decrease by about 1 cm. The depth of the solution in the HCl tube will not change appreciably.

HAZARDS

Because concentrated sulfuric acid is a strong acid and a powerful dehydrating agent, it must be handled with great care. Likewise, concentrated nitric acid is a powerful oxidizing agent which reacts very quickly with skin tissue. Rubber or plastic gloves and a protective apron should be worn for these demonstrations. Spills should be neutralized with an appropriate agent such as sodium bicarbonate ($NaHCO_3$) and then wiped up.

Concentrated aqueous ammonia can irritate the skin, and its vapors are harmful to the eyes and mucous membranes.

Hydrochloric acid can cause severe burns. Hydrochloric acid vapors are extremely irritating to the skin, eyes, and respiratory system.

Phosphoric acid can cause severe burns to the eyes and mucous membranes and is irritating to the skin.

Concentrated nitric acid is both a strong acid and a powerful oxidizing agent. Contact with combustible materials can cause fires. Contact with the skin can result in severe burns. The vapor irritates the respiratory system, eyes, and other mucous membranes.

The brown nitrogen dioxide, the white fuming sulfur trioxide gas, and the fumes of hydrogen chloride are all extremely toxic and should not be allowed to escape into the room. These gases are irritating to the respiratory system; inhaling any one of them can result in severe pulmonary effects which may not be apparent until several hours after exposure.

Do not allow anyone to eat the hard-boiled eggs that have been used in Procedure C.

In Procedure E, do not use tubes with a diameter larger than 14 mm. Larger tubes of sulfuric acid will become dangerously hot.

DISPOSAL

The slightly basic waste solutions from Procedure A should be flushed down the drain with water. The waste solutions from the other procedures should be combined in a 1-liter beaker containing 500 mL of water. Neutralize the mixture by adding sodium bicarbonate ($NaHCO_3$) until there is no longer any fizzing, and flush the mixture down the drain with water.

DISCUSSION

Acids are categorized as acids because they possess a common set of properties: they change the colors of certain vegetable pigments, they dissolve metals, and they release carbon dioxide gas from solid carbonates. However, although they do possess these common properties, each still has properties unique to itself. In this demonstration the differing properties of four common mineral acids are highlighted. These four mineral acids are hydrochloric acid, nitric acid, phosphoric acid, and sulfuric acid. The term *mineral acid* is generally synonymous with *inorganic acid* and is used to distinguish these acids from organic acids. The derivation of these terms is from the original sources of these acids. The mineral acids were prepared from minerals, and organic acids were drived from products of living organisms.

The four acids used in this demonstration are among the top 25 chemicals produced in the United States. The production of these acids is listed in the table. Sulfuric acid is produced in greater quantity than any other chemical—nearly twice as much as the next greatest, nitrogen.

Mineral Acid Production in 1985 [3]

Acid	Pounds produced (in billions)
Sulfuric	79.23
Phosphoric	20.98
Nitric	15.56
Hydrochloric	5.56

Each of the procedures in this demonstration shows how several properties differ among the acids, but each procedure highlights the differences in one particular property. Procedure A emphasizes their differing oxidizing abilities. Procedures B and E highlight their differing affinities for water. Procedure C illustrates differences in the way they react with the protein of an egg, and Procedure D contrasts their effects on paper.

In Procedure A, the same set of steps is performed with each acid. The differences in the behavior of the acids are dramatic. Each acid is combined with a piece of copper in a sealed Erlenmeyer flask that is connected to a beaker of water with glass tubing whose free end is immersed in the water. With nitric acid, the flask quickly fills with red-brown nitrogen dioxide, which is produced when the nitric acid oxidizes the copper.

$$4 \, HNO_3(aq) + Cu(s) \longrightarrow Cu^{2+}(aq) + 2 \, NO_3^-(aq) + 2 \, NO_2(g) + 2 \, H_2O(l)$$

(Associated with the action of nitric acid upon copper is an amusing anecdote from Ira Remsen about his early experiences with chemistry. This story, in Remsen's own words, is on pages 90–91 in Demonstration 8.9. The reaction of nitric acid with copper is also used in Demonstrations 6.15 and 6.16 in Volume 2 as a source for nitrogen dioxide gas.) As gas is produced by the reaction, some of it escapes through the tubing and bubbles through the water. At this point the flask is cooled to slow the reaction and to reduce the pressure of the gas inside, and the water flows through the tubing and into the flask. Water is drawn into the flask because the NO_2 produced in the reaction is very soluble in water, and as it dissolves the pressure in the flask falls, allowing the atmosphere to push the water in. When the water enters the flask, it dilutes the acid, slows the reaction, and forms a solution which is blue because of the hydrated copper ions $(Cu(H_2O)_4^{2+})$ produced in the reaction. Then concentrated ammonia is added to the solution to neutralize the acid, and the solution turns deep blue as a result of the formation of the deeply colored complex between copper ions and ammonia $(Cu(NH_3)_4^{2+})$.

The same steps are performed with hydrochloric acid in place of nitric acid. There is no evidence of a reaction between hydrochloric acid and copper, so the mixture is heated. When the flask is cooled, water flows quickly into the flask, because, during heating, hydrogen chloride gas was driven from the hydrochloric acid solution, filling the flask with highly soluble HCl gas. As the HCl gas dissolves in the water, the pressure inside the flask decreases and the atmosphere pushes the water into the flask. However, the solution formed when the water enters the flask is colorless. When ammonia is added to the mixture, the solution remains colorless, meaning that very few, if any, copper ions have been formed. Hydrochloric acid, unlike nitric acid, is not an oxidizing agent, and it does not react with copper.

When these steps are repeated with phosphoric acid, very little water enters the flask, and no blue color appears. Phosphoric acid is not decomposed by heating, and it is not an oxidizing agent.

The last of the mineral acids, sulfuric acid, produces quite a show when it is treated in the same fashion. Without heating, the copper does not appear to react. When the acid is heated it produces dense white fumes in the flask, and bubbles escape through the tubing. After the heating stops, water slowly flows into the tubing, because when the acid was heated, water-soluble sulfur trioxide gas was formed in the flask. (This gas is responsible for the dense white fumes that form when the acid is heated.) When the first few drops of water fall onto the acid, the rest of the water is expelled from the tube followed by a blast of bubbles. Sulfuric acid has a high affinity for water, and when it combines with water, it liberates a great deal of heat. This heat causes some of the water to vaporize, increasing the pressure in the flask and driving water from the tube. (The spattering that results when a small amount of water comes in contact with sulfuric acid illustrates why this acid is always added to water, and not vice versa, when the acid is diluted.) As the flask cools once more, water flows slowly back through the tube, and when it falls on the acid, it is expelled again. This repeats several times until the acid in the flask becomes so dilute that insufficient heat is liberated to vaporize the water, at which point the water flows into the flask and forms a solution. This solution is pale blue, indicating that some copper has dissolved: hot, concentrated sulfuric acid is an oxidizing agent.

$$2\ H_2SO_4(aq) + Cu(s) \longrightarrow Cu^{2+}(aq) + SO_4^{2-}(aq) + SO_2(g) + 2\ H_2O(l)$$

Not much of the copper dissolves, however, because it soon forms a coating of copper sulfate, which is not very soluble in concentrated sulfuric acid. This coating protects the copper from further reaction with the sulfuric acid.

Procedure B investigates the long-term behavior of the acids in containers open to the air. The beaker containing sulfuric acid will always gain weight, because sulfuric acid is very hygroscopic and absorbs water from the air. The beaker of phosphoric acid will not gain weight because it is much less hygroscopic than sulfuric acid. The other two acids are volatile and will evaporate, so their beakers will lose weight. In fact, hydrochloric acid and nitric acid will evaporate completely if their containers are not sealed.

Procedure C shows that nitric acid reacts with egg white to form a yellow compound. Nitric acid reacts with proteins producing a yellow compound in a reaction called the xanthoproteic reaction [4]. In this reaction, the nitric acid nitrates the aromatic portions of the amino acids phenylalanine, tyrosine, and tryptophan, which are constituents of the protein, forming bright yellow compounds.

$$R{-}CH_2C_6H_5 + HNO_3 \longrightarrow R{-}CH_2C_6H_4NO_2 + H_2O$$

In this equation, R represents the remainder of a phenylalanine-containing protein. The product contains $-CH_2C_6H_4NO_2$, which is similar to nitrotoluene ($CH_3C_6H_4NO_2$), a yellow compound.

Procedure D shows how these mineral acids differ in their behavior toward paper. Hydrochloric and phosphoric acids have no effect on paper, but concentrated nitric acid turns it yellow, and sulfuric acid turns it black. Most paper is made from wood pulp, which is a mixture of cellulose and lignin. The major constituent of paper is cellulose, a polysaccharide. Polysaccharides are polymers of carbohydrates. Carbohydrates contain the elements carbon, hydrogen, and oxygen. The hydrogen and oxygen atoms are in the same ratio as in water—hence the name carbohydrate. Sulfuric acid has such a great affinity for water that it extracts the elements of water from carbohydrates, leaving behind the carbon. For this reason, paper turns black when it comes in contact with sulfuric acid. Most paper made from wood pulp also contains small amounts of lignin, the exact chemical nature of which is not known. However, lignin is known to be a polymeric material containing phenyl groups. These phenyl groups are nitrated by nitric acid, forming yellow nitrophenyl groups.

Procedure E contrasts the amount of heat liberated when each of these acids is diluted with water. Nitric, hydrochloric, and phosphoric acids become only warm when they are diluted, but sulfuric acid becomes extremely hot—hot enough to vaporize water. This procedure also shows the volume changes that occur when the acids are diluted. In each case, about 30 mL of acid is combined with 20 mL of water. The density of concentrated sulfuric acid is 1.84 g/mL and that of water is 1.0 g/mL [5], which indicates that 55 g of H_2SO_4 is combined with 20 g of water, producing a solution that is 73% sulfuric acid by weight. The density of a 73% H_2SO_4 solution is 1.67 g/mL [5]. Therefore, the volume of 75 g of 73% H_2SO_4 is 45 mL. This is noticeably less than the combined volume of the two components, which is 50 mL. With the other acids, the change in volume upon mixing is much less significant, although with nitric acid, it may be noticeable.

REFERENCES

1. H. N. Alyea and F. B. Dutton, Eds., *Tested Demonstrations in Chemistry,* 6th ed., Journal of Chemical Education: Easton, Pennsylvania (1965), p. 42.
2. G. S. Newth, *Chemical Lecture Experiments,* Longmans, Green and Co.: London (1928), p. 274.
3. D. Webber, *Chem. Eng. News* 64(16): 12 (1986).
4. H. R. Mahler and E. H. Cordes, *Biological Chemistry,* 2d ed., Harper and Row: New York (1971).
5. R. C. Weast, Ed., *CRC Handbook of Chemistry and Physics,* 66th ed., CRC Press: Boca Raton, Florida (1985).

8.8

Etching Glass with Hydrogen Fluoride

A gray powder and a colorless liquid are placed in a dish, which is then covered with a watch glass and gently heated. After about 5 minutes the watch glass is no longer transparent: the undersurface of the glass has been attacked and roughened [1].

MATERIALS

10 g calcium fluoride, CaF_2

25 mL concentrated (18M) sulfuric acid, H_2SO_4

lead dish (see Procedure for description), or porcelain evaporating dish, with diameter of 10 cm and depth of 4 cm

piece of duct tape, ca. 5 cm square (optional)

watch glass, with diameter of 15 cm, or 15-cm square of window glass

single-edged razor blade (optional)

ring stand, with ring and wire gauze to support dish

gloves, plastic or rubber

Bunsen burner

tongs

paper towel

PROCEDURE

Preparation

A lead dish can be made from lead sheet available at plumbing supply shops. Cut a disk with a diameter of 15 cm from the lead sheet. Place the disk on a sandbag and hammer the sheet into the form of a dish about 4 cm deep and 10 cm in diameter.

If the watch glass is to be etched with a pattern, firmly press the duct tape to the convex side of the watch glass. Using a single-edged razor blade, cut the pattern into the tape, and peel off that portion of the tape to expose the glass.

Place 10 g CaF_2 in the lead dish, and set it on a ring stand in a fume hood, positioning the burner under the dish.

Presentation

This demonstration should be presented only in a fume hood. Wearing gloves, pour 25 mL of concentrated H_2SO_4 onto the CaF_2 in the dish, and immediately cover the mixture with the watch glass with the convex side (taped side) down. Gently heat the mixture until it begins to bubble. Remove the burner, and allow the dish to stand covered for about 5 minutes. Using tongs, lift the watch glass from the dish and rinse it with cool water. Dry it, peel off the tape, and examine the undersurface.

HAZARDS

Avoid heating the mixture vigorously. Foaming can occur and overflow the dish.

Hydrogen fluoride gas is very irritating to the respiratory system and can cause burns to the eyes.

Because sulfuric acid is both a strong acid and a powerful dehydrating agent, it must be handled with great care. Spills should be neutralized with an appropriate agent, such as sodium bicarbonate ($NaHCO_3$), and then wiped up. The dilution of concentrated sulfuric acid is a highly exothermic process and releases sufficient heat to cause burns.

DISPOSAL

Allow the dish and contents to cool. Because a fourfold excess of sulfuric acid over CaF_2 is used in this procedure, the solid residue in the dish contains concentrated H_2SO_4. To dilute this acid, stand back from the dish and carefully fill it with crushed ice. Allow the ice to melt, and neutralize the mixture by adding sodium bicarbonate ($NaHCO_3$) until there is no longer any fizzing. Then flush the mixture down the drain with water.

DISCUSSION

This demonstration shows an interesting property of hydrogen fluoride: it reacts with glass. The ability of HF to dissolve glass does not indicate that hydrofluoric acid is a particularly strong acid. In fact, as acids go, HF is rather weak; its pK is 3.46 [2]. The reactivity of HF with glass is due, not to its acidity, but instead to the fluorine it contains. Glass is a mixture whose major component is silica (SiO_2). Silica reacts with gaseous HF to form SiF_4:

$$4\ HF(g) + SiO_2(s) \longrightarrow SiF_4(g) + 2\ H_2O(l) \qquad \Delta H° = -194\ kJ/mol$$

and with aqueous HF to produce H_2SiF_6, which is soluble in water [2, 3]:

$$6\ HF(aq) + SiO_2(s) \longrightarrow H_2SiF_6(aq) + 2\ H_2O(l) \qquad \Delta H° = -61.6\ kJ/mol$$

Gaseous HF is used in this reaction because the etched surface produced on the glass is translucent, whereas the etched surface produced by aqueous HF is transparent.

If a porcelain evaporating dish is used instead of a lead bowl, the porcelain, which contains silica, will be attacked by hydrogen fluoride. However, it will not be seriously damaged in a single demonstration, and it can be used many times before it is seriously affected. In place of duct tape, other types of tape can be used, but they are not as effective at protecting the glass surface. A coating of wax can also be used.

REFERENCES

1. H. N. Alyea and F. B. Dutton, Eds., *Tested Demonstrations in Chemistry,* 6th ed., Journal of Chemical Education: Easton, Pennsylvania (1965), p. 34.
2. J. A. Dean, *Lange's Handbook of Chemistry,* 13th ed., McGraw-Hill Book Co.: New York (1985).
3. F. A. Cotton and G. Wilkinson, *Advanced Inorganic Chemistry,* 4th ed., John Wiley and Sons: New York (1980).

8.9

"Coin-Operated
Red, White, and Blue Demonstration":
Fountain Effect
with Nitric Acid and Copper

Three flasks are arranged side by side and connected with glass tubing. All three contain colorless liquids; the one on the left is half filled, the one in the middle is completely filled, and the one on the right contains only a small amount of liquid. When two copper pieces are added to the flask on the right, a sequence of visible changes begins. The sequence ends after 10–20 minutes with each flask half filled, one with a blue liquid, one with a colorless liquid, and the last with a red liquid (Procedure A). A similar sequence of events causes a flag to appear and glowing liquid to flow through a spiral tube (Procedure B). Nitric acid dissolves brass but not gold; aqua regia dissolves both (Procedure C).

MATERIALS FOR PROCEDURE A

For preparation of indicator solutions, see pages 27–29.
For preparation of stock solutions of acids and bases, see pages 30–32.

400 mL 0.35M sodium hydroxide, NaOH (To prepare 1 liter of solution, dissolve 14 g of NaOH in 600 mL of distilled water and dilute the resulting solution to 1.0 liter.)

1 mL phenolphthalein indicator solution

1140 mL 0.10M nitric acid, HNO_3

25 mL concentrated (16M) nitric acid, HNO_3

2 3-g pieces of copper (Copper U.S. cents dated prior to 1983 can be used. Do *not* use a post-1982 copper-clad zinc cent, because it will cause the reaction to become too vigorous and difficult to control.)

3 1-liter round-bottomed flasks

3 stands with clamps

4 right-angle bends of glass tubing, with outside diameter of 1 cm and the length of each arm ca. 15 cm

1-holed stopper to fit flask

2-holed stopper to fit flask

15 cm rubber tubing to fit right-angle bends

83

MATERIALS FOR PROCEDURE B

For preparation of indicator solutions, see pages 27–29.
For preparation of stock solutions of acids and bases, see pages 30–32.

Cyalume lightstick

400 mL 0.35M sodium hydroxide, NaOH (For preparation, see Materials for Procedure A.)

1 mL phenolphthalein indicator solution

1140 mL 0.10M nitric acid, HNO_3

25 mL concentrated (16M) nitric acid, HNO_3

2 3-g pieces of copper (Copper U.S. cents dated prior to 1983 can be used. Do not use a post-1982 copper-clad zinc cent, because it will cause the reaction to become too vigorous and difficult to control.)

Tinkertoy building set

string, ca. 60 cm

2 tall ring stands

6 clamps and holders for ring stands

4 rubber bands

2 test tubes, 13 mm × 100 mm

small flag on a staff about 12 inches long (Staff should fit into Tinkertoy connector spool, or flag can be mounted on Tinkertoy rod.)

white poster board, ca. 30 cm × 60 cm

adhesive tape, ca. 20 cm

sheet of transparency material, ca. 20 cm square

2 thumbtacks

spiral condenser

3 100-mL beakers

3 1-liter round-bottomed flasks

2 short ring stands

4 right-angle bends of glass tubing, with outside diameter of 1 cm and the length of each arm ca. 15 cm

1-holed stopper to fit flask

2-holed stopper to fit flask

15 cm rubber tubing to fit right-angle bends

2.5-inch adjustable hose clamp (available at automotive supply stores)

soft packing foam, 10 cm thick, or plastic bag filled with cotton balls

gloves, plastic or rubber

single-edged razor blade

masking tape, ca. 3 cm

triangular file

MATERIALS FOR PROCEDURE C

50 mL water

50 mL concentrated (16M) nitric acid, HNO$_3$

piece of brass foil, ca. 1 cm square

piece of gold foil, ca. 1 cm square (e.g., gold leaf [obtainable from sign-painting shops])

75 mL concentrated (12M) hydrochloric acid, HCl (optional)

600-mL beaker

2 150-mL beakers

white backdrop (e.g., 30-cm × 60-cm poster board), or overhead projector

250-mL beaker

PROCEDURE A [1]

Preparation

Assemble the apparatus shown in Figure 1. Clamp the three 1-liter round-bottomed flasks to stands, and set them side by side. Insert one of the right-angle bends through the 1-holed stopper and two of the bends through the 2-holed stopper. Adjust the bends of flask 2 so their tips are near the center of the flask when the stopper is seated in the mouth of the flask. Connect the free ends of the bends with rubber tubing. Hang the last right-angle bend in the mouth of the unstoppered flask; the tip of the bend should be near the center of the flask.

Pour 400 mL of 0.35M NaOH into the open flask (flask 1). Unstopper the other two flasks and pour 1 mL of phenolphthalein indicator solution and 1140 mL of 0.10M HNO$_3$ into flask 2. Pour 25 mL of concentrated HNO$_3$ into flask 3. Reseat the stopper securely in flask 2.

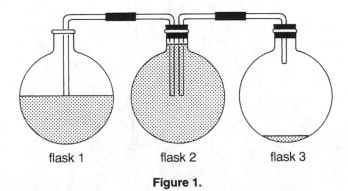

flask 1   flask 2   flask 3

Figure 1.

Presentation

Drop two copper pieces into the flask containing the concentrated HNO_3 (flask 3), and reseat the stopper firmly in its mouth. Immediately, a red-brown gas is formed in flask 3, and liquid flows from flask 2 into flask 1, turning red. After several minutes, red-brown gas fills flask 3 and liquid stops flowing from flask 2 to flask 1. Eventually, liquid begins to flow from flask 2 into the concentrated HNO_3 flask (flask 3), and liquid flows from flask 1 into flask 2. As the red liquid from flask 1 enters flask 2, it becomes colorless. The solution in flask 3 is blue.

PROCEDURE B

Preparation

Using Tinkertoys, assemble the apparatus shown in Figure 2. The string should be wrapped counterclockwise twice around the upper horizontal shaft. This shaft should rotate freely in the bearings when the free end of the string is pulled, but the rotation should be stopped when the flag is up.

Prepare the assembly as shown in Figure 3. Clamp the connector spool at the bottom of the Tinkertoy assembly near the top of a tall ring stand. Rotate the upper shaft so the flag is down. Use two rubber bands to fasten the two test tubes in an upright position to the front spool. Use adhesive tape to fasten the white poster board to the ring stand so that it hides the flag when it is inverted but reveals the flag when it is upright. Roll the piece of transparency material into a cone, secure it with tape, and tack it with thumbtacks to the connector spool below the front bearing, to form a funnel into which

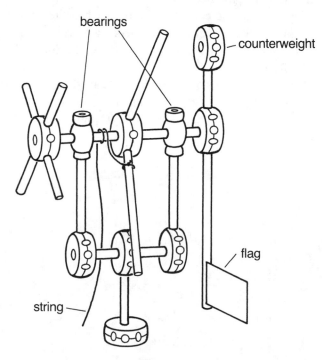

Figure 2.

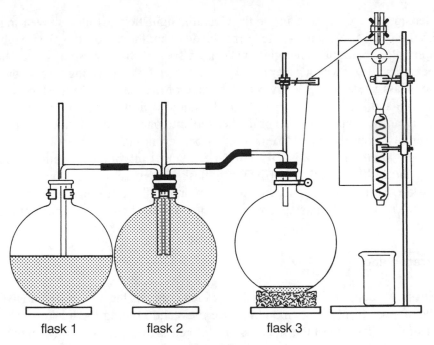

flask 1   flask 2   flask 3

Figure 3.

contents of the test tubes will spill when the rotating shaft is turned. Clamp the spiral condenser to the ring stand in such a position that liquid poured into the cone will flow into the condenser. Place a 100-mL beaker under the condenser.

   Clamp two of the 1-liter round-bottomed flasks to the short ring stands, and clamp the third flask to the remaining tall stand. Set the stands side by side. Insert one of the right-angle bends through the 1-holed stopper and two of the bends through the 2-holed stopper. Adjust the bends of flask 2 so their tips are near the center of the flask when the stopper is seated in the mouth of a flask. Connect the free ends of the bends with rubber tubing.

   Place the stand with the Tinkertoy assembly next to the stand holding flask 3; the rotating shaft should be perpendicular to the row of flasks. Adjust the clamp on flask 3 so that it is held quite loosely. Securely attach the 2.5-inch adjustable hose clamp to the top of flask 3 and place the piece of foam under the flask. Attach a clamp to the top of the stand. Thread the string from the Tinkertoy assembly over the clamp at the top of the stand holding the flask 3, and tie the free end to the hose clamp. Position the stand holding the Tinkertoy assembly so the string is taut.

   Test the mechanical assembly in this fashion. Fill the two test tubes with water. Slowly pour water into the flask 3. As the flask fills, it should sink into the foam. (If it doesn't, the foam is too stiff and a softer piece is needed.) As the flask sinks, it pulls the string, which causes the shaft of the Tinkertoy assembly to rotate counterclockwise. (If the shaft doesn't rotate, check to be sure the string is taut, that the assembly is not binding, and that the string is wrapped in the proper direction around the assembly.) By the time flask 3 is half filled with water, the contents of the test tubes should have spilled into the cone and run through the condenser into the beaker. At the same time, the flag should have appeared from behind the poster board. Make any necessary adjustments in the apparatus to assure that it functions properly. Empty the water from the flask and from the beaker. Return the apparatus to its original position.

Separate the two liquids inside the Cyalume lightstick as follows. Wearing plastic or rubber gloves, use a single-edged razor blade to cut the top off the plastic tube. Pour the liquid along with the inner glass vial into a 100-mL beaker. Remove the vial from the beaker and dry it with a paper towel. Wrap the vial with masking tape so only about 1 cm of glass is exposed. Gently score the side of the glass above the tape with a triangular file and snap the vial open. Pour the liquid from the vial into another 100-mL beaker. Pour the liquid from one of the beakers into one of the test tubes on the Tinkertoy assembly; pour the liquid from the other beaker into the other test tube.

Pour 400 mL of 0.35M NaOH into flask 1. Pour 1 mL of phenolphthalein indicator solution and 1140 mL of 0.10M HNO_3 into flask 2. Pour 25 mL of concentrated HNO_3 into flask 3. Seat the stopper securely in the mouth of flask 2. Hang the last right-angle bend in the mouth of the unstoppered flask; the tip of the bend should be near the center of the flask.

Presentation

Drop two copper pieces into flask 3, which contains the concentrated HNO_3, and seat the stopper firmly in its mouth. Move the stand holding the Tinkertoy assembly away from this flask until the string is taut. A red-brown gas is formed in the flask of HNO_3, and liquid flows from flask 2 to flask 1, turning red. After several minutes, red-brown gas fills flask 3, and liquid stops flowing from flask 2. Eventually, liquid begins to flow from flask 2 into flask 3, and liquid flows from flask 1 into flask 2. As the red liquid from flask 1 enters flask 2, it becomes colorless. The solution in flask 3 is blue. As liquid flows into flask 3, the upper shaft of the Tinkertoy assembly will rotate. Eventually, the flag will flip up and appear, and the contents of the test tubes will spill into the cone. When the contents of the test tubes mix, they will glow as they flow through the spiral condenser into the beaker.

PROCEDURE C

Preparation

Pour 50 mL of water into the 600-mL beaker. In a fume hood, carefully pour 25 mL concentrated HNO_3 into each of the two 150-mL beakers.

Presentation

This demonstration should be presented in a fume hood or in a well-ventilated area.

Place the 600-mL beaker containing water before the white backdrop or on an overhead projector. Set one of the 150-mL beakers in the 600-mL beaker. Drop the piece of brass foil into this 150-mL beaker and cover it by inverting the 250-mL beaker over it inside the 600-mL beaker (see Figure 4). Immediately, brown fumes forms above the liquid, and the solution in the small beaker turns blue. All of the metal dissolves within a minute.

Remove the 250-mL and 150-mL beakers from the large beaker. Place the other 150-mL beaker of HNO_3 in the 600-mL beaker. Drop the piece of gold foil into the beaker. No reaction occurs.

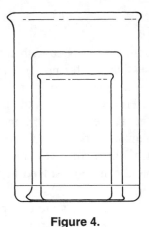

Figure 4.

Carefully pour 75 mL of 12M HCl into the mixture of gold and nitric acid. Cover the 150-mL beaker with the inverted 250-mL beaker. After a few seconds the liquid begins to turn a yellowish color and the gold dissolves.

HAZARDS

Concentrated nitric acid is both a strong acid and a powerful oxidizing agent. Contact with combustible materials can cause fires. Contact with the skin can result in severe burns. The vapor irritates the respiratory system, eyes, and other mucous membranes.

Hydrochloric acid can cause severe burns. Hydrochloric acid vapors are extremely irritating to the respiratory system, eyes, and skin.

Aqua regia, formed by mixing concentrated nitric and concentrated hydrochloric acids, is a strong oxidizing agent. Contact with combustible materials can cause fires. Contact with the skin can result in severe burns.

Sodium hydroxide can cause severe burns of the eyes, skin, and mucous membranes. Dust from solid sodium hydroxide is very irritating to the eyes and respiratory system.

The brown gas produced in the reactions is nitrogen dioxide. Nitrogen dioxide is irritating to the respiratory system; inhaling it can result in severe pulmonary irritation which is not apparent until several hours after exposure. A concentration of 100 ppm is dangerous for even a short period of time, and exposure to concentrations of 200 ppm or more can be fatal.

DISPOSAL

The waste solutions should be neutralized by adding sodium bicarbonate (NaHCO$_3$) until fizzing stops, and the neutralized solutions should be flushed down the drain with water.

DISCUSSION

Procedure A uses the reaction of copper with nitric acid to initiate a series of chemical and physical processes that last for nearly 20 minutes. It provides an opportunity for students to make careful observations and to postulate the identity of the original solutions in each flask. Procedure B uses the physical process of Procedure A to drive a simple mechanism that causes a flag to appear near the end of the process, and two liquids that glow when mixed to be poured into a spiral glass tube. Procedure C shows that brass and gold can be distinguished chemically by their reactivity toward nitric acid. It also shows that gold can be dissolved in a 3:1 mixture of nitric and hydrochloric acids, called aqua regia.

In Procedure A, when the copper pieces are dropped into a flask containing concentrated nitric acid, the flask fills with a red-brown gas, nitrogen dioxide. As gas is produced by the reaction, the pressure in the flask builds. The increased pressure forces half of the 0.10M HNO_3 solution containing phenolphthalein into the flask containing 0.35M NaOH. As the phenolphthalein mixes with the NaOH solution, it turns red. As the reaction of copper with nitric acid nears completion, the reaction slows and the flask cools. Furthermore, the very soluble NO_2 produced by the reaction has entered the middle flask, where it dissolves in the 0.10M HNO_3. The combined effect of the cooling flask and the dissolving gas is to lower the pressure in the stoppered flasks. This causes the liquid from the open flask to be forced back into the middle flask. As the red liquid mixes with the acid, the phenolphthalein returns to colorless. When the middle flask is filled with liquid, the liquid flows into the flask that contained the copper pieces. This liquid dilutes the solution of copper nitrate formed by the reaction of copper with nitric acid. The copper nitrate colors this solution blue.

Associated with the action of nitric acid upon copper is an amusing anecdote from Ira Remsen about his early experiences with chemistry. Remsen reminisces:

> While reading a textbook of chemistry I came upon the statement, "nitric acid acts upon copper." I was getting tired of reading such absurd stuff and I was determined to see what this meant. Copper was more or less familiar to me, for copper cents were then in use. I had seen a bottle marked nitric acid on a table in the doctor's office where I was then "doing time." I did not know its peculiarities, but the spirit of adventure was upon me. Having nitric acid and copper, I had only to learn what the words "act upon" meant. The statement "nitric acid acts upon copper" would be something more than mere words. All was still. In the interest of knowledge I was even willing to sacrifice one of the few copper cents then in my possession. I put one of them on the table, opened the bottle marked nitric acid, poured some of the liquid on the copper and prepared to make an observation. But what was this wonderful thing which I beheld? The cent was already changed and it was no small change either. A green-blue liquid foamed and fumed over the cent and over the table. The air in the neighborhood of the performance became colored dark red. A great colored cloud arose. This was disagreeable and suffocating. How should I stop this? I tried to get rid of the objectionable mess by picking it up and throwing it out of the window. I learned another fact. Nitric acid not only acts upon copper, but it acts upon fingers. The pain led to another unpremeditated experiment. I drew my fingers across my trousers and another fact was discovered. Nitric acid acts upon trousers. Taking everything into consideration, that was the most impressive experiment and relatively

probably the most costly experiment I have ever performed. . . . It was a revelation to me. It resulted in a desire on my part to learn more about that remarkable kind of action. Plainly, the only way to learn about it was to see its results, to experiment, to work in a laboratory. [2]

The reaction of nitric acid with copper is also used in Demonstrations 6.15 and 6.16 in Volume 2 as a source for nitrogen dioxide gas.

Procedure C shows that two yellow metals, brass and gold, can be distinguished by their reaction with nitric acid. Brass reacts vigorously, liberating a red-brown gas, and eventually dissolves completely. Gold, on the other hand, appears inert to nitric acid.

Brass is an alloy of zinc and copper. Both of these metals react with nitric acid:

$$Zn(s) + 4\ HNO_3(aq) \longrightarrow Zn(NO_3)_2(aq) + 2\ NO_2(g) + 2\ H_2O(l)$$

$$Cu(s) + 4\ HNO_3(aq) \longrightarrow Cu(NO_3)_2(aq) + 2\ NO_2(g) + 2\ H_2O(l)$$

The solution formed when brass reacts with nitric acid is blue because of the blue $[Cu(H_2O)_4]^{2+}$ ion in solution.

Gold does not dissolve in concentrated nitric acid. However, it does dissolve in a mixture of 3 parts concentrated hydrochloric acid to 1 part concentrated nitric acid. The reaction can be represented in a somewhat simplified form by the equation

$$Au(s) + 4\ HCl(aq) + 3\ HNO_3(aq) \longrightarrow H[AuCl_4](aq) + 3\ NO_2(g) + 3\ H_2O(l)$$

Nitric acid oxidizes hydrochloric acid, forming chlorine and chlorine oxides in solution. Gold reacts readily with chlorine, which explains why it dissolves in aqua regia [3]. The complexing action of chloride ions in the formation of $AuCl_4^-$ also makes the mixture of hydrochloric and nitric acids more effective in dissolving gold than nitric acid alone. It is because of its ability to dissolve gold that the 3:1 mixture of hydrochloric acid and nitric acid is called aqua regia (royal water). The 3:1 ratio has been found empirically to be the most effective, perhaps because it provides the best combination of oxidizing and complexing power.

REFERENCES

1. R. I. Perkins, *J. Chem. Educ.* 63:781 (1986).
2. F. H. Getman, *The Life of Ira Remsen,* Journal of Chemical Education: Easton, Pennsylvania (1940), pp. 9–10.
3. F. A. Cotton and G. Wilkinson, *Advanced Inorganic Chemistry,* 3d ed., John Wiley and Sons: New York (1972).

8.10

Fountain Effect with Ammonia, Hydrogen Chloride, and Indicators

Four inverted flasks are connected by a series of glass tubes, and a tube extends from the lowest flask into a beaker of colored water. When a small amount of water is injected into the highest flask, water begins to flow through the tube and into the lowest flask. The four flasks are sequentially filled with water, and the water in each flask is a different color. (Described in Demonstration 6.24 in Volume 2 of this series is the injection of a small amount of water into an inverted round-bottomed flask connected by a glass tube to a reservoir of water below it. Soon after the injection, the water from the reservoir rushes into the flask, turning red as it enters and forming a fountain inside the flask.)

MATERIALS

For preparation of indicator solutions, see pages 27–29.
For preparation of stock solutions of acids and bases, see pages 30–32.

cylinder of ammonia, NH_3, with valve and rubber tubing

6 mL universal indicator or phenolphthalein indicator solution

20 mL 2M acetic acid, $HC_2H_3O_2$

1 liter tap water

30-cm straight glass tube, with outside diameter of 6 mm or 7 mm

3 2-holed rubber stoppers to fit 125-mL round-bottomed flasks

3 U-bends of glass tubing, with outside diameter of 6 mm or 7 mm, each arm ca. 18 cm, and the distance between arms ca. 18 cm

3 125-mL round-bottomed flasks

250-mL separatory funnel

1-holed rubber stopper to fit mouth of 250-mL separatory funnel

2 ring stands, with four clamps

1-liter beaker

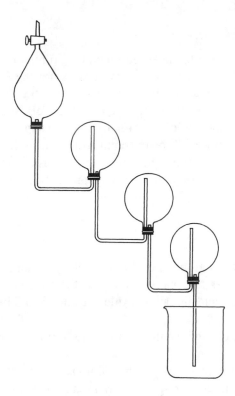

PROCEDURE[†]

Preparation

Assemble the apparatus as illustrated in the figure. Insert the 30-cm straight glass tube in one of the 2-holed stoppers so it just clears the narrow end of the stopper. Insert one arm of a U-bend through the other hole so the tip of the arm reaches to within 1 cm of the bottom of one of the round-bottomed flasks when the stopper is seated in the mouth of the flask. Insert the other arm of the U-bend in another 2-holed stopper so it just clears the narrow end of the stopper. Assemble the other two U-bends and 2-holed stoppers in a fashion similar to that just described. Insert the free arm of the last U-bend through the 1-holed stopper for the funnel so its end just clears the narrow end of the stopper. Clamp the flasks and funnel to the two ring stands so the stopper assembly will fit in their mouths and the free end of the straight tube will reach to within 1 cm of the bottom of the 1-liter beaker when the apparatus is set next to the beaker.

Set the apparatus in a fume hood and attach the ammonia cylinder to the tip of the separatory funnel. Open the stopcock on the separatory funnel and the valve on the ammonia cylinder. Allow ammonia gas to flow through the apparatus for about 1 minute, filling the flasks with NH_3. Close the valve on the ammonia cylinder, and then close the funnel stopcock.

Pour 6 mL of universal indicator and 20 mL of 2M acetic acid into the 1-liter beaker. Fill the beaker with tap water. Set the ammonia-filled apparatus next to the beaker so the glass tube is immersed in the solution in the beaker.

†This demonstration was developed by Robert Becker of Greenwich High School, Greenwich, Connecticut.

Presentation

Fill the tube at the top of the inverted separatory funnel with water. Open the stopcock momentarily to allow a few drops of water to enter the funnel, and quickly close the stopcock. The solution will rise from the beaker into the first flask. As it enters, the color of the solution changes. When the first flask is full, the solution will flow into the second flask, then into the third, and finally into the separatory funnel. The color of the solution in each container will be different from that in the others.

When the solution has stopped flowing into the flasks, open the funnel stopcock, and the solution will slowly drain out of the apparatus back into the beaker.

HAZARDS

Ammonia gas irritates all parts of the respiratory system and is severely irritating to the eyes. Ammonia gas also causes burns to the skin and is toxic when inhaled. Mixtures of ammonia gas and air can be explosive and should be kept away from sparks or open flame.

Glacial acetic acid can irritate the skin, and its vapors are irritating to the eyes and respiratory system.

Use only round-bottomed flasks in this demonstration. Flat-bottomed flasks (e.g., Erlenmeyer flasks) are not as strong and may break under the stress developed in this procedure.

DISPOSAL

The waste solution should be flushed down the drain with water.

DISCUSSION

The procedure in this demonstration is a variation of the ammonia fountain described in Demonstration 6.24 in Volume 2 of this series. However, rather than using a single ammonia-filled flask, this demonstration uses four containers filled with ammonia gas. These containers are connected in such a way that, as each one is filled with water, the water begins to flow into the next, until all four are filled. A universal indicator is used to produce a series of color changes as the water flows through the containers. Phenolphthalein can be used if universal indicator is unavailable; however, in this case, when the liquid flows into each flask, it turns red-purple, but when the flask is nearly filled, the color fades away.

In the procedure, when the acidified water begins to enter the lowest container, virtually all of the ammonia gas dissolves, neutralizing the acid and forming a basic aqueous ammonia solution. The pH of this solution is high, and the universal indicator assumes a basic color. As the acid solution continues to flow into the lowest container, the concentration of the ammonia solution decreases and so does its pH. The decrease in pH causes the universal indicator to change to a less basic color. Eventually, the solution flows from the lowest container into the next, until all four containers are filled. As acid solution flows through each container, the solution in the container becomes progressively less basic, and the color of the indicator changes. Because more acid solution flows through the first container than through the others, the solution it contains is the least basic. When all of the containers have been filled, the most basic solution is in the highest container, and the least basic is in the lowest container. Because each container is at a different pH, the universal indicator in each container is a different color.

8.11

Fizzing and Foaming: Reactions of Acids with Carbonates

When a colorless liquid is poured into a glass containing an opaque yellow liquid, the mixture turns pink and foams up over the top of the glass (Procedure A). When baking soda is added to a mixture of lemon juice and dishwashing detergent in a soda glass, the glass fills with white foam (Procedure B). When a powdered carbonate and an acid are mixed in a stoppered test tube, within a few seconds the stopper shoots several feet from the mouth of the tube (Procedure C). A gas is prepared by the reaction of marble chips with an acid. This is described in Demonstration 6.1 of Volume 2 of this series.

MATERIALS FOR PROCEDURE A

For preparation of indicator solutions, see pages 27–29.
For preparation of stock solutions of acids and bases, see pages 30–32.

20 g sodium hydrogen carbonate (baking soda), $NaHCO_3$

5 mL methyl red indicator solution

50 mL water

20 g powdered, sudsing laundry detergent (e.g., Tide)

50 mL 6M hydrochloric acid, HCl

1-liter tall-form beaker, or large soda glass

stirring rod

cafeteria tray or crystallizing dish, with diameter 15 cm or greater

MATERIALS FOR PROCEDURE B

fresh lemon or lime, or 25 mL reconstituted lemon juice, orange juice, or vinegar

10 drops liquid dishwashing detergent (Do not use liquid detergent made for automatic dishwashers; it will not foam.)

ca. 20 cm³ (heaping tablespoon) baking soda (sodium hydrogen carbonate), $NaHCO_3$

knife (if fresh fruit is used)

soda glass

shallow pan

tablespoon

MATERIALS FOR PROCEDURE C

10 mL vinegar (5% w/w acetic acid, $HC_2H_3O_2$, in water)

6 g sodium carbonate, Na_2CO_3, or sodium bicarbonate, $NaHCO_3$

long-stemmed funnel

test tube, 25 mm × 200 mm, with cork

stand, with clamp to hold test tube

spatula

PROCEDURE A [1]

Preparation

Place 20 g $NaHCO_3$, 5 mL methyl red indicator solution, and 50 mL of water in the 1-liter tall-form beaker. Stir 20 g of powdered laundry detergent into the mixture. The mixture is yellow. Set the beaker on the cafeteria tray or in the crystallizing dish.

Presentation

Pour 50 mL of 6M HCl into the beaker and quickly stir the mixture. The mixture will turn pink and foam up over the sides of the beaker and into the crystallizing dish.

PROCEDURE B [2]

Preparation and Presentation

Cut the lemon or lime in half and squeeze its juice into the soda glass. (Or pour about 25 mL of reconstituted lemon juice, orange juice, or vinegar into the glass.) Add 10 drops of liquid dishwashing detergent to the glass. Holding the glass over a shallow pan, add a heaping tablespoon of $NaHCO_3$ to the glass and stir the mixture. The glass will fill with a thick white foam and some may spill over into the pan.

PROCEDURE C [3]

Preparation

Pour 10 mL of vinegar through the long-stemmed funnel into the test tube, taking care to avoid getting any liquid on the sides of the tube near its mouth. Clamp the tube to the stand at an angle so that the liquid comes to within 6 cm of the mouth of the tube.

Presentation

With a spatula, carefully position 6 g of Na_2CO_3 in the upper, dry part of the test tube, and firmly seat the cork in the mouth of the tube. Remove the tube from the clamp, invert it once to mix its contents, and quickly replace it in the clamp. Aim the tube to avoid hitting anyone or anything breakable. Within 10 seconds the cork will be ejected from the mouth of the tube with a loud pop and travel up to 15 feet from the mouth of the tube.

HAZARDS

Do not allow anyone to eat what appears to be an ice-cream soda produced in Procedures A and B.

Because the cork is ejected from the test tube with considerable force, care should be taken that no person or breakable object is in the line of fire.

Hydrochloric acid can irritate the skin. Its vapors are extremely irritating to the eyes and respiratory system.

DISPOSAL

Any residue should be dissolved in vinegar or acetic acid and flushed down the drain with water.

DISCUSSION

One of the characteristic properties of an acid is that it liberates a gas when it is combined with a carbonate or bicarbonate. This gas is carbon dioxide, and the reactions that produce it are represented by these equations:

$$CO_3^{2-}(aq) + 2 H^+(aq) \longrightarrow CO_2(g) + H_2O(l)$$

$$HCO_3^-(aq) + H^+(aq) \longrightarrow CO_2(g) + H_2O(l)$$

In Procedure A, the reaction is between hydrochloric acid and sodium hydrogen carbonate (sodium bicarbonate, baking soda). The carbon dioxide gas formed in the reaction is trapped in tiny soap bubbles formed by the laundry detergent mixture. The color change of yellow to pink is a result of the change of the methyl red indicator from its basic color of yellow to its acid color of red when the hydrochloric acid is added to the mixture.

In Procedure B, all of the materials used are common household products, which makes this an easy demonstration to present, suitable for an audience of all ages and backgrounds. The reaction is between the acids in the products used—acetic acid in the case of vinegar and citric acid in the case of the fruit juices—and baking soda. Citric acid reacts with sodium hydrogen carbonate as indicated in the equation

$$C_3H_4(OH)(COOH)_3(aq) + 3 NaHCO_3(aq) \longrightarrow$$

$$C_3H_4(OH)(COONa)_3(aq) + 3 H_2O(l) + 3 CO_2(g)$$

In Procedure C, a reaction between a carbonate and an acid is carried out in a corked test tube. As the reaction proceeds, the pressure of the gas in the tube increases. Initially, friction between the cork and the glass prevents the cork from releasing the gas. However, eventually the pressure becomes great enough so that the friction is overcome, and the cork is ejected from the tube.

REFERENCES

1. H. N. Alyea and F. B. Dutton, Eds., *Tested Demonstrations in Chemistry,* 6th ed., Journal of Chemical Education: Easton, New Jersey (1965).
2. B. George, *J. College Sci. Teach.* 16:75 (1986).
3. L. A. Ford, *Chemical Magic,* T. S. Denison and Co.: Minneapolis, Minnesota (1959), p. 29.

8.12

Sealed-Bag Reactions
with Acids and Bases

Two substances are mixed in a sealed plastic bag on the pan of a balance, and the mass of the bag and its contents is recorded to the nearest centigram every minute for a period of 10 minutes. The mass, as measured with the balance, will change.

MATERIALS FOR PROCEDURE A

For preparation of stock solutions of acids and bases, see pages 30–32.

15 mL 2M sodium hydroxide, NaOH

5 g dry ice, CO_2 (solid carbon dioxide), or cylinder of CO_2 gas with valve and 1 meter of rubber tubing

test tube, 15 mm × 125 mm, with stopper

ca. 1-liter plastic bag (e.g., sandwich bag)

rubber band

balance accurate to 0.01 gram (an electronic top-loading model is most convenient)

MATERIALS FOR PROCEDURE B

For preparation of stock solutions of acids and bases, see pages 30–32.

15 mL 2M hydrochloric acid, HCl

5 g mossy zinc

test tube, 15 mm × 125 mm, with stopper

ca. 1-liter plastic bag (e.g., sandwich bag)

rubber band

balance accurate to 0.01 gram (an electronic top-loading model is most convenient)

PROCEDURE A

Preparation

Fill the test tube to within a centimeter of the top with 2M NaOH and stopper it. Place the test tube inside the 1-liter plastic bag. Fill the bag with carbon dioxide by subliming 5 g of dry ice inside the bag, or by flushing and filling the bag with CO_2 from a cylinder. Tightly seal the bag with a rubber band.

Presentation

Place the bag assembly on the balance, weigh it to the nearest centigram, and record the mass. Through the bag and without breaking the seal, hold the test tube in one hand and the stopper with the other hand. Carefully, withdraw the stopper from the test tube allowing the CO_2 gas to mix with the sodium hydroxide. Return the bag to the balance and record the mass every minute for a period of 10 minutes or longer. As the CO_2 dissolves in the NaOH, the mass of the sealed bag and its contents appears to increase.

PROCEDURE B

Preparation

Fill the test tube to within a centimeter of the top with 2M HCl and stopper it. Place the test tube along with 5 grams of mossy zinc inside the 1-liter plastic bag. Flatten the bag to squeeze out the air and tightly seal it with a rubber band.

Presentation

Place the bag assembly on the balance, weigh it to the nearest centigram, and record the mass. Through the bag and without breaking the seal, hold the test tube in one hand and the stopper with the other hand. Cautiously, withdraw the stopper from the test tube and allow the HCl to contact the zinc. Record the mass every minute for a period of 10 minutes or longer. The bag gradually expands, and the mass of the bag and its contents appears to decrease.

HAZARDS

Sodium hydroxide solutions can cause severe burns to the eyes, skin, and mucous membranes.

Hydrochloric acid can irritate the skin. Its vapors are extremely irritating to the eyes and respiratory system.

DISPOSAL

The waste solutions should be flushed down the drain with water.

DISCUSSION

This demonstration uses acid-base reactions in sealed containers to demonstrate that the buoying effect of air on an object depends on the volume of the object. In both cases a reaction in a closed system appears to change the mass of the system, seemingly violating the law of conservation of mass. However, the apparent change in mass is a result of the buoying effect of air on the weighing process. The apparent mass determined by weighing is the true mass minus the mass of fluid (air) displaced. These procedures illustrate the difference between mass and weight. The mass of the contents of the bag does not change. However, the weight, which is the downward force due to gravity, does change.

In Procedure A, a plastic bag that contains a tube of sodium hydroxide solution is filled with carbon dioxide gas and weighed. The tube of NaOH solution is opened and the apparent mass of the bag is recorded periodically. The mass of the bag as determined with the balance increases gradually. This seeming violation of conservation of mass is actually a result of the decreasing buoying effect of air on the bag as the volume of the bag decreases. The CO_2 gas in the bag slowly dissolves in the NaOH solution, a reaction between an acidic oxide and a basic solution. (See Demonstrations 8.14 through 8.17 for more reactions of acidic and basic oxides.)

$$CO_2(g) + 2\, NaOH(aq) \longrightarrow Na_2CO_3(aq) + H_2O(l)$$

Because the amount of gas in the bag decreases, the volume of the bag diminishes. As the volume of the bag decreases, the volume of air it displaces also decreases. According to Archimedes' principle, an object is buoyed (i.e., its apparent mass is reduced) by an amount equal to the mass of the fluid it displaces. Because the volume of air displaced by the bag decreases, the buoying effect of the displaced air decreases, and the mass of the bag appears to increase.

In Procedure B, a tube containing hydrochloric acid is sealed in a plastic bag with pieces of zinc, and the bag is weighed. The tube of acid is opened, the acid mixes with the zinc, and the apparent mass of the bag is recorded periodically. The mass of the bag as determined with the balance decreases steadily. The apparent decrease in mass is a result of the increasing volume of the bag, and the resulting increase in the buoying effect of air. The volume of the bag increases because a gas is produced by the reaction of hydrochloric acid with zinc:

$$Zn(s) + 2\, HCl(aq) \longrightarrow ZnCl_2(aq) + H_2(g)$$

This reaction is an example of a characteristic property of acids: they react with active metals releasing hydrogen gas.

The apparent change in mass observed in this demonstration depends on the volume change of the bag, which depends on the capacity of the particular bag employed. An estimate of the observed change in mass can be made by assuming a volume change of 1 liter. A volume change of 1 liter will result in a change in displacement of 1 liter of air. The mass of 1 liter of air at a temperature of 25°C and 1 atm of pressure can be calculated from the ideal gas equation, using an average molar mass for air of 29 g/mol (the average of 80% N_2 and 20% O_2):

$$m = \frac{(29\ \text{g/mol})(1.0\ \text{atm})(1.0\ \text{liter})}{(0.0821\ \text{liter} \cdot \text{atm/mol} \cdot \text{K})(298\ \text{K})} = 1.2\ \text{g}$$

In this demonstration, a top-loading automatic electronic balance is more convenient than a balance with mechanical weights. With the electronic balance the mass of the bag can be monitored continuously, and the steady change in the apparent mass can be easily seen.

8.13

Hydrolysis:
Acidic and Basic Properties of Salts

Comparing the color of universal indicator in a variety of 0.1M salt solutions with the color in pH-standard solutions yields an estimate of the pH of the salt solution (Procedure A). The colors can be displayed by overhead projection (Procedure B). The pH of the salt solution can also be determined with a pH meter (Procedure C).

MATERIALS FOR PROCEDURE A

For preparation of indicator solutions, see pages 27–29.

100 mL each of buffer solutions with pH of 1, 3, 5, 7, 9, and 11 (For preparation, see Procedure A of Demonstration 8.1.)

15 mL universal indicator, pH range 1–13

one (or both) of the following strongly acidic salt solutions:

100 mL 0.1M aluminum nitrate, $Al(NO_3)_3$ (To prepare 1 liter of solution, dissolve 38 g of $Al(NO_3)_3 \cdot 9H_2O$ in 600 mL of distilled water, and dilute the resulting solution to 1.0 liter.)

100 mL 0.1M tin(IV) chloride, $SnCl_4$ (To prepare 1 liter of stock solution, dissolve 26 g of $SnCl_4$ in 600 mL of distilled water and dilute the resulting solution to 1.0 liter.)

one (or more) of the following mildly acidic salt solutions:

100 mL 0.1M calcium nitrate, $Ca(NO_3)_2$ (To prepare 1 liter of stock solution, dissolve 16 g of $Ca(NO_3)_2$ in 600 mL of distilled water and dilute the resulting solution to 1.0 liter.)

100 mL 0.1M ammonium chloride, NH_4Cl (To prepare 1 liter of stock solution, dissolve 5.4 g of NH_4Cl in 600 mL of distilled water and dilute the resulting solution to 1.0 liter.)

100 mL 0.1M potassium dihydrogen phosphate, KH_2PO_4 (To prepare 1 liter of stock solution, dissolve 14 g of KH_2PO_4 in 600 mL of distilled water and dilute the resulting solution to 1.0 liter.)

100 mL 0.1M magnesium nitrate, $Mg(NO_3)_2$ (To prepare 1 liter of stock solution, dissolve 26 g of $Mg(NO_3)_2 \cdot 6H_2O$ in 600 mL of distilled water and dilute the resulting solution to 1.0 liter.)

100 mL 0.1M ammonium nitrate, NH_4NO_3 (To prepare 1 liter of stock solution, dissolve 8.0 g of NH_4NO_3 in 600 mL of distilled water and dilute the resulting solution to 1.0 liter.)

one (or more) of the following neutral salt solutions:

100 mL 0.1M sodium chloride, NaCl (To prepare 1 liter of stock solution, dissolve 5.8 g of NaCl in 600 mL of distilled water and dilute the resulting solution to 1.0 liter.)

100 mL 0.1M sodium nitrate, $NaNO_3$ (To prepare 1 liter of stock solution, dissolve 8.5 g of $NaNO_3$ in 600 mL of distilled water and dilute the resulting solution to 1.0 liter.)

100 mL 0.1M potassium chloride, KCl (To prepare 1 liter of stock solution, dissolve 7.5 g of KCl in 600 mL of distilled water and dilute the resulting solution to 1.0 liter.)

100 mL 0.1M ammonium acetate, $NH_4C_2H_3O_2$ (To prepare 1 liter of stock solution, dissolve 7.7 g of $NH_4C_2H_3O_2$ in 600 mL of distilled water and dilute the resulting solution to 1.0 liter.)

one (or more) of the following mildly basic salt solutions:

100 mL 0.1M sodium acetate, $NaC_2H_3O_2$ (To prepare 1 liter of stock solution, dissolve 23 g of $NaC_2H_3O_2 \cdot 3H_2O$ in 600 mL of distilled water and dilute the resulting solution to 1.0 liter.)

100 mL 0.1M sodium fluoride, NaF (To prepare 1 liter of stock solution, dissolve 4.2 g of NaF in 600 mL of distilled water and dilute the resulting solution to 1.0 liter.)

100 mL 0.1M potassium nitrite, KNO_2 (To prepare 1 liter of stock solution, dissolve 8.5 g of KNO_2 in 600 mL of distilled water and dilute the resulting solution to 1.0 liter.)

100 mL 0.1M sodium sulfite, Na_2SO_3 (To prepare 1 liter of stock solution, dissolve 13 g Na_2SO_3 in 600 mL of distilled water and dilute the resulting solution to 1.0 liter.)

100 mL 0.1M sodium hydrogen carbonate, $NaHCO_3$ (To prepare 1 liter of stock solution, dissolve 8.4 g of $NaHCO_3$ in 600 mL of distilled water and dilute the resulting solution to 1.0 liter.)

100 mL 0.1M dipotassium hydrogen phosphate, K_2HPO_4 (To prepare 1 liter of stock solution, dissolve 17 g of K_2HPO_4 in 600 mL of distilled water and dilute the resulting solution to 1.0 liter.)

100 mL 0.1M sodium hydrogen sulfite, $NaHSO_3$ (To prepare 1 liter of stock solution, dissolve 10 g of $NaHSO_3$ in 600 mL of distilled water and dilute the resulting solution to 1.0 liter.)

one (or both) of the following strongly basic salt solutions:

100 mL 0.1M trisodium phosphate, Na_3PO_4 (To prepare 1 liter of stock solution, dissolve 16 g of Na_3PO_4 in 600 mL of distilled water and dilute the resulting solution to 1.0 liter.)

100 mL 0.1M sodium carbonate, Na_2CO_3 (To prepare 1 liter of stock solution, dissolve 11 g of Na_2CO_3 in 600 mL of distilled water and dilute the resulting solution to 1.0 liter.)

11 or more 250-mL beakers (six for buffer solutions plus one for each salt solution)

11 or more labels for beakers

dropper

6 glass stirring rods (one for each buffer solution; to be cleaned and reused for salt solutions)

lighted background (For assembly instructions, see Procedure A of Demonstration 8.1), or 20-cm × 50-cm white background

MATERIALS FOR PROCEDURE B

For preparation of indicator solutions, see pages 27–29.

overhead projector, with transparency and marker

15 mL each of buffer solutions with pH of 1, 3, 5, 7, 9, and 11 (For preparation see Procedure A of Demonstration 8.1.)

15 mL universal indicator, pH range 1–13

15 mL each of one (or more) of the strongly acidic salt solutions listed in Materials for Procedure A

15 mL each of one (or more) of the mildly acidic salt solutions listed in Materials for Procedure A

15 mL each of one (or more) of the neutral salt solutions listed in Materials for Procedure A

15 mL each of one (or more) of the mildly basic salt solutions listed in Materials for Procedure A

15 mL each of one (or more) of the strongly basic salt solutions listed in Materials for Procedure A

11 or more 50-mL beakers (six for buffer solutions plus one for each salt solution)

dropper

6 glass stirring rods (one for each buffer solution; to be cleaned and reused for salt solutions)

MATERIALS FOR PROCEDURE C

100 mL each of one (or more) of the strongly acidic salt solutions listed in Materials for Procedure A

100 mL each of one (or more) of the mildly acidic salt solutions listed in Materials for Procedure A

100 mL each of one (or more) of the neutral salt solutions listed in Materials for Procedure A

100 mL each of one (or more) of the mildly basic salt solutions listed in Materials for Procedure A

100 mL each of one (or more) of the strongly basic salt solutions listed in Materials for Procedure A

5 or more 250-mL beakers (one for each salt solution)

5 or more labels for beakers

calibrated pH meter with large display

PROCEDURE A

Preparation

Place one each of the following labels on six 250-mL beakers: "pH 1," "pH 3," "pH 5," "pH 7," "pH 9," and "pH 11." Pour 100 mL of the appropriate buffer solution into each beaker. To each beaker add 10 drops of universal indicator and stir each mixture. Arrange the beakers in order of increasing pH in front of a lighted or white background.

Label separate 250-mL beakers with the name or formula of each salt solution to be used. Pour 100 mL of the appropriate 0.1M salt solution into each beaker.

Presentation

Add 10 drops of universal indicator to a beaker of salt solution and stir the mixture thoroughly. Compare the color of this solution with the colors of the buffer solutions in order to estimate the pH of the salt solution. Record the estimated pH. (Solutions of neutral salts in distilled water may not be precisely neutral, because even tiny amounts of impurities in the salt or in the distilled water [e.g., dissolved CO_2] will cause deviations from pH 7.)

Repeat the procedure described in the preceding paragraph with each of the beakers of salt solution.

PROCEDURE B

Preparation

Write the labels "pH 1," "pH 3," "pH 5," "pH 7," "pH 9," and "pH 11" on an overhead transparency. Arrange six 50-mL beakers on the overhead projector so each beaker is adjacent to one of the labels. Pour 15 mL of the appropriate buffer solution into each beaker. To each beaker add 10 drops of universal indicator and stir each mixture.

Choose one or more salt solutions for use from each of the five groups, and label separate 50-mL beakers with the name or formula of each salt. Pour 15 mL of the appropriate 0.1M salt solution into each beaker.

Presentation

Place a beaker containing a salt solution on the overhead projector. Add 10 drops of universal indicator to a beaker of salt solution and stir the mixture thoroughly. Compare the color of this solution with the colors of the buffer solutions in order to estimate the pH of the salt solution. Record the name or formula of the salt and the estimated pH of its solution. (Solutions of neutral salts in distilled water may not be precisely neutral, because even tiny amounts of impurities in the salt or in the distilled water [e.g., dissolved CO_2] will cause deviations from pH 7.)

Repeat the procedure described in the preceding paragraph with each of the beakers of salt solution.

PROCEDURE C

Preparation

Choose one or more salt solutions for use from each of the five groups, and label a corresponding number of 250-mL beakers with the name or formula of each salt. Pour 100 mL of the appropriate 0.1M salt solution into each beaker.

Presentation

Insert the pH electrode in one of the salt solutions and record the pH of the solution. Repeat this for each solution.

HAZARDS

Solid ammonium nitrate and potassium nitrite should not be allowed to contact combustible materials, for this can result in fire.

Dust from sodium fluoride irritates the respiratory system, eyes, and skin. Ingestion can result in nausea, diarrhea, and abdominal pains. Chronic effects include shortness of breath, cough, elevated temperature, and cyanosis.

DISPOSAL

The waste solutions should be flushed down the drain with water.

DISCUSSION

As this demonstration shows, a salt solution can have a pH ranging from very acidic to very basic. Procedures A and B use a universal indicator to determine the pH of salt solutions; Procedure C uses a pH meter.

The process by which a salt affects the pH of its solution has been called hydrolysis. Hydrolysis is a reaction involving water and a solute in which ions of water, H^+ or OH^-, are formed. Such a reaction occurs with the anion of a weak acid or the cation of a weak base. For example, the acetate ion (the anion of a weak acid, acetic acid) reacts with water by removing a hydrogen ion from it and forming acetic acid and hydroxide ion, which makes the solution basic:

$$C_2H_3O_2^-(aq) + H_2O(l) \rightleftharpoons HC_2H_3O_2(aq) + OH^-(aq)$$

The anions of other weak acids behave in a similar fashion. The weaker the acid, the farther to the right the hydrolysis reaction proceeds, and the more basic the solution. The salt of a weak base produces an acidic solution. Ammonium chloride is a salt of the weak base ammonia. Ammonium ions react with water by transferring a hydrogen ion to the water, which makes the solution acidic:

$$NH_4^+(aq) + H_2O(l) \rightleftarrows NH_3(aq) + H_3O^+(aq)$$

Cations of other weak bases also turn their solutions acidic, and the weaker the base, the more acidic the salt solution.

These reactions are Brønsted-Lowry acid-base reactions. In the case of the anion of a weak acid, water behaves as an acid, transferring a hydrogen ion to the anion and forming a molecule of the weak acid. The anion of the weak acid behaves as a base, accepting the hydrogen ion from water. Thus, the anion of a weak acid is a base. Similarly, in the case of the cation of a weak base, the cation transfers a hydrogen ion to a water molecule, forming a molecule of the weak base. Water behaves as a base, and the cation of a weak base is an acid.

As this demonstration shows, the anion of a weak acid is itself basic, and the cation of a weak base is itself acidic. The acidity or basicity of these ions can be quantitatively related to the basicity or acidity of their parent bases and acids. When the anion of a weak acid is dissolved in water, some of the anions react with water to form weak acid molecules and hydroxide ions. This is represented for the acetate ion in an equation presented above. The equilibrium constant expression for this hydrolysis reaction is

$$K_h = \frac{[HC_2H_3O_2][OH^-]}{[C_2H_3O_2^-]}$$

Multiplying the right side of the equation by the ratio $[H^+]/[H^+]$ does not change the equality, because the value of this ratio is 1.

$$K_h = \frac{[HC_2H_3O_2][OH^-]}{[C_2H_3O_2^-]} \cdot \frac{[H^+]}{[H^+]} = \frac{[HC_2H_3O_2]}{[C_2H_3O_2^-][H^+]} \cdot [H^+][OH^-]$$

The expression on the right is the product of the inverse of the K_a expression of acetic acid and the K_w expression for water. The equilibrium constant for the hydrolysis reaction is equal to the ratio of the K_w to the K_a of the weak acid, $K_h = K_w/K_a$. This equation shows the inverse relationship between the acidity of a weak acid and the basicity of its anion.

8.14

Acidic and Basic Properties of Oxides

A metal oxide is added to a beaker of water containing universal indicator, and a nonmetal oxide is added to another beaker with like contents. The mixture containing metal oxide becomes blue-green, whereas the mixture containing the nonmetal oxide turns yellow-orange (Procedure A). These phenomena can be observed by overhead projection (Procedure B). A mixture of two indicators can be used in place of universal indicator (Procedure C) and viewed by overhead projection (Procedure D).

MATERIALS FOR PROCEDURE A

For preparation of indicator solutions, see pages 27–29.

1 liter distilled water

1 mL universal indicator

ca. 1 g of *one* of the following metal oxides: calcium oxide, CaO; magnesium oxide, MgO; or lithium oxide, Li_2O

ca. 1 g of *one* of the following nonmetal oxides: phosphorus(V) oxide, P_4O_{10}; or solid carbon dioxide (dry ice), CO_2

2 600-mL beakers

2 glass stirring rods

dropper

spatula

MATERIALS FOR PROCEDURE B

For preparation of indicator solutions, see pages 27–29.

overhead projector

40 mL distilled water

1 mL universal indicator

ca. 0.1 g of *one* of the following metal oxides: calcium oxide, CaO; magnesium oxide, MgO; or lithium oxide, Li_2O

ca. 0.1 g of *one* of the following nonmetal oxides: phosphorus(V) oxide, P_4O_{10}; or solid carbon dioxide (dry ice), CO_2

2 50-mL beakers

2 glass stirring rods

dropper

spatula

MATERIALS FOR PROCEDURE C

For preparation of indicator solutions, see pages 27–29.

1 liter distilled water

0.5 mL bromothymol blue indicator solution

0.5 mL methyl red indicator solution

ca. 1 g of *one* of the following metal oxides: calcium oxide, CaO; magnesium oxide, MgO; or sodium oxide, Na_2O

ca. 1 g of *one* of the following nonmetal oxides: phosphorus(V) oxide, P_4O_{10}; or solid carbon dioxide (dry ice), CO_2

2 600-mL beakers

2 glass stirring rods

dropper

spatula

MATERIALS FOR PROCEDURE D

For preparation of indicator solutions, see pages 27–29.

overhead projector

40 mL distilled water

0.5 mL bromothymol blue indicator solution

0.5 mL methyl red indicator solution

ca. 0.1 g of *one* of the following metal oxides: calcium oxide, CaO; magnesium oxide, MgO; or sodium oxide, Na_2O

ca. 0.1 g of *one* of the following nonmetal oxides: phosphorus(V) oxide, P_4O_{10}; or solid carbon dioxide (dry ice), CO_2

2 50-mL beakers

2 glass stirring rods

dropper

spatula

PROCEDURE A

Preparation

Pour 500 mL of distilled water into each of the two 600-mL beakers, and place a glass stirring rod in each beaker. Add 10 drops of universal indicator to each beaker, and stir both solutions thoroughly.

Presentation

Using the spatula, add about 1 g of the metal oxide to the first beaker and about 1 g of nonmetal oxide to the second beaker. When the mixture in each beaker is stirred, its color will change. The mixture containing the metal oxide will become blue-green, and the one containing the nonmetal oxide will turn yellow-orange.

PROCEDURE B

Preparation

Pour 20 mL of distilled water into each of the two 50-mL beakers, and place a glass stirring rod in each beaker. Add 10 drops of universal indicator to each beaker, and stir both solutions thoroughly.

Presentation

Set the two 50-mL beakers on the overhead projector. Using the spatula, add about 0.1 g of the metal oxide to the first beaker and about 0.1 g of nonmetal oxide to the second beaker. When the mixture in each beaker is stirred, its color will change. The mixture containing the metal oxide will become blue-green, and the one containing the nonmetal oxide will turn yellow-orange.

PROCEDURE C

Preparation

Pour 500 mL of distilled water into each of the two 600-mL beakers, and place a glass stirring rod in each beaker. Add 10 drops of bromothymol blue indicator and 10 drops of methyl red indicator to both beakers. Stir both solutions thoroughly.

Presentation

Using the spatula, add about 1 g of the metal oxide to one beaker and stir the mixture well. The solution will change from yellow to green. Similarly, add about 1 g of nonmetal oxide to the other beaker. When the solution is stirred, the color changes from yellow to orange.

PROCEDURE D

Preparation

Pour 20 mL of distilled water into each of the two 50-mL beakers, and place a glass stirring rod in each beaker. Add 10 drops of bromothymol blue indicator and 10 drops of methyl red indicator to both beakers. Stir both solutions thoroughly.

Presentation

Set the two 50-mL beakers on the overhead projector. Using the spatula, add about 0.1 g of the metal oxide to one beaker and stir the mixture well. The solution will change from yellow to green. Similarly, add about 0.1 g of nonmetal oxide to the other beaker. When the solution is stirred, the color changes from yellow to orange.

HAZARDS

Phosphorus(V) oxide is caustic and can cause burns to the skin.
Dry ice is extremely cold ($-78°C$) and contact with the skin can cause frostbite.

DISPOSAL

The waste solutions should be flushed down the drain with water.

DISCUSSION

This demonstration illustrates the general phenomena that oxides of metals are basic in aqueous solution and oxides of nonmetals are acidic. Procedures A and B show the effect of adding these oxides to water containing a universal indicator. With the metal oxide, the indicator changes to a color in its basic range, whereas with the non-metal oxide, the color shifts to one in the acidic range. Procedures C and D show the same effect, but instead of using a complex universal indicator, they use a mixture of two common indicators to produce a similar result. The colors produced by this mixture are described in the following chart:

Indicator	Acid color	Neutral color	Base color
bromothymol blue	yellow	yellow	blue
methyl red	red	yellow	yellow
mixture	orange	yellow	green

The colors of the mixture result from combining the colors of the individual indicators.

In order for the color of the indicator to change, the oxide must react with water to liberate hydrogen ions or hydroxide ions. Metal oxides react with water to form metal hydroxides:

$$M_xO_y + y\,H_2O \longrightarrow x\,M(OH)_{2y/x}$$

To change the pH of the solution, the metal oxide must react with water at an appreciable rate, and the hydroxide must be at least moderately soluble. The oxides used in the demonstration meet both criteria. Many metal oxides react very slowly, and only a few form soluble hydroxides. Others which will produce a visible change in the color of the indicator include bismuth(III) oxide (Bi_2O_3), zinc oxide (ZnO), and lead(II) oxide (PbO). However, these react slowly, and it takes several minutes for the color change to become apparent. Some of the oxides react vigorously with water, particularly those of the heavier alkali metals (e.g., potassium and rubidium). Because these reactions can be violent, it is not recommended that they be used in this demonstration. In fact, the reaction of calcium oxide with water, although not violent, is quite exothermic, as shown in Demonstration 1.7 in Volume 1 of this series.

Nonmetal oxides react with water to form acids. Most of these acids are soluble in water, so the pH of the water falls when such an oxide is added to it. The nonmetal oxides used in this demonstration react with water as indicated in the following equations:

$$P_4O_{10}(s) + 6\ H_2O(l) \longrightarrow 4\ H_3PO_4(aq)$$

$$CO_2(s) + H_2O(l) \longrightarrow H_2CO_3(aq)$$

Both of the acids produced, phosphoric acid and carbonic acid, are weak acids that are partly ionized in aqueous solution, causing the solution to be acidic. Other reactions of carbon dioxide and water are investigated in Demonstration 6.2 in Volume 2 of this series. Phosphorus(V) oxide and carbon dioxide are used in this demonstration because they are perhaps the easiest of the nonmetal oxides to handle. Most nonmetal oxides are gases and quite noxious. Phosphorus(V) oxide itself is caustic and should be handled with care. Solid carbon dioxide, dry ice, is extremely cold and should not be handled with bare fingers.

8.15

Colors, Bubbles, and Fog: Acidic Properties of Carbon Dioxide in Aqueous Solutions

Dry ice is dropped into tall cylinders filled with colored indicator solutions. Cold carbon dioxide gas bubbles up through the solutions, condensing water vapor into fog at the top of the cylinders. As the carbon dioxide gas bubbles through the solution, it dissolves, gradually making the solutions weakly acidic and causing the colors of the indicator solutions to change. This is described in Demonstration 6.2 in Volume 2 of this series.

8.16

Acidic Properties of Nitrogen(IV) Oxide

A flask is filled with colorless nitrogen(II) oxide gas generated by the reaction of nitric acid with copper. When air enters this flask, reddish brown nitrogen(IV) oxide forms. The reddish brown color disappears when the gas dissolves in a small amount of water in the flask. This solution of nitrogen(IV) oxide in water is acidic, which is revealed by the color of an indicator in the water. This is described in Demonstration 6.15 in Volume 2 of this series.

8.17

Acidic Properties
of Combustion Products of Sulfur,
Nitrogen, and Chlorinated Polymers

Sulfur is burned in air, and the products of the combustion form an acidic solution in water (Procedures A and C). Nitric oxide also forms an acidic solution (Procedures B and D), as do the combustion products of polyvinyl chloride (Procedure E).

MATERIALS FOR PROCEDURE A

See Materials for Procedure A of Demonstration 6.18 in Volume 2 of this series.

MATERIALS FOR PROCEDURE B

For preparation of indicator solutions, see pages 27–29.
For preparation of stock solutions of acids and bases, see pages 30–32.

400 mL tap water

2 mL bromocresol green indicator solution

ca. 10 mL 5M aqueous ammonia, NH_3, in a dropper bottle (Amount needed depends on pH of tap water.)

ca. 2 mL concentrated (16M) nitric acid, HNO_3

2 g copper turnings

5-liter round-bottomed flask, with cork support ring

dropper

10-cm length of glass tubing, with outside diameter of 7 mm

1-holed rubber stopper to fit 5-liter flask

2-holed rubber stopper to fit 25-mm × 200-mm test tube

test tube, 25 mm × 200 mm

28-cm glass rod, with outside diameter small enough to fit through the 10-cm length of glass tubing

rubber pipette bulb

MATERIALS FOR PROCEDURE C

For preparation of indicator solutions, see pages 27–29.
For preparation of stock solutions of acids and bases, see pages 30–32.

300 mL tap water

2 mL bromocresol green indicator solution

ca. 10 mL 5M aqueous ammonia, NH_3, in a dropper bottle (Amount needed depends on pH of tap water.)

3 g powdered sulfur

500-mL filter flask

dropper

right-angle glass bend, with outside diameter of 7 mm, one arm 30 cm long, the other 10 cm

1-holed rubber stopper to fit glass tube with outside diameter of 38 mm

1-holed rubber stopper to fit filter flask

glass tube, 30 cm long with outside diameter of 38 mm

2 stands, each with a clamp

Bunsen burner

water aspirator, with hose

spatula

MATERIALS FOR PROCEDURE D

For preparation of indicator solutions, see pages 27–29.
For preparation of stock solutions of acids and bases, see pages 30–32.

300 mL tap water

2 mL bromocresol green indicator solution

ca. 10 mL 5M aqueous ammonia, NH_3, in a dropper bottle (Amount needed depends on pH of tap water.)

ca. 2 mL concentrated (16M) nitric acid, HNO_3

2 g copper turnings

500-mL filter flask

dropper

40-cm length of glass tubing with outside diameter of 10 mm, bent at 90 degrees about 10 cm from one end

#5 2-holed rubber stopper to fit 25-mm × 200-mm test tube

#7 1-holed rubber stopper to fit filter flask

ring stand and utility clamp to support filter flask

test tube, 25 mm × 200 mm

water aspirator, with hose

30-cm length of ⅛-inch glass, metal, or wooden rod

MATERIALS FOR PROCEDURE E

For preparation of indicator solutions, see pages 27–29.
For preparation of stock solutions of acids and bases, see pages 30–32.

300 mL tap water

2 mL bromocresol green indicator solution

ca. 10 mL 5M aqueous ammonia, NH_3, in a dropper bottle (Amount needed depends on pH of tap water.)

piece of chlorine-containing polymer food wrap, 25 cm × 25 cm (e.g., Saran Wrap)

500-mL filter flask

dropper

right-angle glass bend, with outside diameter of 7 mm, one arm 30 cm long, the other 10 cm

1-holed rubber stopper to fit glass tube with outside diameter of 38 mm

1-holed rubber stopper to fit filter flask

glass tube, 30 cm long with outside diameter of 38 mm

2 stands, each with a clamp

Bunsen burner

water aspirator, with hose

PROCEDURE A

See Procedure A of Demonstration 6.18 in Volume 2 of this series.

PROCEDURE B

Preparation

Pour 400 mL of tap water into the 5-liter flask. Add 2 mL of bromocresol green indicator solution. If the mixture is not blue, add just enough drops of 5M NH_3 to turn it blue.

Assemble the glass tubing and the two rubber stoppers as shown in Figure 1. The wide end of the flask's stopper should be about 1 cm from one end of the glass tubing, and there should be about 5 cm between its narrow end and the wide end of the test tube's stopper.

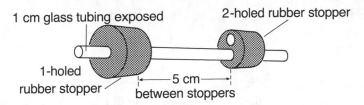

1 cm glass tubing exposed

2-holed rubber stopper

1-holed rubber stopper

—5 cm—
between stoppers

Figure 1.

Assemble the apparatus as illustrated in Figure 2. Pour just enough concentrated nitric acid (HNO$_3$) into the test tube to fill the curved bottom of the test tube (about 2 mL). Roll 2 g of copper turnings between the palms of your hands to form a wad that can be held by friction in the upper part of the test tube. Insert the wad just below the mouth of the test tube, and do not allow the copper to contact the nitric acid. Seat the 2-holed stopper (of the tubing and stopper assembly) snugly in the mouth of the test tube. Insert the test tube in the 5-liter flask and firmly seat the stopper in the mouth of the flask.

Insert one end of the 28-cm glass rod all of the way into the pipette bulb.

Presentation

Insert the glass rod in the glass tubing and push the copper turnings down to the nitric acid. Push the copper into the acid slowly, to prevent a rapid reaction between the copper and the acid. A red-brown gas will fill the test tube and escape into the flask through the open hole in the stopper. Some of the gas may also escape through the glass tubing and into the room; to minimize this, allow the base of the pipette bulb to rest on the mouth of the glass tubing. After the reaction has subsided, shake the flask vigorously to promote the dissolving of the red-brown gas in the indicator solution. The blue solution will turn yellow.

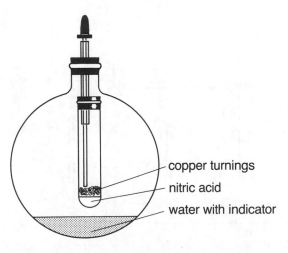

copper turnings
nitric acid
water with indicator

Figure 2.

PROCEDURE C

Preparation

Pour 300 mL of tap water into the 500-mL filter flask. Add 2 mL of bromocresol green indicator solution to the flask. If the mixture is not blue, add just enough drops of 5M NH_3 to turn it blue.

Assemble the apparatus as illustrated in Figure 3. Insert the short arm of the right-angle bend in the glass tube's stopper and the long arm in the flask's stopper. Seat the stopper in the mouth of the flask and adjust the right-angle bend so its tip is within 1 cm of the bottom of the flask. Seat the other stopper in one end of the glass tube and clamp the other end of the tube to the stand so the tube is horizontal. Set the Bunsen burner under the glass tube. Clamp the filter flask to the other stand, and attach the hose from the aspirator to the tubulation on the flask. With a spatula place about 3 g of powdered sulfur about halfway in the glass tube.

Presentation

Turn on the aspirator and adjust it to produce moderate bubbling in the flask. The color of the liquid in the flask does not change. Light the burner and place it under the sulfur in the tube. As the sulfur heats it will melt and begin to burn with a blue flame. When it begins to burn, stop heating the sulfur. After 10–30 seconds the indicator solution becomes yellow.

PROCEDURE D

Preparation

Pour 300 mL of tap water into the 500-mL filter flask. Add 2 mL of bromocresol green indicator solution to the flask. If the mixture is not blue, add just enough drops of 5M NH_3 to turn it blue.

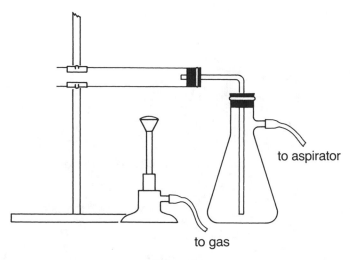

to aspirator

to gas

Figure 3.

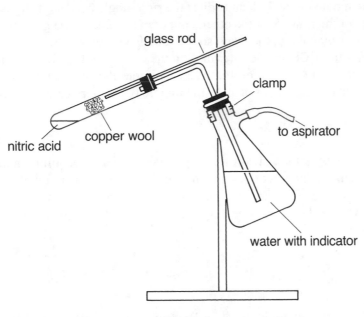

Figure 4.

Assemble the apparatus as illustrated in Figure 4. Insert the short arm of the right-angle bend in the 2-holed test-tube stopper and the long arm in the flask's stopper. Seat the stopper in the mouth of the flask and adjust the right-angle bend so its tip is within 1 cm of the bottom of the flask. Clamp the flask to the stand at an angle of about 30 degrees off vertical. Pour just enough concentrated HNO_3 into the test tube to fill the curved bottom of the test tube (about 2 mL). Roll 2 g of copper turnings between the palms of your hands to form a wad that can be held by friction in the upper part of the test tube. Insert the wad just below the mouth of the test tube, and do not allow the copper to contact the nitric acid. Firmly attach the test tube to the 2-holed stopper on the flask assembly. Attach the hose from the aspirator to the tubulation on the flask.

Presentation

Turn on the aspirator and adjust it to produce moderate bubbling in the flask. The color of the liquid in the flask does not change. Insert the rod through the open hole in the test tube stopper. With this rod push the copper ball down into the nitric acid. A red-brown gas will be produced at the copper and be aspirated into the liquid in the filter flask. After about 30 seconds, the indicator solution turns yellow.

PROCEDURE E

Preparation

Pour 300 mL of tap water into the 500-mL filter flask. Add 2 mL of bromocresol green indicator solution to the flask. If the mixture is not blue, add just enough drops of 5M NH_3 to turn it blue.

Assemble the apparatus as illustrated in Figure 3. Insert the short arm of the right-angle bend in the glass tube's stopper and the long arm in the flask's stopper. Seat the stopper in the mouth of the flask and adjust the right-angle bend so its tip is within 1 cm of the bottom of the flask. Seat the other stopper in one end of the glass tube and clamp the other end of the tube to the stand so the tube is horizontal. Set the Bunsen burner under the glass tube. Clamp the filter flask to the other stand, and attach the hose from the aspirator to the tubulation on the flask. Crumple the food wrap into a small ball about 3 cm in diameter, and insert it to the middle of the horizontal 38-mm glass tube.

Presentation

Turn on the aspirator and adjust it to produce moderate bubbling in the flask. The color of the liquid in the flask does not change. Light the burner and place it under the food wrap in the tube. The wrap will char. Continue to heat the wrap until the indicator solution becomes yellow. Stop heating the tube once the color of the indicator has changed.

HAZARDS

To avoid burns, allow the tubes in Procedures C and E to cool before disassembling the apparatus of each.

Concentrated nitric acid is both a strong acid and a powerful oxidizing agent. Contact with combustible materials can cause fires. Contact with the skin can result in severe burns. The vapor irritates the respiratory system, eyes, and other mucous membranes. Spills should be neutralized with sodium bicarbonate ($NaHCO_3$) before being wiped up.

Concentrated aqueous ammonia can irritate the skin, and its vapors are harmful to the eyes and mucous membranes.

Nitrogen dioxide is irritating to the respiratory system; inhaling it can result in severe pulmonary irritation which is not apparent until several hours after exposure. A concentration of 100 ppm is dangerous for even a short period of time, and exposure to concentrations of 200 ppm or more can be fatal.

Sulfur dioxide is an irritating and toxic gas. At concentrations of 3 ppm, the odor of sulfur dioxide is easily detectable. At concentrations over 8 ppm, SO_2 irritates the eyes and throat and induces coughing. Even brief exposure to concentrations over 400 ppm can be fatal.

Hydrogen chloride gas is severely irritating to the eyes and respiratory system. Inhalation must be prevented.

DISPOSAL

The waste solutions should be flushed down the drain with water.

The residue of the food wrap in the tube from Procedure E can be scraped from the tube after it has cooled if the tube is to be used again.

DISCUSSION

In this demonstration, the products of several combustion reactions are shown to be acidic. When sulfur is burned and the product gases dissolved in water, the solution is acidic. When polyvinyl chloride is burned and its combustion products dissolved, an acid solution is obtained. When nitrogen oxides are dissolved, they too produce an acidic solution.

Procedures A and C involve the combustion of sulfur. In Procedure A, which is part of Demonstration 6.18 of Volume 2 of this series, sulfur is burned in a flask containing water and an indicator. When the reaction is complete and the flask is shaken to dissolve the combustion products, the indicator changes color. In Procedure C, the combustion products of sulfur are aspirated through water containing an indicator, and again the indicator changes color.

When sulfur is burned in air, the product is sulfur dioxide:

$$S(s) + O_2(g) \longrightarrow SO_2(g)$$

Sulfur dioxide dissolves in water to form sulfurous acid:

$$SO_2(g) + H_2O(l) \longrightarrow H_2SO_3(aq)$$

Sulfurous acid is a weak acid that ionizes in two steps:

$$H_2SO_3(aq) \rightleftarrows H^+(aq) + HSO_3^-(aq) \qquad pK_1 = 1.76$$

$$HSO_3^-(aq) \rightleftarrows H^+(aq) + SO_3^{2-}(aq) \qquad pK_2 = 7.20$$

Because the pK_1 of sulfurous acid is rather large, its solutions are fairly acidic, and they change the color of bromocresol green indicator, which has a pK_a of 4.8. Sulfurous acid is slowly oxidized by atmospheric oxygen to sulfuric acid:

$$2\ H_2SO_3(aq) + O_2(aq) \longrightarrow 2\ H_2SO_4(aq)$$

After exposure to the atmosphere for 24 hours, a sulfurous acid solution contains enough sulfate ions to form a precipitate of barium sulfate when $Ba^{2+}(aq)$ is added.

Sulfur dioxide is an industrial by-product resulting from burning coal that contains sulfur and from the smelting of sulfide ores. When sulfur dioxide enters the atmosphere, it combines with water in the air to make sulfurous acid. The sulfurous acid is oxidized by atmospheric oxygen to sulfuric acid, a strong acid. This acid is washed to earth by rain (acid rain) and has a damaging effect on streams, lakes, and marble structures. Some industries employ scrubbers to remove sulfur oxides from effluent gases. These scrubbers work in a fashion similar to the apparatus used in Procedure C, where the products of sulfur combustion are passed through water. The sulfur oxides dissolve in the water and are removed from the effluent.

Procedures B and D show that nitrogen dioxide is acidic. Nitrogen dioxide in the atmosphere is a product of the combustion of nitrogen in air. This combustion occurs only at very high temperatures, such as in lightning bolts and in the cylinders of internal-combustion engines.

$$N_2(g) + 2\ O_2(g) \longrightarrow 2\ NO_2(g)$$

Because it is difficult to produce and capture nitrogen dioxide by burning nitrogen in a classroom setting, nitrogen dioxide is generated instead by the reaction of copper with nitric acid:

$$Cu(s) + 4\ HNO_3(aq) \longrightarrow Cu(NO_3)_2(aq) + 2\ NO_2(g) + 2\ H_2O(l)$$

(The reaction of copper with nitric acid also generates small amounts of NO and traces of N_2.) In Procedure B this reaction is done inside a test tube within a flask containing water and an indicator. When the generation of NO_2 is complete, the flask is shaken to dissolve the NO_2, and the indicator changes color. In Procedure D, the NO_2 is aspirated through water containing an indicator and the indicator changes color.

Procedure E deals with the products of the combustion of polyvinyl chloride. A piece of plastic food wrap is heated in a tube, air is drawn through the tube and into water containing an indicator. The indicator changes to its acid color.

The plastic wrap does not burn immediately in a flame, but it does char. If it is heated enough, fumes will continue to be emitted without flame for some time. It can be seen from this demonstration that these fumes are acid. Their noxiousness is a likely cause of deaths from fires in airplanes, where the seating materials and other internal cabin components are made from other forms of polyvinyl chloride (PVC). Polyvinyl chloride is polymerized from vinyl chloride:

$$x \; H_2C=CHCl \longrightarrow [-CH_2-\underset{\underset{Cl}{|}}{CH}-]_x$$

It is ranked 19th among the top 50 chemicals produced in the United States (1985), with an annual production of about 3.9 million tons. It is used widely for making bottles, phonograph records, and floor tile, as well as other products. As this demonstration shows, incineration in open air is *not* a suitable method for the disposal of chlorine-containing plastics. Incineration leads to the production of acidic air pollutants. For this reason, these plastics are relegated to landfill sites.

Combustion reactions that produce acidic products, such as those illustrated in the procedures of this demonstration, are believed to be responsible for a decrease in the pH of rainwater in several localities throughout the world, especially in the developed nations [1–4]. These demonstrations can be used to show that these combustion products produce acids when dissolved in water [5, 6].

REFERENCES

1. G. E. Likens, R. F. Wright, J. N. Galloway, and T. J. Butler, *Sci. Am.* 241(4):43 (1979).
2. *Acid Rain,* American Chemical Society: Washington, D.C. (1982).
3. N. C. Baird, *Chem13News,* 116:8 (1980).
4. I. Peterson, *Sci. News* 122:138 (1982).
5. O. T. Zajicek, *J. Chem. Educ.* 62:158 (1985).
6. L. H. Barrow, *J. Chem. Educ.* 62:339 (1985).

8.18

Acid-Neutralizing Capacity of Lake Beds

When an acidic solution containing a yellow indicator is poured onto marble chips, the color changes to green. When the same yellow solution is poured over granite chips, the solution remains yellow (Procedure A). The same phenomena can be displayed by overhead projection (Procedure B).

MATERIALS FOR PROCEDURE A

For preparation of indicator solutions, see pages 27–29.
For preparation of stock solutions of acids and bases, see pages 30–32.

100 g granite chips, ca. 5 mesh (largest chips ca. 0.5 cm in diameter)

310 mL 0.1M hydrochloric acid, HCl

500 mL distilled water

100 g marble chips, ca. 5 mesh (largest chips ca. 0.5 cm in diameter)

1 mL bromocresol green indicator solution

3 600-mL beakers

glass stirring rod

2 labels for beakers

dropper

MATERIALS FOR PROCEDURE B

For preparation of indicator solutions, see pages 27–29.
For preparation of stock solutions of acids and bases, see pages 30–32.

overhead projector, with transparency and marker

10 g granite chips, ca. 5 mesh (largest chips ca. 0.5 cm in diameter)

65 mL 0.1M hydrochloric acid, HCl

160 mL distilled water

10 g marble chips, ca. 5 mesh (largest chips ca. 0.5 cm in diameter)

1 mL bromocresol green indicator solution

2 150-mL beakers

glass stirring rod

2 10-cm petri dishes

dropper

PROCEDURE A

Preparation

Put 100 g of granite chips in a 600-mL beaker and add 100 mL of 0.1M HCl. Stir the mixture for about 30 seconds and decant the HCl into a sink flushed with running water. Repeat this washing of the granite chips twice, then rinse them in a similar fashion with distilled water. These washings will remove any carbonaceous contamination of the granite chips.

Keep the granite chips in the 600-mL beaker, and put the marble chips in another. Label the appropriate beakers "granite" and "marble."

Pour 5 mL of 0.1M HCl into a clean 600-mL beaker and dilute the acid to 500 mL with distilled water. Add 1 mL bromocresol green indicator solution to the beaker and stir the mixture. The solution should be yellow; if it isn't, add drops of 0.1M HCl until it becomes yellow.

Presentation

Pour half of the diluted HCl solution into the container of marble chips and the other half into the container of granite chips. Swirl the containers to circulate the acid over the chips. The solution over the marble chips will turn green, but that over the granite will remain yellow.

PROCEDURE B

Preparation

Put 10 g of granite chips in a 150-mL beaker and add 20 mL of 0.1M HCl. Stir the mixture for about 30 seconds and decant the HCl into a sink flushed with running water. Repeat this washing of the granite chips twice, then rinse them in a similar fashion with distilled water. This washing will remove any carbonaceous contamination of the granite chips.

Write the labels "marble" and "granite" on an overhead transparency. Place the transparency on the overhead projector and arrange the petri dishes on the transparency so each is adjacent to one of the labels. Place the granite chips and marble chips in the appropriate petri dishes.

Pour 1 mL of 0.1M HCl into a clean 150-mL beaker and dilute the acid to 100 mL with distilled water. Add 1 mL bromocresol green indicator solution to the beaker and stir the mixture. The solution should be yellow; if it isn't, add drops of 0.1M HCl until it just becomes yellow.

Presentation

Pour half of the diluted HCl solution into the petri dish containing marble chips and the other half into the one containing granite chips. Swirl the dishes to circulate the acid over the chips. The solution over the marble chips will turn green, but that over the granite will remain yellow.

DISPOSAL

The waste solutions should be flushed down the drain with water. The granite and marble chips should be rinsed with water and dried for use in repeated performances of this demonstration.

DISCUSSION

The acidity of some lakes has increased dramatically in the last 30 years, while other nearby lakes have remained nearly neutral [1]. The increase in acidity is attributed to acid rain resulting from industrial air pollution (see Demonstration 8.17). Presumably all of the lakes in a region are exposed to the same acid rain. How is it that some remain neutral while others become increasingly acidic? The neutrality of some lakes has been attributed to the soil through which rainwater flows into the lake.

Some soils are a mixture of clays. These are silicate materials, similar to granite. Other soils are sandy and contain limestone, which is a carbonate, as is marble. This demonstration shows that carbonate minerals, like marble, are able to neutralize acid solutions, whereas silicate minerals, like granite, are not. The chemical differences in soils surrounding lakes can be a factor in determining whether the lake remains neutral or becomes more acidic as a result of acid rain.

Bromocresol green indicator is used in this demonstration because it changes color around pH 5.5, which is the approximate pH of normal rainwater and lakes. Thus, even though normal rainwater is acidic, it will produce the basic or neutral color of bromocresol green. Only if it is more acidic than normal will rainwater produce the acidic color of this indicator.

REFERENCE

1. I. Peterson, *Sci. News* 123:332 (1983).

8.19

Amphoteric Properties of the Hydroxides of Aluminum, Zinc, Chromium, and Lead

When aluminum is added to a sodium hydroxide solution in a bottle, enough hydrogen gas is generated to fill two balloons attached to the mouth of the bottle, but a third balloon, rather than being inflated, is drawn into the bottle (Procedure A). A precipitate of aluminum hydroxide (Procedure B), chromium hydroxide (Procedure C), and zinc or lead hydroxide (Procedure D) dissolves in either strong acid or strong base. Overhead projection can be used to display these properties of chromium hydroxide (Procedure E) and zinc or lead hydroxide (Procedure F).

MATERIALS FOR PROCEDURE A

For preparation of stock solutions of acids and bases, see pages 30–32.

100 mL 6M sodium hydroxide, NaOH

5 g aluminum (granules or 1-cm × 10-cm × 1-mm strips; do not use foil—it reacts too quickly)

1-liter glass soda bottle

ring and ring stand

ca. 1 g glass wool

3 round rubber balloons, with inflated diameter of 25 cm

heat-resistant gloves

2-m length of string (optional)

MATERIALS FOR PROCEDURE B

For preparation of stock solutions of acids and bases, see pages 30–32.

either

>   solution from Procedure A
>   filter funnel
>   filter paper
>   250-mL Erlenmeyer flask
>   10 mL 1M nitric acid, HNO_3

or

>   200 mL 0.1M aluminum nitrate, $Al(NO_3)_3$ (To prepare 1 liter of stock solution, dissolve 38 g $Al(NO_3)_3 \cdot 9H_2O$ in 600 mL of distilled water, and dilute the resulting solution to 1.0 liter.)

>   30 mL 1M sodium hydroxide, NaOH

50 mL 1M nitric acid, HNO_3

50 mL 1M sodium hydroxide, NaOH

2 300-mL tall-form beakers or 250-mL beakers

2 glass stirring rods

black background (e.g., 15-cm $\times$ 25-cm poster board)

2 50-mL graduated cylinders

MATERIALS FOR PROCEDURE C

For preparation of stock solutions of acids and bases, see pages 30–32.

200 mL 0.010M chromium(III) sulfate, $Cr_2(SO_4)_3$ (To prepare 1 liter of stock solution, dissolve 7.2 g $Cr_2(SO_4)_3 \cdot 18H_2O$ in 600 mL of distilled water, and dilute the resulting solution to 1.0 liter.)

200 mL 0.010M iron(III) sulfate, $Fe_2(SO_4)_3$ (To prepare 1 liter of stock solution, dissolve 5.6 g $Fe_2(SO_4)_3 \cdot 9H_2O$ in 600 mL of distilled water, and dilute the resulting solution to 1.0 liter.)

40 mL 1M sodium hydroxide, NaOH

20 mL 1M nitric acid, HNO_3

4 300-mL tall-form beakers or 250-mL beakers

4 glass stirring rods

2 25-mL graduated cylinders

lighted background (optional)

MATERIALS FOR PROCEDURE D

For preparation of stock solutions of acids and bases, see pages 30–32.

100 mL 0.1M zinc(II) nitrate, $Zn(NO_3)_2$ (To prepare 1 liter of stock solution, dissolve 30 g $Zn(NO_3)_2 \cdot 6H_2O$ and 1 mL of 1M nitric acid in 600 mL of distilled water, and dilute the resulting solution to 1.0 liter.)

>   *or*

>   100 mL 0.1M lead(II) nitrate, $Pb(NO_3)_2$ (To prepare 1 liter of stock solution, dissolve 33 g $Pb(NO_3)_2$ and 1 mL of 1M nitric acid in 600 mL of distilled water, and dilute the resulting solution to 1.0 liter.)

75 mL 1M sodium hydroxide, NaOH

40 mL 1M nitric acid, HNO_3

2 300-mL tall-form beakers or 250-mL beakers

2 stirring rods

black background (e.g., 15-cm × 25-cm poster board)

2 50-mL graduated cylinders

MATERIALS FOR PROCEDURE E

For preparation of stock solutions of acids and bases, see pages 30–32.

overhead projector, with transparency and marker

40 mL 0.010M chromium(III) sulfate, $Cr_2(SO_4)_3$ (To prepare 1 liter of stock solution, dissolve 7.2 g $Cr_2(SO_4)_3 \cdot 18H_2O$ in 600 mL of distilled water, and dilute the resulting solution to 1.0 liter.)

40 mL 0.010M iron(III) sulfate, $Fe_2(SO_4)_3$ (To prepare 1 liter of stock solution, dissolve 5.6 g $Fe_2(SO_4)_3 \cdot 9H_2O$ in 600 mL of distilled water, and dilute the resulting solution to 1.0 liter.)

8 mL 1M sodium hydroxide, NaOH

4 mL 1M nitric acid, HNO_3

4 10-cm petri dishes

2 10-mL graduated cylinders

MATERIALS FOR PROCEDURE F

For preparation of stock solutions of acids and bases, see pages 30–32.

overhead projector

20 mL 0.1M zinc(II) nitrate, $Zn(NO_3)_2$ (To prepare 1 liter of stock solution, dissolve 30 g $Zn(NO_3)_2 \cdot 6H_2O$ and 1 mL of 1M nitric acid in 600 mL of distilled water, and dilute the resulting solution to 1.0 liter.)

   or

20 mL 0.1M lead(II) nitrate, $Pb(NO_3)_2$ (To prepare 1 liter of stock solution, dissolve 33 g $Pb(NO_3)_2$ and 1 mL of 1M nitric acid in 600 mL of distilled water, and dilute the resulting solution to 1.0 liter.)

15 mL 1M sodium hydroxide, NaOH

8 mL 1M nitric acid, HNO_3

2 10-cm petri dishes

2 10-mL graduated cylinders

PROCEDURE A

Preparation

Secure a 1-liter glass soda bottle to a ring stand. Pour 100 mL of 6M NaOH into the bottle. Form two plugs of glass wool and lodge one in the mouth of the bottle. These plugs are to minimize the spattering of NaOH into the balloon during the demonstration.

Presentation

Remove the glass-wool plug from the bottle and drop 5 g of aluminum strips into the bottle. Quickly reinsert the glass-wool plug in the neck of the bottle and slip the opening of a balloon over the top.

The mixture in the bottle will begin to fizz, slowly at first, then vigorously. The bottle will become very hot. Wear heat-resistant gloves to handle the bottle. Within 10 minutes the balloon will inflate to a diameter of 18 cm. Remove the balloon and tie off the opening. Push a fresh glass-wool plug in the mouth of the bottle and attach a second balloon to the bottle. If desired, tether the first balloon with a 2-meter length of string, release the balloon, and allow it to float upward. (**Caution: do not detonate the balloon! It may contain concentrated NaOH solution, which can spatter over the room.**) When the second balloon is inflated to a diameter of 18 cm, tie it off and attach a third balloon. In several minutes, the third balloon will be drawn into the bottle. (The turbid mixture in the bottle can be used in Procedure B of this demonstration.)

PROCEDURE B

Preparation and Presentation

Pour 100 mL of 0.1M $Al(NO_3)_3$ into each of the two 300-mL beakers.† Add 15 mL of 1M NaOH to each beaker, and stir the resulting gelatinous mixtures.

Place each of the beakers before a black background. Pour 50 mL of 1M HNO_3 into one of the beakers and stir the mixture. The gelatinous precipitate will dissolve. Pour 50 mL of 1M NaOH into the other beaker and stir the mixture. The gelatinous precipitate will dissolve in this beaker also.

---

† If the solution from Procedure A is available, remove the balloon from the bottle. Place a funnel lined with filter paper in the mouth of a 250-mL Erlenmeyer flask. Pour the turbid mixture from the bottle into the funnel, collecting the filtrate. Discard the filter paper and filtered solid in a waste receptacle. Pour half of the filtrate from the Erlenmeyer flask into each of the two 300-mL beakers and place a stirring rod in each beaker. While stirring each mixture, slowly add 1M HNO_3 to each beaker until a gelatinous precipitate forms in each.

PROCEDURE C

Preparation

Pour 100 mL 0.010M $Cr_2(SO_4)_3$ into each of two 300-mL beakers. Pour 100 mL of 0.010M $Fe_2(SO_4)_3$ into each of the other two 300-mL beakers. Place a stirring rod in each beaker.

Presentation

If desired, set the beakers before a lighted background. Add 5 mL of 1M NaOH to each container and stir the mixtures. In the two containers of Cr(III), a green precipitate of $Cr(OH)_3$ forms. In the two containers holding Fe(III), a red precipitate of $Fe(OH)_3$ forms.

Add 10 mL of 1M HNO_3 to one of the beakers containing the red $Fe(OH)_3$ precipitate and stir the mixture. The red precipitate redissolves. Add 10 mL of 1M HNO_3 to the other container of green $Cr(OH)_3$ precipitate and stir the mixture. The green precipitate redissolves.

Add 10 mL of 1M NaOH to the other container of the red $Fe(OH)_3$ precipitate and stir the mixture. The red precipitate remains. Add 10 mL of 1M NaOH to the other beaker containing the green $Cr(OH)_3$ precipitate and stir the mixture. The green precipitate redissolves.

PROCEDURE D

Preparation

Pour 100 mL of 0.1M $Zn(NO_3)_2$ or $Pb(NO_3)_2$ into each of the two 300-mL beakers. Place a stirring rod in each beaker.

Presentation

Place the beakers before the black background. Add 25 mL of 1M NaOH to each beaker and stir the mixtures. A white precipitate of the metal hydroxide will form in each beaker. To one of the beakers add an additional 25 mL of 1M NaOH and stir the mixture thoroughly. The white precipitate will redissolve after the second addition of sodium hydroxide. To the other beaker, add 40 mL of 1M HNO_3 and stir the mixture thoroughly. The white precipitate in the second beaker will dissolve after the nitric acid is added.

PROCEDURE E

Preparation

Pour 20 mL 0.010M $Cr_2(SO_4)_3$ into each of two 10-cm petri dishes. Pour 20 mL of 0.010M $Fe_2(SO_4)_3$ into each of the other two 10-cm petri dishes.

Presentation

Set the petri dishes on a transparency atop an overhead projector and write the appropriate label ("$Cr_2(SO_4)_3$" or "$Fe_2(SO_4)_3$") next to each dish. Add 1 mL of 1M NaOH to each dish and swirl them to mix their contents. In the two containers of Cr(III), a green precipitate of $Cr(OH)_3$ forms. In the two containers holding Fe(III), a red precipitate of $Fe(OH)_3$ forms. Both of these precipitates are gelatinous and transparent, and their projected images are colored.

Add 2 mL of 1M HNO_3 to one of the petri dishes containing the red $Fe(OH)_3$ precipitate and swirl the dish. The red precipitate redissolves. Add 2 mL of 1M HNO_3 to one of the dishes containing the green $Cr(OH)_3$ precipitate and swirl the dish. The green precipitate redissolves.

Add 2 mL of 1M NaOH to the other dish containing the red $Fe(OH)_3$ precipitate and swirl the dish. The red precipitate remains. Add 2 mL of 1M NaOH to the other dish containing the green $Cr(OH)_3$ precipitate and swirl the dish. The green precipitate redissolves.

PROCEDURE F

Preparation

Pour 20 mL of 0.1M $Zn(NO_3)_2$ or $Pb(NO_3)_2$ into each of the two 10-cm petri dishes.

Presentation

Set the petri dishes on the overhead projector. Add 5 mL of 1M NaOH to each dish and swirl the dishes to mix their contents. A precipitate of the metal hydroxide will form in each dish, darkening the projected image of each. To one of the dishes add an additional 5 mL of 1M NaOH and swirl the dish. The precipitate will redissolve after the second addition of sodium hydroxide, and the projected image will again appear clear. To the other dish add 8 mL of 1M HNO_3 and swirl the dish. The precipitate in the second dish will also dissolve and the projected image will become clear after the nitric acid is added.

HAZARDS

The bottle in Procedure A becomes very hot. Handle the bottle only with insulated gloves.

The hot vapors escaping from the bottle in Procedure A can carry droplets of hot concentrated sodium hydroxide solution. These vapors can be extremely irritating to the eyes and respiratory system. Anyone close to the bottle should protect his or her eyes with safety goggles. Do not breathe these vapors when handling the bottle.

Concentrated sodium hydroxide solutions can cause severe burns of the eyes, skin, and mucous membranes.

Lead nitrate is harmful if taken internally. The dust from lead salts should not be inhaled.

Nitric acid can irritate the skin; its vapors are irritating to the eyes and respiratory system.

DISPOSAL

The waste solutions should be flushed down the drain with large amounts of water. Dissolve any hydroxide precipitates in 1M HNO_3 before flushing the mixture down the drain.

DISCUSSION

A substance that reacts with both acids and bases—acting as an acid in the presence of a base and as a base when in the presence of an acid—is said to be amphoteric. The hydroxides of the aluminum, chromium, zinc, and lead are amphoteric, as demonstrated in these procedures.

In Procedure A, hydrogen gas is liberated by the reaction of aluminum with sodium hydroxide [1]:

$$2\ Al(s) + 2\ OH^-(aq) + 6\ H_2O(l) \longrightarrow 2\ [Al(OH)_4]^-(aq) + 3\ H_2(g)$$

The hydrogen gas generated by the reaction is collected in balloons. The directions in Procedure A produce enough hydrogen to fill two balloons. However, by the time the third balloon is placed over the mouth of the bottle, the reaction is nearly over. As the reagents are consumed, the generation of the gas slows. Eventually the gases in the bottle cool and the third balloon is drawn into the bottle. The hydrogen-filled balloons produced in this procedure are buoyant in air and rise when released. Because they also contain some air and perhaps some droplets of solution, they may not be quite as buoyant as balloons filled from a tank of hydrogen.

It is rather unusual behavior for a metal to dissolve in a strong base with the liberation of hydrogen. This is generally the behavior of a metal in the presence of an acid. However, aluminum is an unusual metal. The only strong acid with which it reacts readily is hydrochloric acid. It is so inert to other strong acids that nitric acid is shipped in aluminum drums. However, aluminum is corroded by strong alkalis such as sodium and potassium hydroxides, sodium carbonate, and lime (calcium oxide).

In Procedure B, the amphoteric properties of aluminum hydroxide are demonstrated. A precipitate of $Al(OH)_3$ dissolves in nitric acid, neutralizing it, and forming hexaquoaluminum(III) ions [1]:

$$Al(OH)_3(s) + 3\ H^+(aq) + 3\ H_2O(l) \longrightarrow [Al(H_2O)_6]^{3+}(aq)$$

Aluminum hydroxide also dissolves in the base NaOH, neutralizing it and forming tetrahydroxoaluminate(III) ions:

$$Al(OH)_3(s) + OH^-(aq) \longrightarrow [Al(OH)_4]^-(aq)$$

These reactions show that aluminum hydroxide behaves as a base in the presence of an acid such as nitric acid, and it behaves as an acid in the presence of a base such as sodium hydroxide.

Procedures C and E show the amphoteric properties of chromium hydroxide in contrast with the purely basic properties of iron hydroxide. Iron hydroxide dissolves in acid but not in base. However, chromium hydroxide dissolves both in nitric acid and in sodium hydroxide. Like aluminum hydroxide, $Cr(OH)_3$ dissolves in the base $NaOH$, neutralizing it and forming a chromite solution [1]. The chromium-containing species in the chromite solution have not been determined; however, both $[Cr(OH)_6]^{3-}$ and $[Cr(OH)_5(H_2O)]^{2-}$ are likely.

$$Cr(OH)_3(s) + 3\ OH^-(aq) \longrightarrow [Cr(OH)_6]^{3-}(aq)$$

$$Cr(OH)_3(s) + 2\ OH^-(aq) + H_2O(l) \longrightarrow [Cr(OH)_5(H_2O)]^{2-}(aq)$$

Chromium hydroxide also dissolves in nitric acid, neutralizing it, and forming hexaaquochromium(III) ions:

$$Cr(OH)_3(s) + 3\ H^+(aq) + 6\ H_2O(l) \longrightarrow [Cr(H_2O)_6]^{3+}(aq) + 3\ H_2O(l)$$

The sulfate salts are used in this demonstration to improve the solubility of Cr(III). In aqueous solution, $[Cr(H_2O)_6]^{3+}$ is not as soluble as $[Cr(H_2O)_5(SO_4)]^+$, which forms in the presence of sulfate ions. The sulfate assures that the solutions are clear rather than turbid as the $Cr(OH)_3$ dissolves.

Procedures D and F show the amphoteric properties of zinc or lead hydroxide. These hydroxides dissolve in nitric acid.

$$Zn(OH)_2(s) + 2\ H^+(aq) + 2\ H_2O(l) \longrightarrow [Zn(H_2O)_4]^{2+}(aq)$$

$$Pb(OH)_2(s) + 2\ H^+(aq) + 4\ H_2O(l) \longrightarrow [Pb(H_2O)_6]^{2+}(aq)$$

They also dissolve in sodium hydroxide, forming hydroxo complexes.

$$Zn(OH)_2(s) + 2\ OH^-(aq) \longrightarrow [Zn(OH)_4]^{2-}(aq)$$

$$Pb(OH)_2(s) + 4\ OH^-(aq) \longrightarrow [Pb(OH)_6]^{4-}(aq)$$

Further demonstrations and discussion of the complex chemistry of the aqueous Pb^{2+} ion are contained in Demonstration 4.3 in Volume 1 of this series.

REFERENCE

1. F. A. Cotton and G. Wilkinson, *Advanced Inorganic Chemistry,* 3d ed., John Wiley and Sons: New York (1972).

8.20

Differences Between
Acid Strength and Concentration

The pH of 0.1M hydrochloric acid is about the same as that of 0.1M sulfuric acid and lower than that of 0.1M acetic acid. All three acids are neutralized with sodium hydroxide solution. Neutralization of the hydrochloric acid requires only half as much sodium hydroxide as does neutralization of the sulfuric acid, although they have nearly the same pH. Neutralization of the hydrochloric acid requires the same amount of sodium hydroxide as does the neutralization of the acetic acid, although they have different pH values (Procedure A) [1, 2]. Similar differences between pH and neutralizing capacity are also demonstrated with bases (Procedure B).

MATERIALS FOR PROCEDURE A

For preparation of indicator solutions, see pages 27–29.
For preparation of stock solutions of acids and bases, see pages 30–32.

40 mL 0.1M hydrochloric acid, HCl

40 mL 0.1M sulfuric acid, H_2SO_4

40 mL 0.1M acetic acid, $HC_2H_3O_2$

2 mL phenolphthalein indicator solution

200 mL 0.1M sodium hydroxide, NaOH

3 250-mL beakers

3 labels for beakers

dropper

3 glass stirring rods

pH meter, standardized, or pH-indicating paper, pH 0–3 range

100-mL graduated cylinder

MATERIALS FOR PROCEDURE B

For preparation of indicator solutions, see pages 27–29.
For preparation of stock solutions of acids and bases, see pages 30–32.

40 mL 0.1M sodium hydroxide, NaOH

40 mL 0.1M barium hydroxide, $Ba(OH)_2$ (To prepare 1 liter of solution, dissolve

17 g of $Ba(OH)_2$ in 600 mL of distilled water, and dilute the resulting solution to 1.0 liter.)

40 mL 0.1M ammonia, NH_3 (To prepare 1 liter of stock solution, pour 7 mL of concentrated [15M] NH_3 into 600 mL of distilled water, and dilute the resulting solution to 1.0 liter.)

2 mL bromocresol green indicator solution

200 mL 0.1M hydrochloric acid, HCl

3 250-mL beakers

3 labels for beakers

dropper

3 glass stirring rods

pH meter, standardized, or pH-indicating paper, pH 11–13 range

100-mL graduated cylinder

PROCEDURE A

Preparation

Label the three 250-mL beakers with the names of the three acids: hydrochloric acid, acetic acid, and sulfuric acid. Pour 40 mL of the appropriate 0.1M acid into each beaker. Add 10 drops of phenolphthalein indicator solution to each beaker and stir the mixtures thoroughly.

Presentation

Use the pH meter or the pH test paper to measure the pH of each of the three acid solutions. The pH of the HCl and H_2SO_4 solutions will be near 1, and that of the $HC_2H_3O_2$ will be about 3. Record the pH values.

While stirring the HCl solution, *slowly* pour 0.1M NaOH from the 100-mL graduated cylinder until the phenolphthalein indicator changes from colorless to pink. The volume of NaOH solution required to do this will be close to 40 mL. Record the volume.

Repeat the procedure described in the preceding paragraph with the beaker of $HC_2H_3O_2$ solution and with the beaker of H_2SO_4 solution. The $HC_2H_3O_2$ will require about 40 mL of NaOH, and the H_2SO_4 will use about 80 mL.

PROCEDURE B

Preparation

Label the three 250-mL beakers with the names of the three bases: sodium hydroxide, ammonia, and barium hydroxide. Pour 40 mL of the appropriate 0.1M base into each beaker. Add 10 drops of bromocresol green indicator solution to each beaker and stir the mixtures thoroughly.

Presentation

Use the pH meter or the pH test paper to measure the pH of each of the three base solutions. The pH of the NaOH and $Ba(OH)_2$ solutions will be near 13, and that of the NH_3 will be about 11. Record the pH values.

While stirring the NaOH solution, *slowly* pour 0.1M HCl from the 100-mL graduated cylinder until the bromocresol green indicator changes from blue to green or yellow. The volume of HCl solution required to do this will be close to 40 mL. Record the volume.

Repeat the procedure described in the preceding paragraph with the beaker of NH_3 solution and with the beaker of $Ba(OH)_2$ solution. The NH_3 will require about 40 mL of HCl, while the $Ba(OH)_2$ will use about 80 mL.

HAZARDS

Soluble barium compounds are toxic and, if ingested, cause nausea, vomiting, stomach pains, and diarrhea.

DISPOSAL

The waste solutions should be flushed down the drain with water.

DISCUSSION

The notions of acid concentration (as expressed by molarity), acid equivalence (as determined by a titration), and acid strength (as expressed by pH) are sometimes confused. Procedure A of this demonstration illustrates that these are three different properties of acids. The concentration (molarity) of an acid is an operational expression of how the solution can be prepared. That is, it indicates how much of the acid must be used to prepare the solution. The equivalence of an acid solution expresses how much base is required to neutralize it. The equivalence depends on the concentration of the solution and on the number of acidic hydrogens in each molecule of acid. The pH of an acid solution depends not only on the concentration and equivalence of the solution of the acid, but is also related to the degree of ionization of the particular acid in the solution. In Procedure B, these three concepts—strength, concentration, and equivalence—are also applied to bases.

The differences between strength, concentration, and equivalence are illustrated for acids in Procedure A. The procedure uses solutions of hydrochloric acid, sulfuric acid, and acetic acid that have the same concentration, namely 0.1M. This means that each solution can be prepared by dissolving 0.1 mole of its acid in 1 liter of solution. Although the solutions of these three acids have the same concentration, they do not have the same pH. The pH of 0.1M HCl is about the same as that of 0.1M H_2SO_4, namely 1, and both of these have lower pH values than 0.1M $HC_2H_3O_2$, which has a pH of about 3. This indicates that the acid strength of hydrochloric acid is about the same as that of sulfuric acid, but both of these are stronger acids than acetic acid. When 40-mL samples of these acids are titrated with sodium hydroxide solution to a phenolphthalein end point, the hydrochloric acid requires about the same volume of NaOH as does the acetic acid, and the sulfuric acid requires twice as much as either of the other acids. This titration distinguishes acid equivalence from acid strength and acid concentration. All three have the same molar concentration, yet the amount of NaOH required for neutralization of one is different from that of the other two. The two acids of similar strength, HCl and H_2SO_4, as indicated by their pH values, require different amounts of NaOH for neutralization. Two acids that require the same amount of NaOH for neutralization, HCl and $HC_2H_3O_2$, have different pH values and, therefore, different strengths. The results of the demonstration can be summarized in tabular form to highlight the differences among concentration, strength, and equivalence:

Property:	concentration	strength	equivalence
Expressed through:	molarity	pH	mL of titrant
HCl	0.1	1	40
H_2SO_4	0.1	1	80
$HC_2H_3O_2$	0.1	3	40

The neutralization of sulfuric acid requires twice as much sodium hydroxide as the other two acids, because it is a diprotic acid and the other two are monoprotic. This means that each mole of sulfuric acid contains 2 moles of replaceable hydrogens, whereas each mole of hydrochloric acid and acetic acid has only 1 mole of replaceable hydrogens. Even though sulfuric acid is a diprotic acid, the pH of the sulfuric acid is nearly identical with that of the hydrochloric acid, because the first ionization of sulfuric acid, which is complete, suppresses the second ionization, which has an ionization constant of $K_a = 2 \times 10^{-2}$ [3]. The pH of acetic acid is higher than that of the other monoprotic acid, hydrochloric acid, because it is a weak acid.

$$HC_2H_3O_2(aq) \rightleftarrows H^+(aq) + C_2H_3O_2^-(aq)$$

$$K_a = \frac{[H^+][C_2H_3O_2^-]}{[HC_2H_3O_2]} = 1.8 \times 10^{-5}$$

Acetic acid is only partly ionized in solution, while hydrochloric acid is completely ionized.

For bases, a similar pattern is demonstrated in Procedure B, comparing the amounts of hydrochloric acid required to neutralize volumes of 0.1M NaOH, 0.1M NH_3, and 0.1M $Ba(OH)_2$. In this case, the pH of $Ba(OH)_2$, which requires twice as much acid for neutralization as the NaOH or NH_3, is about 0.3 of a unit higher than that of the NaOH. However, this pH difference is much smaller than that between NaOH and NH_3, which require the same amount of acid for neutralization. Although the demonstration of the differences among concentration, strength, and equivalence with these bases is not as clear-cut as with the acids, it is still effective.

REFERENCES

1. R. Perkins, *Chem13News* 147 (Feb.): 11 (1984).
2. M. J. Webb, *J. Chem. Educ.* 58: 193 (1981).
3. J. A. Dean, Ed., *Lange's Handbook of Chemistry*, 13th ed., McGraw-Hill Book Co.: New York (1985).

8.21

Conductivity and Extent
of Dissociation of Acids
in Aqueous Solution

The addition of a few drops of universal indicator to 0.1M solutions of a variety of strong and weak acids produces solutions of different colors, allowing the acids to be arranged in order of acidity. When the same solutions are tested with a conductivity probe, the weaker acids produce a dim glow of the light bulb, whereas the strong acids cause the bulb to glow brightly (Procedure A). A comparison of the conductivities of various concentrations of hydrochloric acid with that of 2M acetic acid allows an estimation of the value of the ionization constant of acetic acid (Procedure B).

MATERIALS FOR PROCEDURE A

For preparation of indicator solutions, see pages 27–29.
For preparation of stock solutions of acids and bases, see pages 30–32.

100 mL 0.1M hydrochloric acid, HCl

100 mL 0.1M acetic acid, $HC_2H_3O_2$

100 mL 0.1M citric acid, $H_3C_6H_5O_7$ (To prepare 1 liter of solution, dissolve 21 g of $H_3C_6H_5O_7 \cdot H_2O$ in 600 mL of distilled water and dilute the resulting solution to 1.0 liter.)

100 mL 0.1M malonic acid, $H_2C_3H_2O_4$ (To prepare 1 liter of solution, dissolve 10.4 g of $H_2C_3H_2O_4$ in 600 mL of distilled water and dilute the resulting solution to 1.0 liter.)

100 mL 0.1M ascorbic acid, $HC_6H_7O_6$ (To prepare 1 liter of solution, dissolve 17 g of $HC_6H_7O_6$ in 600 mL of distilled water and dilute the resulting solution to 1.0 liter.)

100 mL 0.1M propanoic acid, $HC_3H_5O_2$ (To prepare 1 liter of solution, dissolve 7.4 g of $HC_3H_5O_2$ in 600 mL of distilled water and dilute the resulting solution to 1.0 liter.)

100 mL 0.1M glycine, $HC_2H_4O_2N$ (To prepare 1 liter of solution, dissolve 7.5 g of $HC_2H_4O_2N$ in 600 mL of distilled water and dilute the resulting solution to 1.0 liter.)

100 mL 0.1M alanine, $HC_3H_6O_2N$ (To prepare 1 liter of solution, dissolve 8.9 g of $HC_3H_6O_2N$ in 600 mL of distilled water and dilute the resulting solution to 1.0 liter.)

2 mL universal indicator solution, 1–10 pH range

8 250-mL beakers or 300-mL tall-form beakers

8 labels for beakers

white background (e.g., 20-cm × 1-m poster board)

dropper

8 stirring rods

light-bulb conductivity tester (optional)

250-mL wash bottle, filled with distilled water (optional)

400-mL beaker (optional)

MATERIALS FOR PROCEDURE B

For preparation of stock solutions of acids and bases, see pages 30–32.

80 mL 2M hydrochloric acid, HCl

40 mL 2M acetic acid, $HC_2H_3O_2$

140 mL distilled water

special conductivity tester (See Procedure B for assembly instructions.)

7 100-mL beakers

50-mL graduated cylinder

10-mL pipette

4 stirring rods

3 4-mL pipettes

PROCEDURE A

Preparation

Label the eight 250-mL beakers with one each of these labels: "hydrochloric acid," "acetic acid," "citric acid," "malonic acid," "ascorbic acid," "propanoic acid," "glycine," and "alanine." Pour 100 mL of the appropriate 0.1M acid solution into each beaker. Arrange the beakers in a row before the white background.

Presentation

Add 4 drops of universal indicator to each beaker and stir each mixture. Arrange the beakers by color in spectral order from red to yellow (see the table on the following page).

Solution	Color
hydrochloric acid	bright red
malonic acid	orange-red
citric acid	orange-red
ascorbic acid	orange
acetic acid	orange
propanoic acid	orange
glycine	yellow
alanine	yellow

Turn on the conductivity tester and dip its electrodes to the bottom of the alanine solution. The bulb will glow very dimly. Rinse the electrodes with distilled water from the wash bottle, collecting the rinse in the 400-mL beaker. Dip the electrodes to the bottom of the beaker of glycine solution. Compare the brightness of the bulb with its intensity in the alanine solution. Proceed in this fashion, testing each solution in the order in which they are arranged by color. The brightness of the bulb will increase sequentially from alanine to hydrochloric acid.

PROCEDURE B

Preparation

Construct a special conductivity tester as illustrated in Figure 1. This tester allows a single light bulb to be connected to either of two sets of identical electrodes. Cut the wooden form out of a 30-cm piece of 1 × 6 lumber. Attach a light-bulb socket to the center of the wooden plank. Attach two momentary, normally open, push-button switches to the board near the handle. Fashion four 2-cm × 10-cm electrodes from 22-gauge copper metal sheet. Screw the electrodes to the wood. Use 18-gauge lamp cord to make the electrical connections indicated in the schematic (see Figure 2). Connect one lead from the 110-volt plug to one terminal of the light-bulb socket and the other lead from the plug to one pole of one of the button switches. Connect this pole to one pole of the other button switch. Connect the free pole of button 1 to one of the electrodes in set 1.

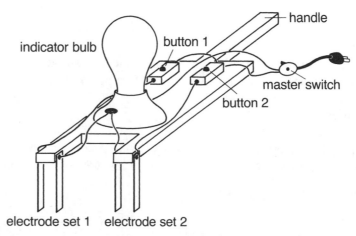

Figure 1. Conductivity tester.

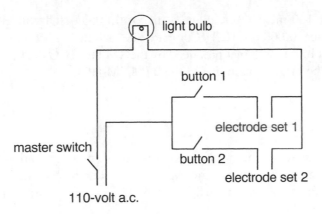

Figure 2. Wiring diagram for conductivity tester.

Likewise, attach the free pole of button 2 to one of the electrodes in set 2. Connect the two remaining electrodes to the free lamp terminal. Be certain that there are no bare electrical connections near the handle end of the tester. If necessary, cover bare connections with electrical tape.

Insert a master line switch in the cord going to the 110-volt plug. Mount a 150- or 100-watt light bulb in the socket.

Presentation

Pour 40 mL of 2M HCl into each of two 100-mL beakers. With the master switch off, insert one set of electrodes on the special conductivity tester in each beaker. Make sure both sets are immersed to the same depth. Turn on the master switch. Press button 1 and the bulb will light. Press button 2 and the bulb will glow just as brightly. Turn off the master switch.

Pour 40 mL of 2M acetic acid into another 100-mL beaker. Rinse set 2 of the electrodes with distilled water and immerse this set in the beaker of acetic acid. Turn on the master switch and alternately press buttons 1 and 2. When button 1 is pressed, the electrodes in the HCl beaker are connected, and the bulb glows brightly. However, when button 2 is pressed and the electrodes in the beaker of acetic acid are in the circuit, the glow from the bulb is much weaker. Turn off the master switch.

With a pipette, remove 10 mL of the 2M HCl solution from the beaker and transfer it to another 100-mL beaker. Add 30 mL of distilled water and stir the mixture. Rinse the electrodes that were in the 2M HCl with distilled water and insert them in the beaker of diluted (0.5M) HCl. Turn on the master switch and alternately press buttons 1 and 2 to compare the brightness of the bulb produced by the 0.5M HCl with the brightness produced by the 2M acetic acid. The 0.5M HCl produces a brighter glow than the 2M $HC_2H_3O_2$.

With a pipette, remove 4 mL of 0.5M HCl from the beaker, transfer it to another 100-mL beaker, add 36 mL of distilled water, and stir the mixture. Compare the brightness of the bulb when this solution of 0.05M HCl is used with that produced by the 2M $HC_2H_3O_2$. Still, the glow with HCl is brighter.

Transfer 4 mL of 0.05M HCl to another 100-mL beaker, add 36 mL of distilled water, and stir the mixture. Test this solution, 0.005M HCl, and the 2M $HC_2H_3O_2$ with the special conductivity tester. Now the glow of the bulb will be nearly the same with both solutions.

In another 100-mL beaker, dilute 4 mL of 0.005M HCl with 36 mL of distilled water to produce 0.0005M HCl. Compare the brightness of the bulb's glow produced by this solution with that produced by the 2M $HC_2H_3O_2$. The glow from the 2M $HC_2H_3O_2$ will be brighter than that of the 0.0005M HCl.

HAZARDS

Hydrochloric acid is harmful to the eyes and skin and irritating to mucous membranes. Its vapors are irritating to the eyes and respiratory system.

Glacial acetic acid can irritate the skin, and its vapors are irritating to the eyes and respiratory system.

Malonic acid is a strong irritant to the skin, eyes, and mucous membranes.

Severe electrical shock can result if the electrodes of the conductivity tester are touched while it is turned on and one of the buttons is pressed.

DISPOSAL

The waste solutions should be flushed down the drain with water.

DISCUSSION

The acidity of a solution depends on the concentration of the acid in the solution, and it is a measure of the extent of ionization of that acid. In Procedure A the acidities of solutions of various acids are compared. All acids are at the same concentration, namely 0.1M, but they have various values of pH, which is revealed by differences in the color of a universal indicator. Each of the solutions is also investigated with a light-bulb conductivity tester. This reveals that the more acidic solutions are more conductive. In Procedure B the conductivity of an acetic acid solution is compared with the conductivity of various concentrations of hydrochloric acid. By assuming that solutions which light the bulb to the same brightness contain the same concentration of ions, an estimate of the dissocation constant, K_a, for acetic acid can be made.

The universal indicator described in Procedure A exhibits the following colors in the acid pH range:

pH 1: deep red
pH 2: red
pH 3: orange-red
pH 4: red-orange
pH 5: orange
pH 6: yellow

Grouping the solutions by color arranges them by pH. The acids fall into one of four groups. The most acidic group contains only hydrochloric acid, whose solution is deep red. The next most acidic group, with an indicator color of orange-red, includes malonic and citric acids. Next comes the orange group: ascorbic, acetic, and propanoic acids. Least acidic are the acids that give a yellow color: glycine and alanine. When the acid solutions are tested with the conductivity tester, they fall into the same groups. The most acidic solution lights the bulb brightest, and the least acidic produces the faintest glow. Because conductivity depends on the concentration of ions in the solution, and each of the solutions has the same nominal concentration, 0.1M, each acid must dissociate or ionize to a different extent. Those that ionize the most are the most acidic; those that ionize the least are the least acidic. The results of the demonstration can be compared with the dissociation constants of the acids in the following table:

pK$_a$ and pH Values for Several Aqueous Weak Acids

Acid	pK$_a$ [1]	pH of 0.1M solution
Malonic	2.83	1.94
Citric	3.14	2.09
Ascorbic	4.10	2.56
Acetic	4.75	2.88
Propanoic	4.87	2.94
Glycine	9.78	5.39
Alanine	9.87	5.44

Solutions of acids with similar pK$_a$ values have similar pH values and conductivities.

The conductivity of a weak acid solution can be used to estimate the value of its K$_a$. In Procedure B the conductivity of a 2M acetic acid solution is compared with that of solutions of hydrochloric acid with a range of concentrations. The comparison is made with a light-bulb conductivity tester. The 2M acetic acid lights the bulb to the same brightness as 0.005M hydrochloric acid and is assumed to contain the same concentration of ions. Because hydrochloric acid is a strong acid, it is completely ionized in solution, and the concentration of hydrogen and chloride ions in 0.005M HCl is 0.005M. Therefore, the concentration of hydrogen and acetate ions in 2M acetic acid is 0.005M, and the concentration of undissociated acetic acid molecules is 2M − 0.005M = 1.995M. Using these values, the value of K$_a$ for acetic acid can be calculated:

$$K_a = \frac{[H^+][C_2H_3O_2^-]}{[HC_2H_3O_2]} = \frac{(0.005)(0.005)}{(1.995)} = 1.3 \times 10^{-5}$$

The calculated value for K$_a$, 1.3×10^{-5}, is within 30% of the literature value of 1.8×10^{-5} at 25°C [1]. Considering the somewhat subjective nature of assessing the brightness of the bulb, this result is certainly in reasonable agreement with the accepted value.

In this demonstration the conductivity of the solution is monitored with a light-bulb conductivity apparatus. This tester uses a light bulb whose power cord is interrupted by a pair of electrodes that are immersed in the solution being tested. In order for the bulb to light, current must flow between the electrodes. A 15-watt bulb requires about 0.13 amps of current to be fully lit, whereas a 150-watt bulb needs 1.3 amps. The more current needed for full brightness, the more ions must be present to carry this current. Therefore, a 150-watt bulb begins to diminish in brightness at a higher ion concentration than does a 15-watt bulb. Thus, a high-wattage bulb will show the decrease in concentration of ions in the solution over a wider range than a low-wattage bulb.

REFERENCE

1. R. C. Weast, Ed., *CRC Handbook of Chemistry and Physics*, 66th ed., CRC Press: Boca Raton, Florida (1985).

8.22

Effects of Ion-Exchange Resins on pH and Solution Conductivity

When a cation-exchange resin is added to a solution of sodium chloride, the pH of the solution decreases by several units, and it remains electrically conductive. When an anion-exchange resin is added to the sodium chloride solution, the pH increases by several units, and it remains conductive. When a mixed-bed ion-exchange resin is used, the pH change is small, and the conductivity of the solution drops significantly [1]. These effects are produced by mixing the solution with the resins in beakers (Procedure A) or by running the solution through a tube containing the resin (Procedure B).

MATERIALS FOR PROCEDURE A

300 mL 0.1M sodium chloride, NaCl (To prepare 1 liter of solution, dissolve 5.8 g of NaCl in 600 mL of distilled water, and dilute the resulting solution to 1.0 liter.)

100 mL distilled water

8 g mixed-bed ion-exchange resin, H^+/OH^- form (e.g., Amberlite MB-1)

> *or*
> an *additional* 4 g of each of the following two resins

4 g strongly acidic cation-exchange resin, H^+ form (e.g., Dowex 50-X8)

4 g strongly basic anion-exchange resin, OH^- form (e.g., Dowex 1-X8)

4 250-mL beakers

light-bulb conductivity tester, with unfrosted light bulb

pH meter, or pH-indicating test paper, or universal indicator

magnetic stirrer, with stir bar

nonmetallic spatula

ring stand

MATERIALS FOR PROCEDURE B

12 g mixed-bed ion-exchange resin, H^+/OH^- form (e.g., Amberlite MB-1)

> *or both of the following:*
> 6 g strongly acidic cation-exchange resin, H^+ form (e.g., Dowex 50-X8)
> 6 g strongly basic anion-exchange resin, OH^- form (e.g., Dowex 1-X8)

75 mL distilled water

25 mL 0.1M sodium chloride, NaCl (For preparation, see Materials for Procedure A.)

glass wool plug to fit inside chromatography column

100-mL chromatography column or buret, with stopcock

buret stand

2 250-mL beakers

nonmetallic spatula

powder funnel to fit chromatography column

2 100-mL beakers

light-bulb conductivity tester, with unfrosted light bulb

PROCEDURE A

Preparation

Pour 100 mL of 0.1M NaCl solution into each of three 250-mL beakers. Pour 100 mL of distilled water into the fourth beaker.

If mixed-bed ion-exchange resin is unavailable, mix 4 g of the cation-exchange resin with 4 g of the anion-exchange resin, using the nonmetallic spatula.

Presentation

Test the conductivity of the 0.1M NaCl solutions and of the distilled water with the conductivity tester. The bulb will glow brightly when the NaCl solutions are tested, but there will be no glow with distilled water. Measure the pH of the distilled water. Set one of the beakers containing NaCl on the stirrer, put the stir bar in the beaker, and turn on the stirrer. Measure the pH of the solution. If you are using a pH meter, leave the electrode in the solution and the meter turned on. With the nonmetallic spatula, add 4 g of the strongly acidic cation-exchange resin. After about 2 minutes the pH of the solution will have dropped into the strongly acidic range. Record the final pH. Test the conductivity of the solution; the lamp will glow brightly.

Set another of the beakers containing NaCl on the stirrer, put the stir bar in the beaker, and turn on the stirrer. Measure the pH of this solution. If you are using a pH meter, leave the electrode in the solution and the meter turned on. Add 4 g of the strongly basic anion-exchange resin. After about 2 minutes the pH of the solution will have risen into the strongly basic range. Record the final pH. Check the conductivity of the solution; the lamp will still glow brightly.

Set the third beaker of 0.1M NaCl on the stirrer, put the stir bar in the beaker, and turn on the stirrer. Measure the pH of this solution. Mount the conductivity tester on the stand so its electrodes are immersed in the solution, and turn on the tester. The lamp will glow brightly. Add 8 g of the mixed-bed ion-exchange resin (or the mixed resins), and stir the mixture. As the mixture is stirred, the conductivity tester's lamp will dim, and after about 2 minutes, there will be no glow. Note the final pH on the meter or with pH test paper. The pH will be near neutral.

PROCEDURE B

Preparation

Loosely pack a small plug of glass wool in the bottom of the chromatography column. Mount the column on the buret stand, place a 250-mL beaker under the column, and open the column's stopcock. In the other 250-mL beaker, make a slurry of 12 g of mixed-bed ion-exchange resin and 75 mL of distilled water. (If mixed-bed ion-exchange resin is unavailable, mix 6 g of the cation-exchange resin with 6 g of the anion-exchange resin, using the nonmetallic spatula.) Pour the slurry through the funnel into the column. Once the last of the slurry has been poured into the column, close the stopcock and allow the resin to settle. Open the stopcock and drain water from the column until about 1 cm of water stands above the top of the settled resin.

Pour 25 mL of 0.1M NaCl solution into one of the 100-mL beakers.

Presentation

Test the NaCl solution with the conductivity tester. The bulb will glow brightly. Pour this solution into the column containing the mixed-bed resin. Place the clean 100-mL beaker under the column, open the stopcock, and allow about 25 mL of liquid to drain into the beaker. Test the collected liquid with the conductivity tester. The bulb will not glow.

HAZARDS

Severe electrical shock can result if the electrodes of the conductivity tester are touched while it is turned on.

DISPOSAL

The spent ion-exchange resins can be regenerated according to the process recommended by the manufacturer for use in repeated performances of this demonstration; otherwise they should be discarded in a solid-waste receptacle. The waste solutions should be flushed down the drain with water.

DISCUSSION

This demonstration shows the functioning of several ion-exchange resins. These resins are capable of removing certain ions from a solution and replacing them with others. In the first part of Procedure A, an acidic ion-exchange resin removes sodium ions from a sodium chloride solution and replaces them with hydrogen ions. As this occurs, the released hydrogen ions make the solution acidic and the pH of the solution decreases. In the second part of Procedure A, a basic ion-exchange resin removes chloride ions from a sodium chloride solution and replaces them with hydroxide ions. As this occurs, the released hydroxide ions make the solution basic and the pH of the solution increases. In the last part of Procedure A, a mixed-bed resin, a combination of acidic and basic resins, removes sodium ions, replacing them with hydrogen ions, and removes chloride ions, replacing them with hydroxide ions. The hydrogen ions and hydroxide ions released into the solution react with each other and form water, neutralizing each other. Therefore, the pH of the solution does not change, and the concentration of ions in the solution decreases as a result of the action of the ion-exchange resin. This decrease in ion concentration is monitored with a light-bulb conductivity tester. Procedure B uses a mixed-bed ion-exchange resin in a tube to remove the ions from a sodium chloride solution. In this procedure, the sodium chloride solution is allowed to flow through the resin in the tube, and the effluent is tested with a light-bulb conductivity tester to show that the ions have been removed.

The phenomenon of ion exchange has been known for over 130 years, even before Svante Arrhenius formulated the ionic theory. In 1850, nine years before the birth of Arrhenius, J. T. Way published a paper entitled "On the Power of Soils to Absorb Manure," in which he examined how soluble fertilizers such as potassium chloride are retained by soils even after heavy rains [2]. In the experiments that Way described, he poured a solution of potassium chloride over soil suspended in a sieve and collected the effluent. He found that the potassium had been removed from the solution, and that it had been replaced by a chemically equivalent amount of calcium and magnesium. However, the effluent solution contained the same amount of chloride as the original solution. Then, he poured rainwater through the soil and collected the effluent. This effluent did not contain potassium, meaning that the soil had retained the potassium.

The phenomena J. T. Way reported can be explained as follows [3]: Potassium ions displace calcium ions and magnesium ions from the soil, but the chloride ions have no effect on the process and pass through. The process can be represented by the chemical equation

$$2 \, K^+(aq) + Ca^{2+}(soil) \rightleftarrows Ca^{2+}(aq) + 2 \, K^+(soil)$$

The exchange process is an equilibrium. In Way's experiment, all of the potassium was replaced because he had a large excess of calcium and magnesium in the soil. In fact, Way observed that the process could be reversed by washing the soil with a solution of calcium chloride.

Soil is able to bind positive ions such as K^+ and Ca^{2+} because it contains clay, a mixture of insoluble silicates. Silicates are composed of large arrays and long chains of alternating silicon and oxygen atoms. At the ends of these chains are negatively charged oxygen atoms to which the positive ions are attracted. The following diagram represents the end of one such chain:

In order to exchange ions, the silicate must be porous enough that the ions can move through it and interact with the charged oxygen atoms. Therefore, there are two requirements for a substance to function as an ion exchanger: it must have fixed ionic groups that can bind oppositely charged ions, and it must be permeable to these oppositely charged ions.

Most modern commercial ion exchangers are copolymers of styrene and divinylbenzene [4]. This polymer, represented below, has large cross-linked molecules with an open structure that allows ions to permeate it.

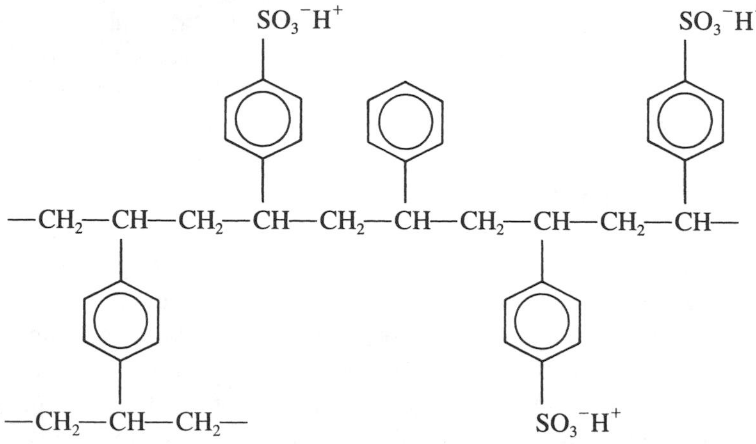

$$SO_3^-H^+ \qquad\qquad SO_3^-H^+$$

$$-CH_2-CH-CH_2-CH-CH_2-CH-CH_2-CH-CH_2-CH-$$

$$-CH_2-CH-CH_2- \qquad\qquad SO_3^-H^+$$

The ionic groups are attached to the benzene rings by chemical reaction. In the structure shown above, the benzene ring on the left comes from divinylbenzene, and the remaining four are from styrene. Divinylbenzene provides the cross-linking which joins polymer strands into a network that can be very open or quite close, depending on the ratio of divinylbenzene to styrene. This ratio can be varied to change the porosity of the polymer, but the usual proportion of divinylbenzene used in commercial resins is 8%. The ionic groups that are chemically attached to the benzene rings may be sulfonic acids ($-SO_3^-H^+$) or quaternary ammonium groups ($-CH_2N(CH_3)_3^+Cl^-$). These two groups are used in over 90% of all commercial ion-exchange resins. The sulfonic acids bind cations, such as H^+ and Na^+, and the ammonium groups bind anions, such as Cl^- and OH^-.

The commercial uses of ion-exchange resins are extremely broad, including desalinization of water, chemical analysis, isolation of trace elements, treatment of wine, and the purification of sugar. One of the foremost commercial uses for ion-exchange resins is in softening water. "Hard" water contains calcium and magnesium ions, which form an insoluble scum when combined with soaps. To prevent the formation of this scum, these ions are removed in a process called softening. A water softener contains a cation-exchange resin. When water containing calcium and magnesium ions flows through the resin, the magnesium and calcium ions are replaced by sodium ions. After a time, all of the sodium ions have been removed from the resin, and the resin must be regenerated by passing a saturated solution of sodium chloride through it. This shifts the ion equilibrium and flushes the magnesium and calcium ions from the resin, replacing them with sodium ions. Then the resin is ready to be used again.

In this demonstration the conductivity of the solution is monitored with a light-bulb conductivity apparatus. This tester uses a light bulb whose power cord is interrupted by a pair of electrodes that are immersed in the solution being tested. In order for the bulb to light, current must flow between the electrodes. A 15-watt bulb requires about 0.13 amps of current to be fully lit, whereas a 150-watt bulb needs 1.3 amps. The more current needed for full brightness, the more ions must be present to carry this current. Therefore, the brightness of a 150-watt bulb begins to diminish at a higher ion concentration than does a 15-watt bulb. Thus, a high-wattage bulb will show the decrease in concentration of ions in the solution over a wider range than a low-wattage bulb.

REFERENCES

1. R. E. Neas, *Western Illinois University Demonstrations Workshop* 33 (1979).
2. J. T. Way, *J. Roy. Agr. Soc. Engl.* 11:313 (1850).
3. F. Helfferich, *Ion Exchange,* McGraw-Hill Book Co.: New York (1962).
4. W. Rieman and H. F. Walton, *Ion Exchange in Analytical Chemistry,* Pergamon Press: Oxford (1970).

8.23

End Point of an Acid-Base Titration Determined by Electrical Conductivity

The probe of a light-bulb electrical-conductivity tester is dipped in a clear pink solution, and the light bulb glows. While the pink solution is stirred, a colorless liquid is added to it from a buret. As the liquids mix, a precipitate forms in the beaker and the glow of the bulb diminishes. The colorless liquid is added until the bulb stops glowing and the pink color disappears. If more of the colorless liquid is added, the bulb begins to glow again, but the pink color does not reappear [1].

MATERIALS

For preparation of indicator solutions, see pages 27–29.
For preparation of stock solutions of acids and bases, see pages 30–32.

200 mL 0.01M barium hydroxide, $Ba(OH)_2$ (To prepare 1 liter of solution, boil 1.2 liters of distilled water to drive out dissolved CO_2, dissolve 3.2 g $Ba(OH)_2 \cdot 8H_2O$ in 600 mL of the boiled water, and dilute the resulting, cooled solution to 1.0 liter with boiled water.)

10 drops phenolphthalein indicator solution

50 mL 0.1M sulfuric acid, H_2SO_4

magnetic stirrer, with stir bar

400-mL beaker

50-mL buret, with stopcock and stand

110-volt light-bulb conductivity tester (e.g., Sargent-Welch S-29761-50) (Use with an incandescent unfrosted light bulb, preferably over 100 watts.)

PROCEDURE

Preparation

Assemble the apparatus as illustrated in the figure. Place the stir bar in the 400-mL beaker and pour 200 mL of 0.01M $Ba(OH)_2$ into the beaker. Add 10 drops of phenolphthalein indicator solution to the beaker and set it on the magnetic stirrer. Fill the buret with 0.1M H_2SO_4 and mount it on the stand over the beaker. Mount the conductivity tester on the stand so its electrodes are immersed in the solution to a depth of at least 4 cm. Turn on the tester to be sure that the bulb lights.

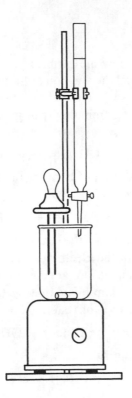

Presentation

Turn on the magnetic stirrer and the conductivity tester. Begin adding H_2SO_4 from the buret while watching the brightness of the bulb. When the acid is added, a precipitate will form in the pink solution. As the equivalence point is approached (after about 17 mL of H_2SO_4 has been added), reduce the rate of acid addition. When the light bulb glows only faintly, add the acid drop by drop. When the bulb no longer glows noticeably, the pink color of the solution will disappear, leaving a milky mixture. Record the volume of acid added up to this point. Add more acid to the beaker, and bulb will glow again.

HAZARDS

Soluble barium compounds are toxic and, if ingested, cause nausea, vomiting, stomach pains, and diarrhea.

The conductivity tester uses 110 volt a.c. Severe electrical shock can result if the electrodes of the conductivity tester are touched while it is turned on.

DISPOSAL

The solid should be filtered from the titration mixture and discarded in a solid-waste receptacle. The filtrate should be neutralized by adding sodium bicarbonate ($NaHCO_3$) until fizzing stops, then be flushed down the drain with water.

DISCUSSION

This demonstration shows how an intrinsic property of a solution, its electrical conductivity, can be used to signal the end point of a titration. As sulfuric acid is added to barium hydroxide, two processes occur that reduce the conductivity of the acid solution. One is the formation of water from the hydrogen ions in the acid and the hydroxide ions in the base:

$$H^+(aq) + OH^-(aq) \longrightarrow H_2O(l)$$

The other reaction is the precipitation of barium sulfate as the sulfate ions in the acid react with the barium ions in the base:

$$SO_4^{2-}(aq) + Ba^{2+}(aq) \longrightarrow BaSO_4(s)$$

The net result of these two processes is the removal of the ions from the solution. Because the ions are responsible for the electrical conductivity of the solution, the electrical conductivity decreases when the ions are removed. When a stoichiometric amount of sulfuric acid is added to the barium hydroxide, virtually all of the ions are removed from the solution. The self-ionization of water,

$$H_2O(l) \rightleftarrows H^+(aq) + OH^-(aq) \qquad K_w = 1.0 \times 10^{-14}$$

and the slight dissolving of barium sulfate,

$$BaSO_4(s) \rightleftarrows Ba^{2+}(aq) + SO_4^{2-}(aq) \qquad K_{sp} = 1.1 \times 10^{-10}$$

are the only sources of ions [2]. At this point the conductivity of the solution is very low. As more sulfuric acid is added, however, the concentration of ions increases and so does the conductivity.

In this demonstration the conductivity of the solution is monitored with a light-bulb conductivity apparatus. This tester uses a light bulb whose power cord is interrupted by a pair of electrodes that are immersed in the solution being tested. In order for the bulb to light, current must flow between the electrodes. A 15-watt bulb requires about 0.13 amps of current to be fully lit, whereas a 150-watt bulb needs 1.3 amps. The more current needed for full brightness, the more ions must be present to carry this current. Therefore, a 150-watt bulb diminishes in brightness farther from the end point of the titration than does a 15-watt bulb. Thus, a high-wattage bulb will show the decrease in concentration of ions in the solution over a wider range than a low-wattage bulb.

REFERENCES

1. H. N. Alyea and F. B. Dutton, Eds., *Tested Demonstrations in Chemistry*, 6th ed., Journal of Chemical Education Press: Easton, Pennsylvania (1965).
2. J. A. Dean, Ed., *Lange's Handbook of Chemistry*, 13th ed., McGraw-Hill Book Co.: New York (1985).

8.24

Effect of Acetate Ion
on the Acidity of Acetic Acid:
Common Ion Effect

Three colorless liquids are simultaneously poured into three cylinders containing a white powder. In one cylinder, frothing begins immediately, and the froth quickly fills the cylinder. In another cylinder, frothing is slightly slower, and in the third, it takes 10–15 seconds to fill the cylinder [1].

MATERIALS

For preparation of indicator solutions, see pages 27–29.
For preparation of stock solutions of acids and bases, see pages 30–32.

2 mL universal indicator (optional)

75 g calcium carbonate powder, $CaCO_3$

180 mL 2M acetic acid, $HC_2H_3O_2$

27 g sodium acetate trihydrate, $NaC_2H_3O_2 \cdot 3H_2O$

ca. 20 mL distilled water

100 mL 2M hydrochloric acid, HCl

dropper

3 250-mL glass cylinders

3 100-mL graduated cylinders

PROCEDURE

Preparation

Place 10–20 drops of universal indicator in the bottom of each of the three 250-mL cylinders, and cover the bottom of each with 25 g of $CaCO_3$ powder. Arrange the cylinders in a row.

Dissolve 27 g of sodium acetate trihydrate in 80 mL of 2M acetic acid, and dilute the resulting solution to 100 mL with distilled water. This solution is approximately 2M in sodium acetate and 2M in acetic acid.

Presentation

With the aid of an assistant, simultaneously pour 100 mL of 2M HCl into one of the 250-mL cylinders, 100 mL of 2M $HC_2H_3O_2$ into the second cylinder, and the 100 mL of $HC_2H_3O_2$–$NaC_2H_3O_2$ mixture into the other. Frothing will occur in all three cylinders, but it will be nearly instantaneous in the cylinder with hydrochloric acid, somewhat slower in the cylinder with acetic acid, and significantly slower in the cylinder containing the sodium acetate and acetic acid mixture. In addition, the color produced by universal indicator in the froth shows that the cylinder with hydrochloric acid is more acidic than the one containing acetic acid, which is more acidic than the one containing the mixture of acetate and acetic acid.

DISPOSAL

The waste solutions should be flushed down the drain with water.

DISCUSSION

The rate of the reaction between calcium carbonate and an acid depends upon the concentration of hydrogen ions in the acid. This demonstration shows that 2M hydrochloric acid has a higher concentration of hydrogen ions than does 2M acetic acid, which has a higher concentration of hydrogen ions than a 2M acetic acid solution also containing 2M acetate ions.

The foam produced in the reaction is the result of the carbon dioxide gas liberated by the reaction of calcium carbonate with acid mixing with water and the calcium carbonate powder:

$$CaCO_3(s) + 2 H^+(aq) \longrightarrow CO_2(g) + H_2O(l) + Ca^{2+}(aq)$$

The rate at which the foam rises depends on the rate of CO_2 production, which depends on the concentration of $H^+(aq)$. The concentrations of $H^+(aq)$ in the solutions used in the demonstration are quite different. The hydrogen ion concentration is greater in 2M hydrochloric acid than in 2M acetic acid, indicating that HCl is more highly ionized in aqueous solution than is $HC_2H_3O_2$. Hydrochloric acid is a strong acid and is completely ionized in aqueous solution, whereas acetic acid is a weak acid that is only partly ionized. The concentration of hydrogen ions is greater in acetic acid than in a mixture of acetic acid and sodium acetate. This fact may be attributed to the effect of the acetate ion from sodium acetate on the ionization of acetic acid. The acetate ion suppresses the ionization of the acid by shifting its ionization equilibrium in the direction of the undissociated molecules:

$$HC_2H_3O_2(aq) \leftrightarrows H^+(aq) + C_2H_3O_2^-(aq)$$

This dissociation is an equilibrium, with a constant of $K_a = 1.8 \times 10^{-5}$. In 2M acetic acid, the concentration of H^+ is 0.006M. When acetate ions are added to a solution of acetic acid, they shift the equilibrium toward the undissociated acid and reduce the concentration of H^+. In 2M acetic acid that also contains 2M acetate ions, the concentration of H^+ is only 0.00002M, which is only 1/300 of that in pure 2M acetic acid. Therefore, the mixed acetic acid–acetate solution liberates CO_2 from $CaCO_3$ more slowly than does pure 2M acetic acid.

Calcium carbonate is used in this demonstration, rather than some other carbonate (e.g., sodium carbonate), because it is insoluble in water. A soluble carbonate would dissolve in the acid solutions, and the reactions would proceed much too fast for any discernible difference in the rates. Furthermore, the insoluble calcium carbonate particles cling to the bubbles of carbon dioxide that are produced by the reactions and form a foam in the cylinders. This foam makes the progress of the reaction more visible.

REFERENCE

1. C. H. Sorum, *J. Chem. Educ.* 25:489 (1948).

8.25

Effect of Molecular Structure on the Strength of Organic Acids and Bases in Aqueous Solutions

Methyl red indicator is red in an aqueous phenol solution, but it is yellow in an aqueous solution of ethanol, showing that phenol is more acidic than ethanol [1, 2]. Thymol blue indicator is yellow in aqueous aniline, but is blue in aqueous ammonia, showing that ammonia is more basic than aniline (Procedure A). Overhead projection can be used to display these phenomena (Procedure B).

MATERIALS FOR PROCEDURE A

For preparation of indicator solutions, see pages 27–29.

- 25 mL 0.3M ethanol, C_2H_5OH (To prepare 100 mL of solution, dissolve 1.5 mL of 95% ethanol in 98.5 mL of distilled water.)

- 25 mL 0.3M phenol, C_6H_5OH (To prepare 100 mL of solution, dissolve 3 g C_6H_5OH in 100 mL distilled water.) **(See the Hazards section before handling phenol.)**

- 25 mL 0.3M ammonia, NH_3 (To prepare 1 liter of solution, dissolve 20 mL of concentrated [15M] NH_3 in 600 mL of distilled water, and dilute the resulting mixture to 1.0 liter.)

- 25 mL 0.3M aniline, $C_6H_5NH_2$ (To prepare 100 mL of solution, dissolve 2.8 g of freshly distilled $C_6H_5NH_2$ in 100 mL of distilled water.)

- 1 mL methyl red indicator solution

- 1 mL thymol blue indicator solution

- 4 test tubes, 25 mm × 200 mm, with stoppers

- 4 labels for test tubes

- gloves, plastic or rubber

- rack for test tubes

- white background (e.g., 20-cm × 30-cm poster board)

- dropper

MATERIALS FOR PROCEDURE B

For preparation of indicator solutions, see pages 27–29.

overhead projector, with transparency and marker

25 mL 0.3M ethanol, C_2H_5OH (To prepare 100 mL of solution, dissolve 1.5 mL of 95% ethanol in 98.5 mL of distilled water.)

25 mL 0.3M phenol, C_6H_5OH (To prepare 100 mL of solution, dissolve 3 g C_6H_5OH in 100 mL distilled water.) **(See the Hazards section before handling phenol.)**

25 mL 0.3M ammonia, NH_3 (To prepare 1 liter of solution, dissolve 20 mL of concentrated [15M] NH_3 in 600 mL of distilled water, and dilute the resulting mixture to 1.0 liter.)

25 mL 0.3M aniline, $C_6H_5NH_2$ (To prepare 100 mL of solution, dissolve 2.8 g of freshly distilled $C_6H_5NH_2$ in 100 mL of distilled water.)

1 mL methyl red indicator solution

1 mL thymol blue indicator solution

4 10-cm petri dishes

gloves, plastic or rubber

dropper

PROCEDURE A

Preparation

Label the four test tubes with the labels "ethanol," "phenol," "ammonia," and "aniline." Wearing gloves, pour 25 mL of 0.3M aqueous solutions of ethanol, phenol, ammonia, and aniline into the appropriate test tubes. Stopper the tubes and place them in a rack before a white background.

Presentation

Add 10 drops of methyl red solution to each of the test tubes containing ethanol and phenol. Stopper the tubes and invert them several times to mix their contents. The phenol solution will be deep red, and the ethanol solution will be orange, indicating that the phenol solution is more acidic than the ethanol.

Add 10 drops of thymol blue solution to each of the test tubes containing ammonia and aniline. Stopper the tubes and invert them several times to mix their contents. The aniline solution will be yellow, and the ammonia solution will be blue, showing that ammonia is more basic.

PROCEDURE B

Preparation and Presentation

Place the four petri dishes on the transparency atop the overhead projector. Write the four labels "ethanol," "phenol," "ammonia," and "aniline" on the transparency, with one label adjacent to each dish. Pour 25 mL of 0.3M aqueous solutions of ethanol, phenol, ammonia, and aniline into the appropriate petri dishes.

Add 10 drops of methyl red solution to the petri dish containing ethanol and 10 drops to the dish containing phenol. Stir the petri dishes to mix their contents. The phenol solution will be deep red, and the ethanol solution will be orange, indicating that the phenol solution is more acidic than the ethanol solution.

Add 10 drops of thymol blue solution to the petri dish containing ammonia and 10 drops to the dish containing aniline. Swirl the petri dishes to mix their contents. The aniline solution will be yellow, and the ammonia solution will be blue, showing that ammonia is more basic than aniline.

HAZARDS

Solid phenol should be handled only with plastic or rubber gloves. It is toxic and causes burns. It can be absorbed rapidly through the skin. Prolonged inhalation of the vapor can have chronic effects. If any solid phenol should contact the skin, the area should be thoroughly washed with large amounts of water.

Pure aniline should be handled only with plastic or rubber gloves. The vapor and liquid are rapidly absorbed through the skin. If any aniline is spilled on the skin, wash it off with plenty of water. It should be used only with adequate ventilation. Long exposure to the vapor can have chronic effects. When even slightly impure, aniline will be orange to dark red. To purify, distill it under reduced pressure. The boiling point at water aspirator pressure (about 30 torr) will be lower than 100°C.

Concentrated aqueous ammonia can irritate the skin, and its vapors are harmful to the eyes and mucous membranes.

Ethanol is flammable and should be kept away from open flames.

DISPOSAL

The waste solutions should be flushed down the drain with water.

DISCUSSION

This demonstration shows that the strength of an acid is affected by the nature of the group to which the acidic or basic part of a molecule is bonded. Both ethanol and phenol contain an —OH group bonded to a hydrocarbon group. In ethanol, the carbon to which the —OH group is bonded is saturated (bonded to four other atoms), whereas in phenol, that carbon is unsaturated (bonded to fewer than four other atoms, namely three). This demonstration shows that the —OH bonded to an unsaturated carbon atom is more acidic than one bonded to a saturated carbon atom.

Both ammonia and aniline contain an —NH$_2$ group. In ammonia it is bonded to a hydrogen atom, and in aniline it is bonded to a benzene ring. The demonstration shows that ammonia, where the —NH$_2$ is bonded to hydrogen, is more basic than aniline, where the —NH$_2$ is bonded to a phenyl group.

REFERENCES

1. David A. Humphreys, *Demonstrating Chemistry*, McMaster University: Hamilton, Ontario (1983).
2. A. J. Gordon and R. A. Ford, *The Chemist's Companion*, John Wiley and Sons: New York (1972).

8.26

Determination of Neutralizing Capacity of Antacids

The neutralizing capacity of an antacid tablet or milk of magnesia is determined by titration with hydrochloric acid. The titration is performed accurately with a buret in Procedure A and semiquantitatively with a graduated cylinder in Procedure B.

MATERIALS FOR PROCEDURE A

For preparation of indicator solutions, see pages 27–29.
For preparation of stock solutions of acids and bases, see pages 30–32.

50 mL 0.50M hydrochloric acid, HCl

either

      antacid tablet (e.g., Tums, Alka-Mints, Rolaids, or Maalox)

      10 mL distilled water

      balance accurate to 0.01 gram

      mortar and pestle

or

      5 mL milk of magnesia

      teaspoon measure

200 mL distilled water

5–10 drops bromophenol blue indicator solution

50-mL buret, with stopcock and stand

400-mL beaker

dropper

magnetic stirrer and stir bar

MATERIALS FOR PROCEDURE B

For preparation of indicator solutions, see pages 27–29.
For preparation of stock solutions of acids and bases, see pages 30–32.

either

162   antacid tablet (e.g., Tums, Alka-Mints, Rolaids, or Maalox)

10 mL distilled water

mortar and pestle

or

5 mL milk of magnesia

teaspoon measure

200 mL distilled water

5–10 drops bromophenol blue indicator solution

50 mL 0.50M hydrochloric acid, HCl

400-mL beaker

dropper

50-mL graduated cylinder

stirring rod

PROCEDURE A

Preparation

Fill the 50-mL buret with 0.50M HCl and mount the buret in its stand.

Presentation

If an antacid tablet is used, weigh one antacid tablet to the nearest 0.01 gram and record the mass. Grind the tablet with a mortar and pestle. Pour the pulverized tablet into a 400 mL beaker. Rinse the powder remaining in the mortar into the beaker with 5–10 mL of distilled water. Pour 200 mL of distilled water into the beaker.

If milk of magnesia is used, pour about 200 mL of distilled water into the 400-mL beaker. Shake the milk of magnesia bottle and fill the teaspoon measure with milk of magnesia. Pour the spoonful of milk of magnesia into the water in the beaker, and dip the spoon in the water to rinse it.

Add 5–10 drops of bromophenol blue indicator solution to the beaker. Place the magnetic stir bar in the beaker, set the beaker on the stirrer, and begin stirring. The mixture will be blue and cloudy.

If a Rolaids tablet is being used, allow about 13 mL of 0.50M HCl to drain from the buret into the beaker. If a Tums or Maalox tablet, drain about 17 mL, and if milk of magnesia or an Alka-Mints tablet, about 24 mL. The color of the mixture will change to yellow near the end of the addition, but when the addition of HCl is stopped, the mixture will return to blue. Then titrate the solution carefully to the bromophenol-blue end point, which coincides with a green color. Note the volume of 0.50M HCl neutralized by the antacid.

PROCEDURE B

Preparation and Presentation

If an antacid tablet is used, grind the tablet with a mortar and pestle. Pour the pulverized tablet into a 400-mL beaker. Rinse the powder remaining in the mortar into the beaker with 5–10 mL of distilled water. Pour 200 mL of distilled water into the beaker.

If milk of magnesia is used, pour about 200 mL of distilled water into the 400-mL beaker. Shake the milk of magnesia bottle and fill the teaspoon measure with milk of magnesia. Pour the spoonful of milk of magnesia into the water in the beaker, and dip the spoon in the water to rinse it.

Add 5–10 drops of bromophenol blue indicator solution. Stir the mixture. The mixture will be blue and cloudy. Fill the graduated cylinder with 50 mL of 0.50M HCl. Pour 5 mL of HCl from the cylinder into the beaker. Stir the mixture in the beaker. Add several more 5-mL aliquots and stir the mixture. With the later aliquots, the mixture will turn yellow upon addition of the acid, then return to blue when the mixture is stirred. When this happens, reduce the size of the aliquot to about 2 mL. Continue to add 2-mL aliquots until the mixture remains green or yellow upon stirring. Note the volume of acid required to neutralize the milk of magnesia.

DISPOSAL

The waste solutions should be flushed down the drain with water.

DISCUSSION

In this demonstration the acid-neutralizing capacity of a typical dose of antacid is determined. In Procedure A, the titration is performed accurately ($\pm 0.1\%$) with a buret; in Procedure B, the titration is performed more qualitatively ($\pm 10\%$) with graduated cylinders. The titrant used in both procedures is hydrochloric acid, which is present in gastric juices at a concentration of about 0.1M.

Antacid tablets are formulated from a number of chemical compounds. Many tablets (e.g., Tums, Alka-Mints) contain carbonates, usually either sodium or calcium carbonate. Others (e.g., Rolaids, Maalox tablets) contain hydroxides, usually magnesium or aluminum hydroxide. Carbonate ions neutralize acid by the process represented in the following equation:

$$CO_3^{2-}(aq) + 2\ H^+(aq) \longrightarrow H_2O(l) + CO_2(g)$$

Because carbon dioxide gas is produced by the reaction, gas pressure will build in the stomach, leading eventually to a "burp of relief" [1]. The amount of calcium carbonate may vary from one brand of tablet to another. For example, one Tums tablet contains 0.5 g of $CaCO_3$, which can neutralize 0.01 mole of HCl, but one Alka-Mints tablet contains 0.85 g of $CaCO_3$, which neutralizes 0.016 mole of HCl [2]. Those antacid tablets that contain aluminum and magnesium hydroxides neutralize acid by this process:

$$OH^-(aq) + H^+(aq) \longrightarrow H_2O(l)$$

Again, different brands of tablets have different formulations and different acid-neutralizing capacities. Maalox tablets contain 0.2 g $Mg(OH)_2$ and 0.2 g $Al(OH)_3$, and they have an acid-neutralizing capacity of about 0.01 mole of HCl per tablet [2]. The antacid action of Rolaids tablets results from the combination of hydroxide and carbonate; they contain dihydroxyaluminum sodium carbonate ($NaAl(OH)_2CO_3$). Each tablet can neutralize 0.008 mole of HCl [2]. A thorough survey of antacid formulations and their neutralizing capacities has been performed by Consumers' Union [3].

Milk of magnesia is a suspension of $Mg(OH)_2$ in water, which may also contain flavoring and coloring agents. The dissolved hydroxide ions neutralize stomach acid:

$$OH^-(aq) + H^+(aq) \longrightarrow H_2O(l)$$

Because the solubility product of $Mg(OH)_2$ is quite small (1.2×10^{-11}), only a small amount of $Mg(OH)_2$ dissolves in the water. The concentration of magnesium ions in a saturated solution of magnesium hydroxide in water is only $1.4 \times 10^{-4}M$, and the hydroxide ion concentration is only $2.8 \times 10^{-4}M$. Therefore, the pH of a saturated solution of $Mg(OH)_2$ is 10.5. Milk of magnesia is only moderately basic, and can be ingested without irritating the lining of the mouth and esophagus. However, because milk of magnesia contains a large excess of undissolved $Mg(OH)_2$, its acid-neutralizing capacity is much greater than its pH of 10.5 would indicate. As the hydroxide ions in solution are consumed in neutralizing stomach acid, more $Mg(OH)_2$ dissolves to replace them. Because $Mg(OH)_2$ dissolves slowly, the titration of milk of magnesia must be done slowly to allow time for the $Mg(OH)_2$ to dissolve.

Acid indigestion, heartburn, and sour stomach are synonyms for a condition whose symptoms are a burning sensation just behind the lower part of the sternum [4]. The condition is caused by a backflow of acidic stomach contents into the lower part of the esophagus. The condition occurs when the contractions of the stomach muscles become sufficiently strenuous to force open the sphincter muscle between the stomach and the bottom of the esophagus. The stomach is lined with cells that secrete a fatty coating that protects the stomach from its acidic contents. However, the esophagus has no such coating, and when the stomach contents enter the esophagus, its lining is irritated and a burning sensation occurs.

The volume of the human stomach varies with its contents, but its maximum volume is about 1 liter. It is not likely that the entire maximum volume would be occupied by gastric juices; much of it would be occupied by undigested food. However, for the sake of illustration, the effect of a standard dose of antacid on the pH of 1 liter of 0.1M HCl can be considered. The standard dose of Tums is one or two tablets. Two tablets contain 1 g of $CaCO_3$, which can neutralize 0.02 mole of HCl. One liter of 0.1M HCl has a pH of 1.0 and contains 0.1 mole of HCl. After the addition of two Tums tablets, 1 liter of gastric juice contains $0.1 - 0.02 = 0.08$ mole of HCl. The pH of this solution is about 1.1. This is an almost negligible change in pH, and alone it cannot account for the relief that the dose provides. Of course, the stomach does contain less than 1 liter of gastric juice, so the pH change caused by the antacid will be greater than that calculated here. Furthermore, when the dose is swallowed, it reaches the irritated portion of the esophagus before it enters the stomach. If the antacid is finely divided, it is likely to linger in the esophagus long enough to neutralize all of the acid in the esophagus. This is why directions for use of antacid tablets advise the consumer to chew the tablets thoroughly.

In addition to being an antacid, milk of magnesia is also a saline laxative. When undissolved magnesium hydroxide enters the intestines, a saturated solution of magnesium hydroxide forms. The osmotic pressure of this solution draws water into the intestines, increasing the volume and mobility of intestinal contents and stimulating the voiding of these contents.

REFERENCES

1. Arm and Hammer Baking Soda label (1980).
2. *Physicians' Desk Reference for Nonprescription Drugs,* 8th ed., Medical Economics Co.: Oradell, New Jersey (1987).
3. *Consumer Reports* 48:412 (1983).
4. D. F. Tapley, R. J. Weiss, and T. Q. Morris, Eds., *The Columbia University College of Physicians and Surgeons Complete Home Medical Guide,* Crown Publishers: New York (1985).

8.27

Instrumental Recording of a Titration Curve

As a basic solution is added at a constant rate to a stirred acidic solution, the pH of the mixture is detected by a pH meter and recorded on a strip-chart recorder [1, 2].

MATERIALS

For preparation of stock solutions of acids and bases, see pages 30–32.

4 liters of 0.1M sodium hydroxide, NaOH

100 mL pH 7 buffer solution (For preparation, see Procedure A of Demonstration 8.1.)

100 mL of one or more of the following acids:

   0.1M hydrochloric acid, HCl

   0.1M nitric acid, HNO_3

   0.1M sulfuric acid, H_2SO_4

   0.1M phosphoric acid, H_3PO_4

   0.1M acetic acid, $HC_2H_3O_2$

   0.1M malonic acid, $CH_2(CO_2H)_2$ (To prepare 1 liter of solution, dissolve 10.4 g of $CH_2(CO_2H)_2$ in 600 mL of distilled water, and dilute the resulting solution to 1 liter.)

   0.1M phenol, C_6H_5OH (To prepare 1 liter of solution, dissolve 9.4 g of C_6H_5OH in 600 mL of distilled water, and dilute the resulting solution to 1 liter.) **(See the Hazards section before handling phenol.)**

200 mL distilled water for each acid

right-angle glass bend, with outside diameter of 7 mm, one arm 5 cm long, the other 35 cm

2-holed rubber stopper to fit 4-liter bottle

4-liter bottle (e.g., empty reagent acid bottle)

10-cm length of glass tubing, with outside diameter of 7 mm

10-cm length of plastic tubing, to fit on 7-mm glass tube

1-meter length of plastic tubing to fit on glass bend

2-way stopcock, with one arm constricted to form a nozzle

500-mL beakers (one for each acid used)

platform or shelf, ca. 75 cm high, to support bottle above table top magnetic stirrer, with stir bar

stand, with clamp to hold stopcock

100-mL graduated cylinder

screw clamp

pH meter, with output terminals for connection to recorder

strip-chart recorder, with speed of at least 6 cm per minute

cables to connect pH meter to recorder

gloves, plastic or rubber

PROCEDURE

Preparation

Assemble the siphon as illustrated in Figure 1. Insert the long arm of the right-angle bend through the 2-holed stopper so the tip of the arm is within 2 cm of the bottom of the 4-liter bottle when the stopper is seated in its mouth. Insert the 10-cm length of glass tubing in the other hole of the stopper. Attach the 10-cm length of plastic tubing to the free end of this straight glass tubing. Attach the 1-meter length of plastic tubing to the free arm of the right-angle bend and the other end to the nonconstricted arm of the stopcock.

Fill the bottle with 0.1M NaOH solution and reseat the stopper of the siphon assembly in the mouth. While holding the open stopcock over a 500-mL beaker, fill the tubing connected to the stopcock with solution by blowing into the short plastic tube until solution begins to flow from the stopcock. Close the stopcock. Set the bottle on a platform about 75 cm above the table top.

Place the 500-mL beaker on the magnetic stirrer, and clamp the stopcock to the stand so NaOH solution will drain into the beaker when the stopcock is opened (see

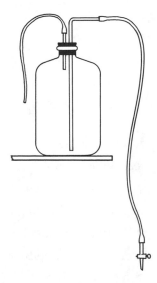

Figure 1.

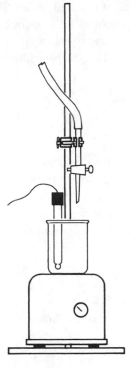

Figure 2.

Figure 2). Without moving the stopcock, remove the beaker and stirrer and place the mouth of a 100-mL graduated cylinder under the stopcock. Determine the flow rate of NaOH solution from the stopcock by measuring the time required for 50 mL of solution to drain into the cylinder. The flow rate should be about 50 mL per minute. If it is more than 70 mL per minute, restrict the flow with a screw clamp on the plastic tubing connected to the stopcock.

Connect the output terminals of the pH meter to the chart recorder. Set the pH meter to "standby," and turn on the recorder. Set the range scale of the recorder to match the full range of the output from the meter (consult the meter manual to determine this). Immerse the pH electrode in the pH 7 buffer and calibrate the meter. At the same time, set the zero adjustment on the recorder so the pen is deflected to half scale when the pH meter reads 7. Make a pH scale for the recorder by dividing the chart from 0 to midrange into seven equal intervals, then extending these intervals from midrange to full scale.

Presentation

Pour about 200 mL of distilled water into the 500-mL beaker with the stirring bar, and replace the beaker on the stirrer so the NaOH solution can drain into it when the stopcock is opened, but do not open the stopcock. Turn on the stirrer and adjust its speed so that a slight vortex is produced in the water. Insert the pH electrode in the water, turn on both the meter and the recorder to display the pH of the water. Add 100 mL of 0.1M acid to the beaker, and observe the change in pH as registered on the meter and on the recorder. Start the chart drive of the recorder and lower its pen to the paper. As the pen crosses a graduation on the paper, open the stopcock. Observe the change in pH as the NaOH solution is added to the acid. After the acid has been neutralized, and the pH of the mixture is above 9, close the stopcock and stop the chart. Remove the chart from the recorder and display the titration curve.

Repeat the procedure described in the preceding paragraph with each of the acid samples at hand.

HAZARDS

Malonic acid is a strong irritant to the eyes, skin, and mucous membranes.

Solid phenol should be handled only with plastic or rubber gloves. It is toxic and causes burns. It can be absorbed rapidly through the skin. Prolonged inhalation of the vapor can have chronic effects. If any solid phenol should contact the skin, the area should be thoroughly washed with large amounts of water.

DISPOSAL

The waste solutions should be flushed down the drain with water.

DISCUSSION

A titration curve for an acid-base titration is a plot of the pH of the analyzed solution versus the volume of titrant added. A titration curve can be determined by adding aliquots of titrant, measuring the pH of the analyzed solution, and plotting the data manually. In this demonstration a titration curve is produced automatically by recording the pH of the analyzed solution on a strip-chart recorder as the titrant is added at a constant rate. The titrant flows at an almost constant rate, because the level of titrant in the large reservoir changes only slightly during the titration. This would not be the case if an ordinary buret were used. Because the titrant is added at a constant rate, the vol-

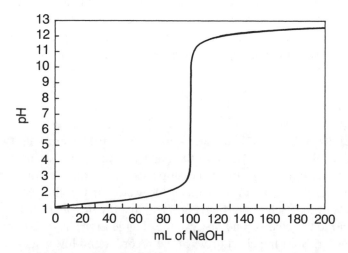

Figure 3. Titration of 100 mL of 0.1M HCl with 0.1M NaOH.

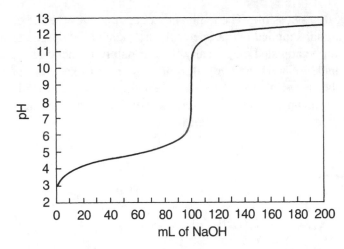

Figure 4. Titration of 100 mL of 0.1M $HC_2H_3O_2$ with 0.1M NaOH.

ume of titrant added is proportional to the time from the start of addition, and this is proportional to the distance the chart has moved under the pen of the recorder.

A titration curve shows how the pH of a solution varies as it is neutralized. Figure 3 contains a titration curve for the addition of NaOH to HCl. Perhaps the most striking feature of such a curve is the steeply rising portion. This steeply rising portion reflects the sudden change in pH that occurs near the equivalence point, the point in the titration where acid and base are in their stoichiometric ratio. Because the change in pH is so dramatic at the equivalence point, it is possible to use an indicator that changes color over a pH range of 2 units and still find the equivalence point accurately.

More subtle features of the shape of titration curves can be illustrated in this demonstration by using a variety of acids and bases. For example, when a weak acid is titrated with a strong base (e.g., acetic acid titrated with sodium hydroxide), the curve rises sharply at first, when the first few milliliters of NaOH are added (Figure 4). However, then it flattens out, becoming more nearly horizontal. This flattening of the curve is a result of the formation of a buffer solution when NaOH is added to $HC_2H_3O_2$ in a less than stoichiometric ratio [3]. This buffer solution resists changes in the pH of the solution as more NaOH is added, and therefore the titration curve flattens. For this reason, the flattened portion of the curve is called the *buffer region*. When weakly basic ammonia is added to strongly acidic hydrochloric acid, the buffer region of the curve occurs on the basic side of the equivalence point. When a weak acid is titrated with a weak base, there is a buffer region on both sides of the equivalence point. A more extensive discussion of the shape of titration curves can be found in most quantitative analysis textbooks.

In setting up the equipment for this demonstration, one must take care to match the range of the output of the pH meter to that of the input of the chart recorder. There is variation among pH meters in the range of the signal they produce at their recorder output. Most produce a signal that changes by 60 mV per pH unit and has a value of 0 volts at pH 7. (This is the voltage produced by the glass electrode relative to the Ag/AgCl reference electrode.) Some meters produce a small signal (e.g., from −25 mV to +25 mV) as the pH varies from 0 to 14. Others produce larger signals, varying perhaps from 0 V to +1 V, or more, across the pH range. The recorder must be adjusted to accept the full range of the signal from the meter, preferably with a pen deflection as close to full scale as possible. The manual that accompanies the meter will indicate the nature of the signal supplied by the meter. If the manual for the meter is not available, the recorder can be adjusted empirically by alternately immersing the pH electrode in 1M HCl and in 1M NaOH, and adjusting the recorder to keep the pen deflections on scale. It may be necessary to set the zero adjustment of the recorder along with the range in order to accomplish this. The zero adjustment of the recorder should be used to set the pen to midrange when the meter reads a pH of 7.

REFERENCES

1. C. W. Schultz and S. L. Spannuth, *J. Chem. Educ.* 56:194 (1979).
2. M. McClendon, *J. Chem. Educ.* 61:1022 (1984).
3. E. G. Meek, *School Sci. Review* 64:528 (1983).

8.28

Buffering Action and Capacity

When a small amount of acid or base is added to distilled water, the pH of the mixture changes dramatically. When a similar amount of acid or base is added to a buffer solution, the pH changes by only a small amount, and a much larger amount of acid or base must be added to produce a large change in the pH (Procedure A). These changes are observed with a pH meter in an acetic acid–acetate buffer (Procedure B) and an ammonium-ammonia buffer (Procedure C). They are also observed with indicators in acetic acid–acetate (Procedure D), ammonium-ammonia (Procedure E), and mixed phosphate buffers (Procedure F). Overhead projection is used to display the colors of the indicators in the acetic acid–acetate buffer (Procedure G) and the ammonium-ammonia buffer (Procedure H).

MATERIALS FOR PROCEDURE A

For preparation of indicator solutions, see pages 27–29.
For preparation of stock solutions of acids and bases, see pages 30–32.

100 mL solution A1 (To prepare 1 liter of solution, dissolve 2.4 g of $NaC_2H_3O_2$ $\cdot 3H_2O$ in 180 mL of 1.0M $HC_2H_3O_2$, and dilute the resulting solution to 1.0 liter with distilled water.)

200 mL solution A2 (To prepare 1 liter of solution, dissolve 13.6 g of $NaC_2H_3O_2$ $\cdot 3H_2O$ in 100 mL of 1.0M $HC_2H_3O_2$ and dilute the resulting solution to 1.0 liter with distilled water.)

200 mL solution A3 (To prepare 1 liter of solution, dissolve 136.1 g of $NaC_2H_3O_2$ $\cdot 3H_2O$ and 57 mL of glacial [17.5M] $HC_2H_3O_2$ in 500 mL of distilled water, and dilute the resulting solution to 1.0 liter.)

100 mL solution A4 (To prepare 1 liter of solution, dissolve 24.5 g of $NaC_2H_3O_2$ $\cdot 3H_2O$ and 18.0 mL of 1.0M $HC_2H_3O_2$ in 500 mL of distilled water, and dilute the resulting solution to 1.0 liter.)

150 mL 1.0M hydrochloric acid, HCl

150 mL 1.0M sodium hydroxide, NaOH

3 mL bromophenol blue indicator solution

3 mL bromothymol blue indicator solution

6 250-mL beakers

labels for beakers

6 magnetic stir bars

2 50-mL burets, with stand

2 100-mL graduated cylinders

2 droppers

magnetic stirrer

MATERIALS FOR PROCEDURE B

For preparation of stock solutions of acids and bases, see pages 30–32.

200 mL distilled water

60 mL 1.0M hydrochloric acid, HCl

60 mL 1.0M sodium hydroxide, NaOH

100 mL 1.0M acetic acid, $HC_2H_3O_2$

100 mL 1.0M sodium acetate, $NaC_2H_3O_2$ (To prepare 1 liter of solution, dissolve 136 g of $NaC_2H_3O_2 \cdot 3H_2O$ in 600 mL of distilled water and dilute to 1 liter.)

2 droppers

2 10-mL graduated cylinders

waterproof marker

4 2.5-cm magnetic stir bars

4 250-mL beakers

3 100-mL graduated cylinders

magnetic stirrer

pH meter with large display readable to ±0.01 units, standardized

250-mL wash bottle filled with distilled water.

MATERIALS FOR PROCEDURE C

For preparation of stock solutions of acids and bases, see pages 30–32.

200 mL distilled water

60 mL 1.0M hydrochloric acid, HCl

60 mL 1.0M sodium hydroxide, NaOH

100 mL 1.0M aqueous ammonia, NH_3

100 mL 1.0M ammonium chloride, NH_4Cl (To prepare 1 liter of solution, dissolve 53.5 g of NH_4Cl in 600 mL of distilled water and dilute the resulting solution to 1.0 liter.)

2 droppers

2 10-mL graduated cylinders

waterproof marker

4 2.5-cm magnetic stir bars

4 250-mL beakers

3 100-mL graduated cylinders

magnetic stirrer

pH meter with large display readable to ±0.01 units, standardized

250-mL wash bottle filled with distilled water

MATERIALS FOR PROCEDURE D

For preparation of indicator solutions, see pages 27–29.
For preparation of stock solutions of acids and bases, see pages 30–32.

600 mL distilled water

2 mL bromocresol green indicator solution

30 mL 6M hydrochloric acid, HCl

30 mL 6M sodium hydroxide, NaOH

900 mL 0.90M ammonium chloride, NH_4Cl (To prepare 1 liter of solution, dissolve 48 g of NH_4Cl in 600 mL of distilled water and dilute the resulting solution to 1.0 liter.)

450 mL 1.0M acetic acid, $HC_2H_3O_2$

450 mL 1.0M sodium acetate, $NaC_2H_3O_2$ (For preparation, see Materials for Procedure B.)

8 500-mL beakers

8 glass stirring rods

white background (e.g., 25-cm × 80-cm poster board)

3 droppers

MATERIALS FOR PROCEDURE E

For preparation of indicator solutions, see pages 27–29.
For preparation of stock solutions of acids and bases, see pages 30–32.

300 mL distilled water

2 mL phenolphthalein indicator solution

1 mL 6M sodium hydroxide, NaOH

300 mL 0.15M sodium acetate, $NaC_2H_3O_2$ (To prepare 1 liter of solution, dissolve 20.4 g of $NaC_2H_3O_2 \cdot 3H_2O$ in 600 mL of distilled water and dilute the resulting solution to 1.0 liter.)

150 mL 1.0M ammonium chloride, NH_4Cl (For preparation, see Materials for Procedure C.)

150 mL 1.0M aqueous ammonia, NH_3

30 mL 6M hydrochloric acid, HCl

3 500-mL beakers

white background (e.g., 25-cm × 40-cm poster board)

3 glass stirring rods

3 droppers

MATERIALS FOR PROCEDURE F

For preparation of indicator solutions, see pages 27–29.
For preparation of stock solutions of acids and bases, see pages 30–32.

750 mL distilled water

30 mL 1.0M hydrochloric acid, HCl

30 mL 1.0M sodium hydroxide, NaOH

2 mL bromothymol blue indicator solution

2 g potassium dihydrogen phosphate, KH_2PO_4

2 g dipotassium hydrogen phosphate, K_2HPO_4

3 500-mL beakers

3 glass stirring rods

2 25-mL graduated cylinders

3 droppers

MATERIALS FOR PROCEDURE G

For preparation of indicator solutions, see pages 27–29.
For preparation of stock solutions of acids and bases, see pages 30–32.

overhead projector, with transparency and marker

60 mL distilled water

2 mL bromocresol green indicator solution

10 mL 6M hydrochloric acid, HCl

10 mL 6M sodium hydroxide, NaOH

90 mL 0.90M ammonium chloride, NH_4Cl (To prepare 1 liter of solution, dissolve 48 g of NH_4Cl in 600 mL of distilled water and dilute the resulting solution to 1.0 liter.)

45 mL 1.0M acetic acid, $HC_2H_3O_2$

45 mL 1.0M sodium acetate, $NaC_2H_3O_2$ (For preparation, see Materials for Procedure B.)

8 10-cm petri dishes

3 droppers

2 10-mL graduated cylinders

MATERIALS FOR PROCEDURE H

For preparation of indicator solutions, see pages 27–29.
For preparation of stock solutions of acids and bases, see pages 30–32.

overhead projector

30 mL distilled water

2 mL phenolphthalein indicator solution

1 mL 6M sodium hydroxide, NaOH

30 mL 0.15M sodium acetate, $NaC_2H_3O_2$ (To prepare 1 liter of solution, dissolve 20.4 g of $NaC_2H_3O_2 \cdot 3H_2O$ in 600 mL of distilled water and dilute the resulting solution to 1.0 liter.)

15 mL 1.0M ammonium chloride, NH_4Cl (For preparation, see Materials for Procedure C.)

15 mL 1.0M aqueous ammonia, NH_3

15 mL 6M hydrochloric acid, HCl

3 10-cm petri dishes

3 droppers

10-mL graduated cylinder

PROCEDURE A

Preparation

Label six 250-mL beakers with the numbers 1 through 6. Pour 100 mL of the appropriate solution into each of the beakers as indicated:

Beaker		Solution	pH
1	A1:	0.18M $HC_2H_3O_2$ and	
		0.018M $C_2H_3O_2^-$	5.7
2	A2:	0.10M $HC_2H_3O_2$ and	
		0.10M $C_2H_3O_2^-$	4.7
3	A3:	1.0M $HC_2H_3O_2$ and	
		1.0M $C_2H_3O_2^-$	4.7
4	A4:	0.018M $HC_2H_3O_2$ and	
		0.18M $C_2H_3O_2^-$	3.7
5	A2:	0.10M $HC_2H_3O_2$ and	
		0.10M $C_2H_3O_2^-$	4.7
6	A3:	1.0M $HC_2H_3O_2$ and	
		1.0M $C_2H_3O_2^-$	4.7

Place a magnetic stir bar in each beaker.

Mount two 50-mL burets on a stand. Fill one with 1.0M HCl and the other with 1.0M NaOH. Fill one 100-mL graduated cylinder with 1.0M HCl and the other with 1.0M NaOH.

Presentation

Add 15 drops of bromophenol blue indicator solution to each of beakers 1, 2, and 3. Stir the solutions on the magnetic stirrer until they are a uniform purple. Add 15 drops of bromothymol blue indicator solution to each of beakers 4, 5, and 6. Stir these mixtures until they are a uniform yellow.

Stir the solution in beaker 1 on the magnetic stirrer while adding 1.0M HCl from the buret. Add HCl until the solution in the beaker turns yellow (about 2 mL of HCl are required). Repeat this with beaker 2 (about 10 mL of HCl are required). In a similar fashion, add 10 mL of 1.0M HCl to beaker 3. The solution in beaker 3 remains purple.

Stir the solution in beaker 4 on the magnetic stirrer while adding 1.0M NaOH from the buret. Add NaOH until the solution turns blue (about 1.5 mL are required). Repeat this with beaker 5 (about 10 mL are required). In a similar fashion, add 10 mL of 1.0M NaOH to beaker 6. The solution in beaker 6 remains yellow.

While stirring the purple solution in beaker 3, pour 1.0M HCl from the graduated cylinder into the beaker until the solution turns yellow. This requires about 90 mL of HCl. While stirring the yellow solution in beaker 6, pour 1.0M NaOH from the graduated cylinder into the beaker until the solution turns blue. This requires about 90 mL of NaOH.

PROCEDURE B

Preparation

Calibrate each of the two droppers to deliver about 1 mL by drawing 1.0 mL of water from one of the 10-mL graduated cylinders and marking the level of water in the dropper with a waterproof marker.

Presentation

Place a stir bar in a clean 250-mL beaker and pour 100 mL of distilled water into the beaker. Set the beaker on the magnetic stirrer and start the stirrer. Immerse the electrode of the pH meter in the water and measure its pH. (The value may be different from 7 if there are even traces of impurities in the water, such as dissolved carbon dioxide. The pH reading may not be stable because pure water contains few ions and is not sufficiently conductive for an accurate electrochemical measurement of pH.) Record the pH of the water. Use one of the calibrated droppers to add 1 mL of 1.0M HCl to the beaker. The pH will drop by several units. Record the new pH value. Remove the pH electrode from the solution and rinse the electrode with the distilled water in the wash bottle.

Repeat the procedure described in the preceding paragraph with another 250-mL beaker containing 100 mL of distilled water, but use the other dropper to add 1 mL of 1.0M NaOH instead of the HCl. The pH will rise by several units. Record the value of the pH, and rinse the electrode.

Pour 100 mL of 1.0M $HC_2H_3O_2$ into another 250-mL beaker and measure the pH of the solution. Remove the electrode from the solution and rinse it. Pour 100 mL of 1.0M $NaC_2H_3O_2$ into another 250-mL beaker and measure its pH. Pour the $NaC_2H_3O_2$

solution into the $HC_2H_3O_2$ solution and mix them by pouring them back and forth between the beakers several times. Pour half (100 mL) of the mixture into one of the beakers, put a stir bar in the beaker, and set the beaker on the magnetic stirrer. Measure the pH of the mixture and record the value. While stirring the mixture, add 1 mL of 1.0M HCl. The pH of the mixture will fall by a fraction of a unit. Add another 1 mL of HCl; the pH will fall by a similar amount. Pour 10 mL of HCl from one of the graduated cylinders into the beaker. The pH will again drop by a fraction of a unit. Continue to add 10-mL aliquots of HCl until the pH change increases to several units. This will require a total of about 50 mL of HCl.

Put a stir bar in the beaker containing the remainder of the $NaC_2H_3O_2$–$HC_2H_3O_2$ mixture and put the beaker on the stirrer. Measure the pH of the mixture. While stirring the mixture, add 1 mL of 1.0M NaOH. The pH of the mixture will rise by a fraction of a unit. Add another 1 mL of NaOH; the pH will rise by a similar amount. Pour 10 mL of NaOH from the other graduated cylinder into the beaker. The pH will again increase by a fraction of a unit. Continue to add 10-mL aliquots of NaOH until the pH change increases to several units. This will require a total of about 50 mL of NaOH.

PROCEDURE C

Preparation

Calibrate each of the two droppers to deliver about 1 mL by drawing 1.0 mL of water from one of the 10-mL graduated cylinders, and marking the level of the water in the dropper with a waterproof marker.

Presentation

Place a stir bar in a clean 250-mL beaker and pour 100 mL of distilled water into the beaker. Set the beaker on the magnetic stirrer and start the stirrer. Immerse the electrode of the pH meter in the water and measure its pH. (The value may be different from 7 if there are even traces of impurities in the water, such as dissolved carbon dioxide. The pH reading may not be stable because pure water contains few ions and is not sufficiently conductive for an accurate electrochemical measurement of pH.) Record the pH of the water. Use one of the calibrated droppers to add 1 mL of 1.0M HCl to the beaker. Record the new pH value. Remove the pH electrode from the solution and rinse the electrode with the distilled water in the wash bottle.

Repeat the procedure described in the preceding paragraph with another 250-mL beaker containing 100 mL of distilled water, but use the other dropper to add 1 mL of 1.0M NaOH instead of the HCl. Record the value of the pH, and rinse the electrode.

Pour 100 mL of 1.0M NH_3 into another 250-mL beaker and measure the pH of the solution. Remove the electrode from the solution and rinse it. Pour 100 mL of 1.0M NH_4Cl into another 250-mL beaker and measure its pH. Pour the NH_4Cl solution into the NH_3 solution and mix them by pouring them back and forth between the beakers several times. Pour half, 100 mL, of the mixture into one of the beakers, put a stir bar in the beaker, and set the beaker on the magnetic stirrer. Measure the pH of the mixture and record the value. While stirring the mixture, add 1 mL of 1.0M HCl. The pH of the mixture will fall by a fraction of a unit. Add another 1 mL of HCl; the pH will fall by a similar amount. Pour 10 mL of HCl from a graduated cylinder into the beaker. The pH will again drop by a fraction of a unit. Continue to add 10-mL aliquots of HCl until the pH change increases to several units. This will require a total of about 50 mL of HCl.

Put a stir bar in the beaker containing the remainder of the NH_4Cl–NH_3 mixture and put the beaker on the stirrer. Measure the pH of the mixture. While stirring the mixture, add 1 mL of 1.0M NaOH. The pH of the mixture will rise by only a fraction of a unit. Add another 1 mL of NaOH; the pH will rise by a similar amount. Pour 10 mL of NaOH from a graduated cylinder into the beaker. The pH will again increase by only a fraction of a unit. Continue to add 10-mL aliquots of NaOH until the pH change increases to several units. This will require a total of about 50 mL of NaOH.

PROCEDURE D

Preparation and Presentation

Pour 300 mL of distilled water into each of two 500-mL beakers, and place a stirring rod in each. Set the beakers before the white background, add 4 drops of bromocresol green indicator solution to each container, and stir the mixtures. Add 2 drops of 6M HCl to one of the containers and stir the mixture. The color of the mixture will change from blue to yellow. Add 3 drops of 6M NaOH to this container and stir the mixture. The color will return to blue. Add 3 drops of 6M HCl to this container and stir the mixture. The color will return to yellow. Keep the solutions in the two containers as color standards.

Pour 300 mL of 0.90M NH_4Cl into each of three 500-mL beakers, and place a stirring rod in each. Set the beakers before the white background, add 4 drops of bromocresol green indicator solution to each container, and stir the mixtures. Add 2 drops of 6M HCl to one of the containers and stir the mixture. The color will change from green to yellow. Compare the color of this solution with that of the yellow solution produced earlier. Add 2 drops of 6M NaOH to the second container. The color of the mixture will change from green to blue. Compare the color of this solution with that of the blue solution produced earlier. Keep the three solutions as color standards.

Pour 150 mL of 1.0M $HC_2H_3O_2$ into each of three 500-mL beakers, and place a stirring rod in each. Add 150 mL of 1.0M $NaC_2H_3O_2$ to each container and stir the mixtures. Add 4 drops of bromocresol green to each container and stir the mixtures. Add 2 drops of 6M HCl to one of the containers and stir the mixture. The mixture will remain green. Add 2 more drops and stir the mixture. The mixture still remains green. Add more 6M HCl, 5 mL at a time, and stir after each addition, until the mixture turns yellow. This will require about four such additions. Add 2 drops of 6M NaOH to the second container. The solution will remain green. Add more 6M NaOH, 5 mL at a time, and stir after each addition, until the mixture turns blue. This will require about four such additions.

PROCEDURE E

Preparation and Presentation

Pour 300 mL of distilled water into a 500-mL beaker. Set the beaker before the white background, add 4 drops of phenolphthalein indicator solution to the water, and stir the mixture. Add 2 drops of 6M NaOH to the container and stir the mixture. The mixture will change from colorless to pink.

Pour 300 mL of 0.15M NaC$_2$H$_3$O$_2$ into a second 500-mL beaker, and set the beaker before the white background. Add 4 drops of phenolphthalein indicator solution to the container and stir the mixture. Add 2 drops of 6M NaOH to the second container and stir the mixture. The mixture will change from colorless to pink.

Pour 150 mL of 1.0M NH$_4$Cl and 150 mL of 1.0M NH$_3$ into a third 500-mL beaker, stir the mixture, and place the container before the white background. Add 4 drops of phenolphthalein indicator solution to the solution and stir the mixture. This mixture will be pink.

Add 3 drops of 6M HCl to the first container (distilled water with NaOH) and stir the mixture. The solution will change from pink to colorless. Add 3 drops of 6M HCl to the second container (acetate with NaOH) and stir the mixture. This mixture will also change from pink to colorless. Add 3 drops of 6M HCl to the last container (ammonia-ammonium mixture) and stir the solution. The mixture will remain pink. Add another 3 drops of 6M HCl to this solution. The mixture will still be pink. Add more 6M HCl, 5 mL at a time, and stir after each addition, until the mixture becomes colorless. This will require about four such additions.

PROCEDURE F

Preparation

Pour 250 mL of distilled water into each of three 500-mL beakers, and place a stirring rod in each beaker. Fill a 25-mL graduated cylinder with 1.0M HCl and a second cylinder with 1.0M NaOH.

Presentation

Add 10 drops of bromothymol blue indicator solution to each of the three beakers of distilled water. While stirring the solution, add 1.0M HCl drop by drop to the first beaker until the solution becomes yellow. This will take only several drops. Add 1.0M NaOH drop by drop to this first beaker until the color changes to blue. This will also take only several drops.

Add 2 g of KH$_2$PO$_4$ and 2 g of K$_2$HPO$_4$ to each of the remaining two beakers of water and stir the mixtures until the solids dissolve. While stirring the solution in one of these beakers, add 1.0M HCl drop by drop until twice as many drops have been added as were required to change the color of the solution in the first beaker. While stirring the solution, slowly pour 1.0M HCl from one of the graduated cylinders into the beaker until the solution turns yellow. This will require about 20 mL of HCl. Slowly add 1.0M NaOH from the other graduated cylinder to the last beaker while stirring the solution. Continue to add NaOH until the mixture turns green. This will require about 20 mL of NaOH.

PROCEDURE G

Preparation and Presentation

Set two 10-cm petri dishes on a transparency atop the overhead projector. Pour 30 mL of distilled water into each dish. Add 4 drops of bromocresol green indicator solution to each dish, and swirl the dishes to mix their contents. Add 2 drops of 6M HCl to each of the dishes and swirl the dishes. The color of the mixtures will change from blue to yellow. Add 3 drops of 6M NaOH to each dish and swirl them again. Both will turn blue. Add 3 drops of 6M HCl to one of the dishes and swirl it. The color will return to yellow. Write the label "acid" on the transparency adjacent to the yellow dish, and write "base" adjacent to the blue dish.

Set three more 10-cm petri dishes on the overhead projector. Pour 30 mL of 0.90M NH_4Cl into each dish. Add 4 drops of bromocresol green indicator solution to each petri dish and swirl the dishes to mix their contents. The mixtures will be green. Add 2 drops of 6M HCl to one of the dishes and swirl the dish. The color will change to yellow. Compare the color of this solution with that of the yellow solution produced earlier. Add 2 drops of 6M NaOH to the second dish. The color of the mixture will change from green to blue. Compare the color of this solution with that of the blue solution produced earlier. Remove the first two petri dishes, but keep the three new ones on the projector as color standards.

Place three more 10-cm petri dishes on the overhead projector. Pour 15 mL of 1.0M $HC_2H_3O_2$ into each of the dishes. Add 15 mL of 1.0M $NaC_2H_3O_2$ to each dish. Add 4 drops of bromocresol green to each dish and swirl them to mix their contents. The mixtures will be green. Add 2 drops of 6M HCl to one of the dishes and swirl it. The mixture will remain green. Add 2 more drops and swirl the dish; the mixture still remains green. Add more 6M HCl, 1 mL at a time, and swirl the dish after each addition, until the mixture turns yellow. This will require about four such additions. Add 2 drops of 6M NaOH to the second dish and swirl it. The solution will remain green. Add more 6M NaOH, 1 mL at a time, and swirl the dish after each addition, until the mixture turns blue. This will require about four such additions.

PROCEDURE H

Preparation and Presentation

Set a 10-cm petri dish on the overhead projector. Pour 30 mL of distilled water into the dish. Add 4 drops of phenolphthalein indicator solution to the dish, and swirl it to mix its contents. Add 2 drops of 6M NaOH to the dish and swirl it. The mixture will change from colorless to magenta.

Set another 10-cm petri dish on the overhead projector, and pour 30 mL of 0.15M $NaC_2H_3O_2$ into it. Add 4 drops of phenolphthalein indicator solution to the dish and swirl it to mix its contents. Add 2 drops of 6M NaOH to the second dish and swirl it. The mixture will change from colorless to magenta.

Set a third 10-cm petri dish on the overhead projector. Pour 15 mL of 1.0M NH_4Cl and 15 mL of 1.0M NH_3 into it, and swirl it to mix its contents. Add 4 drops of phenolphthalein indicator solution to the dish and swirl it. This mixture will be pink.

Add 3 drops of 6M HCl to the first dish (distilled water with NaOH) and swirl the mixture. The solution will change from magenta to colorless. Add 3 drops of 6M HCl to the second dish ($NaC_2H_3O_2$ with NaOH) and swirl the mixture. This mixture will also change from magenta to colorless. Add 3 drops of 6M HCl to the last container (NH_4Cl and NH_3 mixture) and swirl the solution. The mixture will remain pink. Add another 3 drops of 6M HCl to this solution. The mixture will still be pink. Add more 6M HCl, 1 mL at a time, and swirl the mixture after each addition, until the mixture becomes colorless. This will require about four such additions.

HAZARDS

Concentrated solutions of acetic acid and hydrochloric acid can irritate the skin. Their vapors are extremely irritating to the eyes and respiratory system.

Concentrated sodium hydroxide solutions can cause severe burns to the eyes, skin, and mucous membranes.

Concentrated aqueous ammonia can irritate the skin, and its vapors are harmful to the eyes and mucous membranes.

DISPOSAL

The waste solutions should be flushed down the drain with water.

DISCUSSION

A buffer solution is one that resists changes in its pH when an acid or a base is added to it. Buffer solutions typically contain a weak acid and a salt of the weak acid, or a weak base and a salt of the weak base. Procedure A shows with indicators that a solution containing both acetic acid and sodium acetate resists changes in pH when an acid or a base is added. It also shows that the amount of acid or base that can be added before a large pH change occurs depends on the concentration of the acid and salt in the solution [1]. Procedures B and C use a pH meter to demonstrate buffering action, the first using a mixture of acetic acid and sodium acetate, the other using a mixture of ammonia and ammonium chloride [2]. The remaining five procedures use indicator solutions to show buffering action. Procedures D and G demonstrate that a solution containing a mixture of acetic acid and sodium acetate resists changes in its pH; Procedures E and H do the same with a mixture of ammonia and ammonium chloride; and Procedure F shows that a buffer can be made from two salts of a polyprotic acid.

In Procedure A, the solutions in beakers 2, 3, 5, and 6 all have the same pH value, namely 4.7. These solutions all contain acetic acid and acetate ions in a 1:1 molar ratio. Therefore, the pH is equal to the pK_a of acetic acid. The solutions in beakers 2 and 5 differ from those in 3 and 6 in that the concentrations of acetic acid and acetate ions in 3 and 6 are 10 times the concentrations of these solutes in 2 and 5. As the demonstration shows, more acid or base is required to cause a large pH change in the solutions with the higher concentrations (beakers 3 and 6) than is required for a similar change in the solutions with lower concentrations (beakers 2 and 5). Therefore, the concentrations of the acid and salt in the buffer determine its buffering capacity, that is, the amount of acid or base it can consume before a large pH change occurs. Beaker 1 has a pH value 1 unit higher than beakers 2, 3, 5, and 6, whereas beaker 4 has a pH value 1 unit lower. Although the concentration of $HC_2H_3O_2$ is greater in beaker 1 than it is in beaker 2, it has a smaller capacity for acid than does the solution in beaker 2. This illustrates that the useful range for a buffer is limited to within ± 1 unit of the pK_a of the weak acid.

The first part of Procedure B is a demonstration of the effect on the pH of water when acid or base is added to it. When 1 mL of 1.0M HCl is added to 100 mL of water, the pH changes from near 7 to about 2. When 1 mL of 1.0M NaOH is added to 100 mL of water, the pH rises to 12. Then a buffer solution is prepared by mixing equal volumes of 1.0M acetic acid and 1.0M sodium acetate. The pH of the acid is about 2.4, that of the salt is about 9, and that of the mixture is near 4.7. When 1 mL of 1.0M HCl is added to 100 mL of the mixture, or if even 10 mL are added, the change in the pH of the mixture is only a fraction of a pH unit. In fact it requires about 50 mL of 1.0M HCl to change the pH of the mixture by the same amount that only 1 mL of acid changed the pH of water. A similar result is obtained when sodium hydroxide is added to the mixture—nearly 50 mL are required to change the pH by several units. The mixture resists changes in its pH; it is a buffer solution.

Procedure C is similar to Procedure B, but a mixture of ammonia and ammonium chloride is used in place of the acetate–acetic acid mixture. The mixture of a weak base and its salt proves to be as effective a buffer (although not at the same pH) as the mixture of a weak acid and its salt.

The first part of Procedures D and G is a demonstration of the effect on the color of the bromocresol green indicator when acid or base is added to its solution. When 2 drops of 6M HCl are added to 300 mL of water, the color changes from blue to yellow. When 3 drops of 6M NaOH are added to the solution, the color changes from yellow to blue. Only a few drops of acid or base are required to change the color of the indicator. Next, the effect of adding acid to a salt solution is investigated. (The salt used is ammonium chloride because its pH is about the same as that of the buffer to be investigated later, and the color of the indicator in this solution is the same as in the buffer.) Again, the color is changed when only a few drops of acid or base are added. Then a buffer solution is prepared by mixing equal volumes of 1.0M acetic acid and 1.0M sodium acetate. To change the color of the indicator in this solution, far more acid must be added than was added to water or to the NH_4Cl solution to produce a color change. In fact it requires about 200 times as much! Similarly, nearly 200 times as much base is needed to cause the same color change produced in water or in the NH_4Cl solution. The mixture does resist changes in its pH; it is a buffer solution.

In Procedures E and H the buffering ability of a mixture of ammonia and ammonium chloride is investigated. This is done using phenolphthalein as the indicator. The mixture of a weak base and its salt proves to be as effective a buffer as the mixture of a weak acid and its salt.

Procedure F uses a mixture of KH_2PO_4 and K_2HPO_4 as the buffer solution, with bromothymol blue as the indicator. This shows that a mixture of salts of a polyprotic acid can also form a buffer solution.

A buffer solution made of a weak acid and its salt resists changes in pH, because it contains an acid that neutralizes added base and it contains a base that neutralizes added acid [3]. In the mixture of acetic acid and sodium acetate used in Procedure D, acetic acid neutralizes added base (hydroxide ions):

$$HC_2H_3O_2(aq) + OH^-(aq) \longrightarrow C_2H_3O_2^-(aq) + H_2O(aq)$$

In this mixture the acetate ions function as a base, neutralizing added acid (hydrogen ions):

$$C_2H_3O_2^-(aq) + H^+(aq) \longrightarrow HC_2H_3O_2(aq)$$

In the mixture of ammonia and ammonium chloride used in Procedure E, ammonia acts as a base, neutralizing added acid:

$$NH_3(aq) + H^+(aq) \longrightarrow NH_4^+(aq)$$

Ammonium ions function as an acid, neutralizing added base:

$$NH_4^+(aq) + OH^-(aq) \longrightarrow NH_3(aq) + H_2O(l)$$

A more quantitative description of the functioning of buffer solutions can be found in virtually any general chemistry textbook.

All of the procedures show that a buffer solution has a limited capacity to resist changes in pH. When a sufficiently large amount of acid or base is added to a buffer solution, the pH of the solution will change drastically. The capacity of a buffer solution is determined by the amount of weak acid (or weak base) and salt that the buffer contains. If the amount of base added to a buffer solution is a stoichiometric excess over the amount of acid contained in the buffer, the buffer will be overwhelmed by the base. Similarly, if the amount of acid exceeds the amount of base in the buffer, again the buffer's capacity will be exceeded. Therefore, the higher the concentrations of weak acid (or weak base) and salt in the solution, the greater the capacity of the buffer solution to resist changes in pH.

REFERENCES

1. C. J. Donahue and M. G. Panek, *J. Chem. Educ.* 62:337 (1985).
2. W. L. Felty, *J. Chem. Educ.* 53:229 (1976).
3. J. C. Chang, *J. Chem. Educ.* 53:228 (1976).

8.29

Buffering Action of Alka-Seltzer

When hydrochloric acid is added to both water and a solution containing an Alka-Seltzer tablet, the color of universal indicator shows that the pH change is greater in water than in the Alka-Seltzer solution. A similar effect is produced when the base sodium hydroxide is used in place of acid [1].

MATERIALS

For preparation of indicator solutions, see pages 27–29.
For preparation of stock solutions of acids and bases, see pages 30–32.

1 liter distilled water

2 mL universal indicator, 1–13 pH range

5 mL pH 7 standard solution (For preparation, see Procedure A of Demonstration 8.1.)

2 Alka-Seltzer tablets

50 mL 0.1M hydrochloric acid, HCl

50 mL 0.1M sodium hydroxide, NaOH

5 250-mL beakers

2 droppers

5 glass stirring rods

white background (e.g., 25-cm × 50-cm poster board)

2 50-mL burets, with stopcocks and stands

PROCEDURE

Preparation

Pour 200 mL of distilled water into each of five 250-mL beakers and add 5 drops of universal indicator to each. While stirring the solutions, add about 10 drops of pH 7 buffer to each beaker to adjust the colors of the solution to the same shade of yellow-green. Arrange the beakers in a row in front of a white background. Fill one of the burets with 0.1M HCl and the other with 0.1M NaOH.

Presentation

Add one Alka-Seltzer tablet each to the second and fourth beakers, stir the mixtures, and note their color.

From the buret, add 5 mL of 0.1M HCl to the first and second beakers, stir the mixtures, and note their color. Add another 5 mL of HCl to the second beaker, the one containing Alka-Seltzer. Continue adding 5-mL aliquots to the second beaker until the color of the solution is the same as that in the first beaker. Record how much HCl was added to the second beaker.

From the buret, add 5 mL of 0.1M NaOH to the fourth and fifth beakers, stir the mixtures, and note their color. Add another 5 mL of NaOH to the fourth beaker, the one containing Alka-Seltzer. Continue adding 5-mL aliquots to the fourth beaker until the color of the solution is the same as that in the fifth beaker. Record how much NaOH was added to the fourth beaker.

DISPOSAL

The waste solutions should be flushed down the drain with water.

DISCUSSION

An Alka-Seltzer tablet contains 1.9 g (0.022 mole) of sodium bicarbonate and 1.0 g (0.005 mole) of citric acid. When the tablet is placed in water, the citric acid reacts with the sodium bicarbonate, producing sodium citrate in solution and releasing carbon dioxide gas:

$$3\ HCO_3^-(aq)\ +\ H_3C_6H_5O_7(aq)\ \longrightarrow\ C_6H_5O_7^{3-}(aq)\ +\ 3\ CO_2(g)\ +\ 3\ H_2O(l)$$

Sodium citrate behaves like an antacid, because citrate is the anion of a weak acid. It neutralizes excess acid by forming citric acid:

$$C_6H_5O_7^{3-}(aq)\ +\ 3\ H^+(aq)\ \longrightarrow\ H_3C_6H_5O_7(aq)$$

The excess sodium bicarbonate in the tablet also behaves like an antacid, liberating CO_2:

$$HCO_3^-(aq)\ +\ H^+(aq)\ \longrightarrow\ CO_2(g)\ +\ H_2O(l)$$

The excess sodium bicarbonate is also responsible for the buffering action observed in this demonstration. Bicarbonate ions neutralize the added base, resisting a change in the pH of the solution:

$$HCO_3^-(aq)\ +\ OH^-(aq)\ \longrightarrow\ CO_3^{2-}(aq)\ +\ H_2O(l)$$

REFERENCE

1. N. Friedman, *J. Chem. Educ.* 52:605 (1975).

8.30

Effect of pH
on Protein Solubility†

When a concentrated solution of hydrochloric acid is added drop by drop to a stirred, almost clear, solution of casein in dilute base, the mixture develops a transient cloudiness which persists longer and longer as more acid is added, until suddenly the entire solution turns milky. An additional 2–3 drops of acid cause the solution to clear once more. The process can be reversed by adding concentrated sodium hydroxide drop by drop to the now acidic solution.

MATERIALS

For preparation of stock solutions of acids and bases, see pages 30–32.

50 mL 2M hydrochloric acid, HCl

50 mL 2M sodium hydroxide, NaOH

250 mL 0.5% by weight casein in dilute base (To prepare 1 liter of stock solution, dissolve 0.4 g NaOH in 1 liter of distilled water, and dissolve 5.0 g of purified casein in the resulting solution. The casein solution may be stored for several months if it is refrigerated.)

2 250-mL polyethylene wash bottles

magnetic stirrer and stir bar

400-mL beaker

black backdrop (e.g., 20-cm × 30-cm poster board)

pH meter (optional)

PROCEDURE

Preparation

Put about 50 mL of 2M HCl in one wash bottle and 50 mL of 2M NaOH in the other.

Place the stir bar in the 400-mL beaker, and set the beaker on the magnetic stirrer. Pour 250 mL of casein solution into the beaker. Place the black backdrop behind the beaker.

If a pH meter is to be used, insert the electrode in the casein solution.

---

† This demonstration was developed by Professor Jerry A. Bell of Simmons College in Boston, Massachusetts.

Presentation

Turn on the magnetic stirrer and adjust it to a vigorous stirring rate. Add several drops of 2M HCl from the wash bottle to the outer edge of the stirred solution in the beaker. The first few drops will produce no visible effect. Continue adding drops of HCl until a turbid mixture results. The turbidity will gradually diminish as the mixture is stirred. After the mixture has cleared, add another drop of HCl. Allow the mixture to clear again. Continue to add HCl drop by drop and to allow the solution to clear between additions. Eventually, the addition of a drop of HCl will produce a persistent milky mixture. If a pH meter is used, note the pH at which the milky mixture forms. Without pausing for more than 15 seconds, add another drop or two of HCl. The turbidity will disappear, and the mixture will become even clearer than it was initially.

Add 2M NaOH drop by drop from the other wash bottle. The addition of drops of NaOH will reverse the process observed when the HCl was added. Note the pH at which the milky mixture forms. Do not allow the mixture to remain cloudy for more than 15 seconds, or the precipitated casein may not redissolve upon addition of more NaOH and may coagulate on the pH electrode. Coagulated protein can be dissolved by soaking it in 1M hydrochloric acid.

HAZARDS

Concentrated sodium hydroxide solutions can cause severe burns to the eyes, skin, and mucous membranes.

DISPOSAL

If it is stored refrigerated at an alkaline pH, the casein solution can be reused several times in repeat performances of this demonstration. Otherwise it should be flushed down the drain with water.

The waste solutions should be flushed down the drain with water.

DISCUSSION

This demonstration shows that the solubility of certain proteins is sensitive to the pH of their solution. This indicates that proteins are affected by acid or base and, therefore, have properties of acids or bases. These properties originate in the structure of the protein molecules.

Proteins are polymers of alpha-amino acids bonded together by peptide bonds formed between the carboxyl group on one acid and the amine group on the next. There are about 20 common amino acids found in proteins, and they differ in the side chain attached to their central carbon atom. A general structure of an amino acid may be represented by

$$^+H_3N - \overset{\overset{\displaystyle H}{|}}{\underset{\underset{\displaystyle R}{|}}{C}} - CO_2^-$$

where R is one of a group of about 20 different structures called side chains.
Several of the amino acids have side chains that are acidic or basic. These are listed below:

Amino acid	Side chain (R)
aspartic acid	$-CH_2CO_2H$
glutamic acid	$-CH_2CH_2CO_2H$
lysine	$-(CH_2)_4NH_2$
arginine	$-(CH_2)_3NH(C=NH)NH_2$
histidine	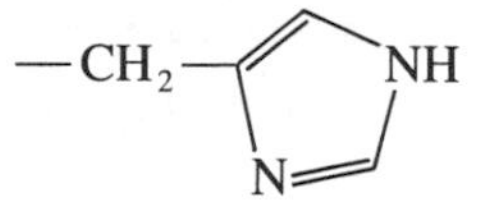 $-CH_2-$ (imidazole ring)

Because proteins contain amino acids that have acidic or basic side chains, they are sensitive to the pH of the solution in which they are dissolved. As the pH of such a solution is changed, the degree of ionization of the groups changes. In order for the protein to remain in solution, it must have a net charge. If the pH is at a value such that few of the acidic groups are ionized and few of the basic groups are protonated, then the protein will come out of solution. This phenomenon is illustrated in this demonstration with a protein found in milk, casein. Casein is a mixture of proteins that contain all of the common amino acids. It is obtained by removing the cream from whole milk and acidifying the resulting skimmed milk, which causes the casein to precipitate.

A protein is a high molecular weight, very polar polymer of alpha-amino acids having many amine and carboxyl groups that can undergo acid-base reactions with the surroundings. In the explanation of how these acid-base reactions affect the solubility of the protein, a few simplifications will be made. First, suppose that the protein has only one acidic group and one basic group, namely a carboxyl group (as in aspartic acid) and an amine group (as in lysine). Second, suppose the amine group is its protonated form ($-NH_3^+$), which behaves as an acid. Then, both side chains can be treated like acids, and their acidities compared directly, rather than comparing the acidity of the carbonyl with the basicity of the amine.

The carboxyl group is an acid and ionizes in solution as indicated in the following equation:

$$R-COOH + H_2O \rightleftarrows R-COO^- + H_3O^+ \qquad pK_a \approx 3$$

The indicated pK_a is a composite of the pK_a values of aspartic acid and glutamic acid. The amine group is a base which can react with a hydrogen ion in solution to form a substituted ammonium ion. This ammonium ion can behave as an acid, as indicated in the following equation:

$$R-NH_3^+ + H_2O \rightleftarrows R-NH_2 + H_3O^+ \qquad pK_a \approx 9$$

The indicated pK_a is based on the pK_a values of the protonated forms of the three basic amino acids—lysine, arginine, and histidine. These pK_a values for the carboxylic acid and protonated amine indicate that the carboxylic acid is the stronger acid.

Suppose the protein is in a solution of high pH (e.g., pH of 11). Because the pH is greater than the pK_a of either the carboxylic acid or the protonated amine, the equilibrium position of both of the above reactions will lie far to the right. Therefore, the amine is in its neutral form ($R-NH_2$), and the carboxylic acid is in its ionic form ($R-COO^-$). The protein can be represented under these conditions as

$$H_2N-protein-COO^-$$

The net charge on this species is -1. Molecules with a net negative charge will repel each other, and the protein will remain in solution.

Next, suppose the pH of the solution is lowered, say, to a pH of 6, by the addition of acid. Now the pH is below the pK_a of the protonated amine but above the pK_a of the carboxylic acid. In this situation, the equilibrium position of the carboxylic-acid ionization will be on the right, but the equilibrium position of the amine ionization will be on the left. The carboxylic acid is in its ionic form ($-COO^-$), and the amine is also ionized as $-NH_3^+$. Then the protein can be represented by

$$^+H_3N-protein-COO^-$$

The net charge on this species is 0. Molecules without a net charge no longer repel one another. In fact, because there are negative and positive charges on the molecule, the negative charge on one molecule can be attracted to the positive charge on another. Therefore, the molecules will be attracted to each other and come together to form aggregates, forcing the water away from the molecules. This leads to the precipitation of these aggregates of protein molecules, and accounts for the milky mixture produced in the demonstration when acid is added to the basic casein solution.

The pH at which the net charge on a protein is 0 is called its *isoelectric point*. Most proteins have their minimum solubility in solutions at their isoelectric pH. The pH at which the maximum cloudiness occurs in the precipitation of casein in this demonstration is an estimate of the isoelectric point for casein. It is a crude estimate, because the maximum cloudiness is difficult to perceive.

Finally, suppose more acid is added to the solution, and the pH falls to about 2. Under these conditions, the pH of the solution is lower than either of the pK_a's, and the equilibrium positions of both ionization reactions are on the left. The molecule can then be represented as

$$^+H_3N-protein-COOH$$

The net charge on this species is positive. Again, molecules which carry a net positive charge repel one another, and the protein precipitate formed at the isoelectric pH redissolves when each molecule gains a net positive charge.

Proteins that have an isoelectric pH below 7 are said to be acidic, and those with an isoelectric pH above 7 are called basic. The isoelectric pH for casein as determined in this demonstration is in the pH range of 4–5. This makes casein quite an acidic protein. The isoelectric pH of casein is this low because casein is a phosphoprotein. This means that there are phosphate groups bonded to the $-OH$ groups on some of the amino acid side chains that have alcohol functional groups. These phosphate esters also have acid-base properties and have an even lower pK_a than the carboxylic acid groups:

$$R-O-PO_3H_2 + H_2O \longrightarrow R-O-PO_3H^- + H_3O^+ \qquad pK_a = 2$$

The presence of the phosphate esters means that a lower pH is required to neutralize enough of these negative charges to yield a protein molecule with a net charge of 0. The result is a low isoelectric pH and an acidic protein.

The biological role (or roles) played by casein is not altogether clear at present. About a quarter of milk solids—which are sold as nonfat dry milk powder—is casein, which provides a good deal of nourishment to the infant animal drinking its mother's milk. Casein also seems to be a carrier of calcium ions from the mother to her offspring. A phosphoprotein would be good for this purpose, because phosphate groups bind calcium ions tightly. Although casein is a major component of dry milk, dry milk is not a successful substitute for casein in this demonstration.

8.31

Lewis Acid-Base Properties of Aluminum Chloride and Hydrogen Chloride

Aluminum chloride is dissolved in dichloromethane, and the acid-base indicator methyl violet is added, turning the solution yellow. When one of a number of organic liquids is added to the solution, the solution becomes purple (Procedure A). A similar effect is observed with a solution of hydrogen chloride in chlorobenzene (Procedure B) [1].

MATERIALS FOR PROCEDURE A

5 g anhydrous aluminum chloride, $AlCl_3$ **(See Hazards section before handling anhydrous aluminum chloride.)**

50 mL dichloromethane, CH_2Cl_2 **(See Hazards section before handling dichloromethane.)**

1 mL methyl violet solution in dichloromethane (To prepare 100 mL of solution, dissolve 0.25 g of methyl violet in 100 mL of dichloromethane.)

5 mL pyridine, C_5H_5N

5 mL triethylamine, $(C_2H_5)_3N$

5 mL ethanol, C_2H_5OH

5 mL methanol, CH_3OH

2 125-mL Erlenmeyer flasks

stopper for 125-mL Erlenmeyer flask

filter funnel, with filter paper

4 test tubes, 25 mm × 200 mm, with stoppers

rack for test tubes

dropper

4 10-mL graduated cylinders

MATERIALS FOR PROCEDURE B

1 mL methyl violet solution (To prepare 100 mL of solution, dissolve 0.25 g of methyl violet in 100 mL of the solvent used in the demonstration [chlorobenzene or dichloromethane].)

40 mL chlorobenzene, C_6H_5Cl, or dichloromethane, CH_2Cl_2 **(See Hazards section before handling these compounds.)**

cylinder of hydrogen chloride gas, HCl, with valve

   or

   10 g sodium chloride, NaCl

   25 mL concentrated (18M) sulfuric acid, H_2SO_4

   long-stemmed funnel

   2-holed rubber stopper to fit Erlenmeyer flask

   125-mL Erlenmeyer flask

   right-angle bend of glass tubing, with outside diameter of 7 mm and the
      length of each arm ca. 5 cm

   stand, with clamp

   Bunsen burner

5 mL diethyl ether, $C_2H_5OC_2H_5$

5 mL pyridine, C_5H_5N, or triethylamine, $(C_2H_5)_3N$

5 mL methanol, CH_3OH

5 mL acetone, CH_3COCH_3

dropper

4 test tubes, 25 mm × 200 mm, with stoppers

rack for test tubes

30-cm length of rubber tubing, with inside diameter of ca. 6 mm

glass delivery tube (a 15-cm piece of glass tubing with outside diameter of 7 mm
   and a right-angle bend ca. 4 cm from one end)

4 10-mL graduated cylinders

PROCEDURE A

Preparation

Working under a fume hood, place 5 g of anhydrous $AlCl_3$ and 50 mL of CH_2Cl_2 in a 125-mL Erlenmeyer flask. Stopper the flask and shake it for several minutes. Filter the CH_2Cl_2 and $AlCl_3$ mixture by pouring it into a filter funnel lined with filter paper, and collect the filtrate in another 125-mL Erlenmeyer flask. Pour 10 mL of the clear filtrate into each of the four test tubes, stopper the tubes, and set them in the rack.

Presentation

Remove the stoppers from the test tubes and add 3 drops of methyl violet solution to each tube. Restopper the test tubes and invert them several times to mix their contents. The violet solution will immediately turn yellow in the $AlCl_3$–CH_2Cl_2 solutions.

Remove the stopper from one of the tubes and add 5 mL of pyridine from one of the 10-mL graduated cylinders. Restopper the tube and invert it several times to mix its contents. The mixture will turn violet. In a similar fashion, add 5 mL of triethylamine to the second tube, 5 mL ethanol to the third, and 5 mL of methanol to the last tube. After each addition, the mixture will become violet.

PROCEDURE B

Preparation

Working under a fume hood, put 4 drops of methyl violet solution in each of the four test tubes, and pour 10 mL of chlorobenzene or dichloromethane into each tube. Stopper the tubes and invert them several times to mix the solutions and set the tubes in the rack. The solutions will be violet.

If a cylinder of hydrogen chloride is used, attach one end of the rubber tubing to the cylinder valve and the other end to the short arm of the glass delivery tube.

If a cylinder of hydrogen chloride gas is unavailable, assemble the hydrogen-chloride generator as illustrated in the figure. Insert the funnel through one hole of the 2-holed stopper so that the tip of the funnel is within 1 cm of the bottom of the Erlenmeyer flask when the stopper is seated in the mouth of the flask. Insert one arm of the right-angle bend through the other hole in the stopper. Connect the free arm of the right-angle bend to the short arm of the glass delivery tube with the rubber tubing. Place 10 g of NaCl in the Erlenmeyer flask. Seat the stopper in the mouth of the flask, and clamp the flask to the stand.

Presentation

Unstopper one of the test tubes and insert the free end of the HCl delivery tube in the solution in the tube. If a cylinder of HCl is used, open the valve to allow a gentle flow of gas to bubble through the chlorobenzene. If the HCl-generator is used, slowly pour 25 mL of concentrated H_2SO_4 into the funnel; hydrogen chloride gas will be produced almost immediately, and it will bubble through the chlorobenzene. When the

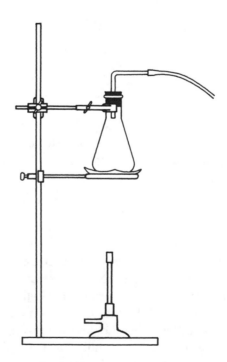

mixture in the tube turns yellow, move the delivery tube to the next test tube. Continue in this manner until the solution in each test tube is yellow. If the production of HCl in the generator should slow, gently warm the generator with the flame of a Bunsen burner.

From one of the 10-mL graduated cylinders, add 5 mL diethyl ether to the first test tube, stopper the tube, and shake it. The mixture will become purple again. Add 5 mL of pyridine or triethylamine to the second tube, stopper and shake it. Again the purple color will return. Repeat this with 5 mL of methanol in the third tube and 5 mL of acetone in the last tube. In each case, the purple color will return.

HAZARDS

Anhydrous aluminum chloride reacts violently with water, liberating a great deal of heat. It is also irritating to the skin and respiratory system. Care should be taken to keep anhydrous aluminum chloride dry.

Dichloromethane can be toxic when inhaled or ingested.

Diethyl ether, acetone, methanol, ethanol, triethylamine, chlorobenzene, and pyridine are flammable organic liquids. They should be handled in a fume hood, and proper precautions should be taken not to inhale or spill material.

Methyl violet is toxic when ingested; its oral LD50† in mice is 105 mg/kg. It also causes persistent stains when it contacts the skin.

DISPOSAL

The filtered solid aluminum chloride should be stirred into 1 liter of water and the mixture flushed down the drain. The dichloromethane or chlorobenzene solutions should be discarded in a receptacle for chlorine-containing organic wastes. Contact local hazardous waste disposal authorities for information on local disposal regulations for chlorine-containing wastes.

DISCUSSION

The most common acid-base concept is that of Brønsted and Lowry: an acid is a substance that releases a hydrogen ion in a reaction, and a base is a substance that receives a hydrogen ion. The reactions in this demonstration do not involve hydrogen ions. Therefore, they do not fall into the category of Brønsted-Lowry acid-base reactions. The Lewis concept of acid-base reactions considers a base to be a substance that forms a bond with another substance by donating a pair of its electrons, and an acid to be a substance that forms a bond by accepting the pair of electrons. Although it is impossible to demonstrate the actual bond formation that occurs in a Lewis acid-base reaction, this demonstration does show that certain reactions in which no hydrogen-ion transfer occurs do produce a color change in a compound that functions as a pH indicator in aqueous solutions.

In Procedure A, aluminum chloride is dissolved in dichloromethane. When methyl violet is added to this solution, the solution turns yellow. In aqueous solutions, methyl violet is a pH indicator which is yellow in very acidic solutions and violet in less acidic and basic solutions. The yellow color can be interpreted as showing that aluminum

† See the note on page 28.

chloride has acidic properties. Because there is no hydrogen in aluminum chloride, it cannot be a Brønsted-Lowry acid. However, it can be a Lewis acid, because the aluminum atom in $AlCl_3$ has an incomplete valence shell that can accept a pair of electrons from a molecule with a free pair. Methyl violet contains such molecules. Methyl violet contains methylated pararosaniline, whose molecular structure is given below. These molecules contain amine groups with nitrogen atoms having nonbonding, or lone, pairs of electrons [2].

Aluminum in $AlCl_3$ forms bonds with these nitrogen atoms by accepting the lone pairs into its valence shell. When it does this, the color of the methyl violet changes from violet to yellow.

When one of the organic liquids—pyridine, triethylamine, ethanol, or methanol—is added to the yellow aluminum chloride solution, the violet color returns. This is a result of a Lewis acid-base reaction between aluminum chloride and the added organic liquid. All of these liquids contain atoms having nonbonding electron pairs: in pyridine and triethylamine, nitrogen atoms, and in ethanol and methanol, oxygen atoms. These form bonds with the aluminum, displacing it from the methyl violet, so the violet color returns. The added organic liquid has acted as a Lewis base reacting with the Lewis acid aluminum chloride. The aluminum chloride has been neutralized by the organic liquid, as indicated by the disappearance of the acid color of methyl violet.

In Procedure B, a similar effect is shown using hydrogen chloride in place of aluminum chloride. This is perhaps not as convincing a demonstration of non–Brønsted-Lowry acidity, because hydrogen chloride contains hydrogen, and its aqueous solution is a well-known laboratory acid, hydrochloric acid. However, it can be argued that in such low-polarity solvents as dichloromethane and chlorobenzene, it is unlikely that any reactions involving ions, such as the hydrogen ion, will occur.

REFERENCES

1. W. F. Luder, W. S. McGuire, S. Zuffanti, *J. Chem. Educ.* 20:344 (1943).
2. J. S. Fritz, "Indicators for Non-Aqueous Acid/Base Titrations," in *Indicators,* E. Bishop, Ed., Pergamon Press: Oxford (1972).

8.32

Reaction Between Ammonia and Hydrogen Chloride

Two test tubes containing hydrochloric acid and aqueous ammonia are held next to each other. When air is blown across the mouths of the tubes, a white smoke is formed. Smoke rings can also be formed. This is described in Demonstration 6.25 in Volume 2 of this series.

9

Liquids, Solutions, and Colloids

Worth E. Vaughan, Rodney Schreiner,
Bassam Z. Shakhashiri, and David B. Shaw

The demonstrations in this chapter illustrate the properties of liquids, solutions, and colloids and are designed to appeal to elementary school students and advanced students in physical chemistry alike. These demonstrations catch the interest of students because the familiar world of common experience lurks behind many of them. We find out such things as, why scale forms in teakettles (9.13), why beer goes flat and fish suffocate in hot water (9.18), why salt melts ice on sidewalks (9.21), why water pipes burst in freezing weather (9.26), what makes dough climb the egg-beater (9.34), how to rid air of suspended solids (9.37), how to keep mayonnaise from separating (9.40), what makes the sky blue and the sunset orange (9.41), and why catsup is hard to get out of the bottle (9.45). Some demonstrations also involve apparent mysteries: the unequal depths of a liquid in the arms of a U-tube (9.30), moving liquids with electricity (9.32), and a stream of liquid that snaps back into its container when the stream is cut (9.35). Others display quite unexpected behavior: a liquid is siphoned without a tube (9.33), a liquid climbs the rod when it is stirred (9.34), and a small amount of powder turns nearly a liter of water to a gel (9.46). A wide variety of properties can be shown, many on a scale large enough to be readily visible to the audience in a lecture hall. In addition, when an overhead projector is used, even rather small displays can be magnified and presented so that they are easily visible. (Demonstrations that involve overhead projection list the projector first under Materials.)

We have undertaken to include demonstrations for which explanations are possible at different levels. Answering "Why?" about phenomena can be challenging. If something happens (liquids boil, freeze, mix, conduct electricity, . . .), explanations are relatively easy to produce. Why the density of water is $1 g/cm^3$ instead of $2 g/cm^3$ is a deeper question. (What *is* the answer?) Much of the background needed for discussing the demonstrations is provided in this introduction. Many of the bulk properties of liquids and solutions can be explained in terms of equilibrium thermodynamics. Much of thermodynamics is expressed in mathematical terms, so the introduction also includes a short description of the concepts and notation of the calculus used in thermodynamics. Some of the bulk properties can be explained in terms of the molecular properties of their constituents. In solutions containing very large solute molecules (i.e., polymer solutions), explanations in terms of molecular properties are very difficult because of the complexity of the solute structure. Much of the current study of polymer solutions is directed toward correlating bulk properties to molecular structures. More extensive discussion of the properties of liquids, solutions, and colloids than presented in this chapter can be found in textbooks of physical chemistry, for example, reference 1.

Technical terms printed in boldface type at their first occurrence in this introduction are explained in the glossary on pages 221–223.

STATES OF MATTER

Most pure substances can exist in any of three different **states of matter**—solid, liquid, and gas—in certain ranges of temperature and pressure. These three states are distinguished by their macroscopic properties. A solid has a definite shape and volume; a liquid has a definite volume, but takes the shape of its container; a gas takes both its shape and its volume from the container. The liquid state is intermediate between the solid and gaseous states. Liquids share some properties with solids and others with gases. Properties in which liquids more closely resemble solids than gases include density, compressibility, and the relationship among pressure, volume, and temperature. Liquids more closely resemble gases in their ability to flow. Liquids also exhibit properties unique to the liquid state, such as surface tension.

As temperature and pressure change, a pure substance can undergo a transition from one state to another. A solid can melt to become a liquid in a process called **fusion,** or a liquid can freeze into a solid. A liquid can evaporate into a gas, a process referred to as **vaporization,** or a gas can **condense** into a liquid. A solid can sublime directly into the gaseous state (**sublimation**), and a gas can condense directly into a solid.

SOME PROPERTIES OF LIQUIDS

The **density** of a substance is the ratio of its mass to its volume: density = mass/volume. Table 1 shows the density of several substances in the three different states. Solids are relatively dense and so are liquids. Gases, on the other hand, are much less dense. If you've ever hefted a milk carton to determine how full it is, you've relied on the difference between the density of milk and the density of air. Both solids and liquids are about 1000 times as dense as gases at atmospheric pressure. Although it is small, there is a difference between the densities of solids and liquids. In general, liquids are about 5–10% less dense than solids (water is a notable exception). Because the densities of liquids and solids are relatively high, the liquid and solid states are referred to as the **condensed states.**

Table 1. Densities (in g/cm^3) of Pure Substances in Different States [2]

	Solid density (at m.p.)	Liquid density (at m.p.)	Liquid density (at b.p.)	Vapor density (at b.p.)
Mercury	14.19	13.69	12.74	0.00388
Sulfur	2.07	1.80	1.48	0.000543
Acetone	0.969	0.918	0.792	0.00215
Water	0.917	1.000	0.958	0.000588
Hydrogen	0.0771	0.0710	0.0709	0.00121

For all three states, density varies with temperature, although the change is not large for liquids and solids. As can be seen from the data of Table 1, the density of liquids generally decreases with increasing temperature (again, water around 4°C is an exception). Because mass is independent of temperature, the volume increases with increasing temperature; that is, substances expand when heated. The density of a substance also depends on pressure: density increases with increasing pressure, because the volume of a substance decreases under those conditions.

In some properties, liquids more closely resemble gases than solids. One of the most obvious differences between solids and liquids is that liquids, like gases, flow. Because both gases and liquids tend to flow, whereas solids do not, the gaseous state and the liquid state are referred to as **fluid states.** Thus the solid state is a condensed nonfluid state, the liquid state is the condensed fluid state, and the gaseous state is the noncondensed fluid state. A measure of the tendency to flow is the **viscosity.** There is a wide range of viscosities among liquids. Water flows readily, glycerine is much slower, and a tar's pace is even more leisurely. Viscosity is one of a group of properties of fluids called **transport properties,** which are related to the flow of matter or energy. Another transport property is **diffusion,** the process by which two substances mix as a result of the motion of their constituent molecules. Transport properties in liquids are determined by intermolecular interactions that depend strongly on the specific molecules involved, and therefore, vary widely from one liquid to another. Transport properties in gases are much less dependent on the identity of the gases.

Liquids have a property that has no counterpart in either gases or solids, namely **surface tension.** Surface tension is the property that makes liquids behave as though they had an elastic skin. It is the property of liquids responsible for the formation of drops, soap bubbles, and meniscuses, and the property that allows water-skimmers to walk on the surface of a pond. Surface tension is defined as the energy required to increase the surface area of the liquid; that is, it is the ratio of energy to area. That liquids exhibit surface tension is consistent with a liquid being a condensed fluid. In order to remain condensed, a liquid must maintain a minimum volume, otherwise it would expand into a gas. Yet to remain a fluid, a liquid must have a surface that is not rigid. Thus surface tension works to maintain a small, nonrigid surface. Surface tension, like viscosity, is determined by the properties of the individual molecules. Molecules in the interior of the liquid experience attractive forces from all sides, but those on the surface are attracted by molecules below it in the bulk of the liquid. Because the attractive forces between molecules depend very much on the identities of the molecules, surface tension varies greatly from one liquid to another.

STRUCTURE OF LIQUIDS

Some of the properties of liquids are like those of solids, some are like those of gases. This suggests that the arrangement of molecules in liquids is in some ways similar to that of solids, and in other ways similar to that of gases. The properties of liquids can be accounted for by the behavior of the individual molecules, just as they can for gases and solids.

The relatively high density and the small values of the compressibility of liquids and solids show that the molecules in liquids and solids are close together. The distance between molecules in liquids and solids is less than the diameter of each particle. This is unlike the situation in gases, where the gas molecules are far apart, and the distance between them may be many times their diameters. The density of solids and liquids is about 1000 times that of gases at atmospheric pressure, so on average the distances between particles differ by roughly the cube root of this factor, namely 10. The molecules in a gas are about 10 times as far apart as those in solids or liquids.

The ability of liquids to flow indicates that their molecules are relatively free to move past one another. This is in distinction to solids, which are rigid, indicating that the positions of molecules in solids are fixed relative to each other. Yet, because liquids are more viscous than gases, the molecules in a liquid must not be so free to move about as those in a gas. Therefore, the forces of attraction between molecules in a liquid must be significantly larger than those in a gas. In a gas, these intermolecular forces are so small that the physical behavior varies only slightly from one gas to another, and there is a universal relationship between pressure, p, volume, V, and temperature, T, that can be applied to all gases, namely the **ideal gas law.** In liquids, the intermolecular forces are much greater, and furthermore, these forces depend on the identity of the molecules. Therefore, there is no universal relationship between p, V, and T for liquids as there is for gases.

Although many properties of liquids are determined by intermolecular forces that depend on the identity of the specific molecules in the liquid, a measure of universality is achieved in the **colligative properties** of solutions. These are properties of liquid solutions which depend on the number of moles of nonvolatile solute relative to total moles of substance in the solution, but which are independent of the identity of the solute. Specifically, the colligative properties are **vapor-pressure lowering, boiling-point elevation, freezing-point depression,** and **osmotic pressure.** Even among these properties, if the solute concentration becomes too large, the universal relationship between the concentration of the solution and values of these properties breaks down.

MATHEMATICAL RELATIONSHIPS AMONG MACROSCOPIC PROPERTIES

Many of the macroscopic properties of liquids and solutions are related to each other. One of the goals of science is to express these relationships mathematically. For each substance, whether pure or a mixture, there is a relationship among the properties of pressure, temperature, volume, and amount. When any three of these properties are specified, there is only one possible value for the fourth. This relationship is expressed in the **equation of state** for the substance. Equations of state are deduced from the experimentally observed properties of a substance. Perhaps the simplest and most well-known equation of state is that for ideal gases: $pV = nRT$, where p is the pressure, V is volume, n is the number of moles (amount), T is absolute temperature, and R is a proportionality constant. The value of R is independent of the identity of the gas and is called the universal gas constant. For most other substances, the equation of state is very complicated.

Much of the study of the properties of matter involves determining how one property changes when another is varied. For example, the variations in the volume of a gas when the pressure is changed at constant temperature has been thoroughly investigated. In a simple experiment, the pressure can be changed from p_1 to p_2, and a volume change from V_1 to V_2 can be observed. In this case, the dependence of V on p is expressed as the ratio of the change in V to the change in p, namely $(V_2 - V_1)/(p_2 - p_1)$. This ratio is usually expressed as $\Delta V/\Delta p$, where the Greek letter Δ represents the

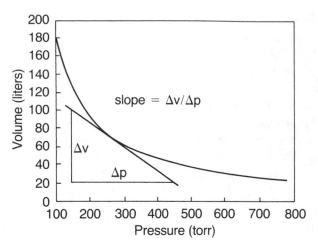

Figure 1. Volume as a function of pressure.

difference between the final state and the initial state; that is, $\Delta V = V_2 - V_1$ and $\Delta p = p_2 - p_1$. If the relationship between p and V were a straight line, then $\Delta V/\Delta p$ would be the slope of the line obtained by plotting V versus p. However, the relationship between p and V is not linear. When V is plotted versus p, as shown in Figure 1, a curve is obtained.

The slope of the curve changes from one value of p to another; that is, the slope is a function of p. The slope of the tangent to the curve at some p_1 can be estimated by $\Delta V/\Delta p$, provided Δp is taken around p_1 and Δp is small. The smaller Δp is, the better the estimate for the value of the slope at p_1. In the branch of mathematics called **calculus,** the symbol dV/dp is used instead of $\Delta V/\Delta p$ to represent the slope of a tangent to the curve. The symbol dV/dp is called the **derivative** of V with respect to p, and dp represents an infinitesimal change in pressure and dV the corresponding change in volume.

According to the ideal gas law, $V = nRT/p$, which indicates that V depends not only on p but also on the amount of gas (n) and on the temperature (T) as well. In other words, V is a function of n, T, and p. This is represented mathematically as $V(n,T,p) = nRT/p$. In such cases, where one variable is a function of more than one other variable, the derivative is expressed as a **partial derivative.** The partial derivative of V with respect to p is expressed with the symbol $\partial V/\partial p$. It is still interpreted as the slope of the V-p curve. In order to indicate explicitly that the other variables are constant, the derivative is often expressed as $(\partial V/\partial p)_{T,n}$.

Mathematicians have developed techniques for finding the derivative of a function. Several of these are quite simple and useful in expressing the properties of substances. Suppose we have the function $y(x) = x^n$, where n is a constant. Then, the formula for finding the derivative of y with respect to x is $\partial y/\partial x = nx^{(n-1)}$. Using this formula with $V(n,T,p) = nRT/p$ gives the following derivatives:

$$(\partial V/\partial n)_{T,p} = RT/p \qquad (\partial V/\partial T)_{n,p} = nR/p \qquad (\partial V/\partial p)_{n,T} = -nRT/p^2$$

These equations give the slopes of the various plots. The plot of V versus n is a straight line whose slope is independent of n, as the equation above on the left shows. The plot of V versus T also has a slope that does not contain the independent variable T and, therefore, is also a straight line. The equation on the right shows that the slope of the plot of V versus p, however, is inversely proportional to p^2 and is not constant.

The derivative of a dependent variable with respect to an independent variable yields the slope of a plot of the dependent variable versus the independent variable. There is a mathematical operation that is the reverse of differentiation, namely **integration.** Integration of a function yields the area under the plot of the dependent variable versus the independent variable. Further information about differentiation and integration can be found in any introductory text on calculus.

Many properties of substances can be expressed in terms of derivatives. For example, the change in the volume of a substance with a change in pressure at constant temperature can be expressed as $(\partial V/\partial p)_T$. The total volume change that occurs when the pressure on a substance is changed depends on, among other things, the amount of the substance. The absolute change in volume of a large amount of substance is greater than the volume change for a small amount of the same substance undergoing the same change in pressure. To make the value independent of the amount of substance—that is, to make it an **intensive property**—the volume change is divided by the total volume of the substance. The property describing the pressure dependence of the volume at constant temperature is the **isothermal compressibility,** β, which is defined as

$$\beta = -\left(\frac{1}{V}\right)\left(\frac{\partial V}{\partial p}\right)_T$$

The minus sign is used to make the value of β positive. For liquids and solids, the values of β are rather small, on the order of 10^{-5} bar^{-1} (atmospheric pressure is about 1 bar). For ideal gases, $\beta = -(1/V)(-nRT/p^2) = (nRT/V)(1/p^2) = 1/p$ and is much larger than for the condensed states. The incompressibility of liquids is dramatically demonstrated whenever a mercury-filled thermometer is heated far beyond the upper limit of its range—the thermometer shatters!

A property describing the temperature dependence of the volume at constant pressure is thermal expansion. Again, the absolute volume change depends on the amount of substance, and therefore, the thermal expansion is divided by the total volume of the substance. This division yields the **coefficient of thermal expansion,** α, which is defined as

$$\alpha = \left(\frac{1}{V}\right)\left(\frac{\partial V}{\partial T}\right)_p$$

The value of α for liquids is about 10^{-3} K^{-1} and varies over an order of magnitude from liquid to liquid. The value for solids is about an order of magnitude smaller than that for liquids. For ideal gases, $\alpha = (1/V)(nR/p) = 1/T$. The property of thermal expansion forms the basis for liquid-in-glass thermometers. As the temperature increases, the volume of the liquid increases and fills more of the thermometer. (Because the coefficient of thermal expansion of glass is much smaller than that of the liquids used in thermometers, the volume change of the glass is negligible compared with that of the liquid.) Graduations on the thermometer translate the volume of the liquid into a temperature.

THE GIBBS FUNCTION

The coefficient of thermal expansion relates the volume change of a substance to the change in its temperature. The coefficient of thermal expansion is only one of many properties related to temperature. The science that deals with the relationships between temperature and other properties of substances is **thermodynamics.** Much of the formulation of thermodynamics in its present shape is due to the work of the American scientist J. Willard Gibbs (1839–1903). His name is attached to a number of thermodynamic relations, including the Gibbs function, the Gibbs phase rule, the Gibbs-Helmholz equation, and the Gibbs-Duhem equation.

The properties of liquids (indeed, of all substances) are related to each other by the **Gibbs function** and its derivatives. The value of the Gibbs function is sometimes called the free energy. For a fixed amount of a pure substance, the Gibbs function relates the temperature and pressure of a substance to other macroscopic properties. In other words, the Gibbs function treats temperature and pressure as the independent variables. The Gibbs function contains complete thermodynamic information; that is, the relationships between all macroscopic properties of a substance are determined by this function. When certain macroscopic properties are measured (or specified), the Gibbs function allows the calculation of the values of other properties. However, the Gibbs function applies only to systems at **equilibrium,** that is, those that undergo no macroscopic change with time. It cannot be used to relate the properties of a system undergoing change.

Because the Gibbs function is a function of p and T, its derivatives with respect to p and T have meaning. The derivative of G with respect to p at constant T is the volume of the substance.

$$\left(\frac{\partial G}{\partial p}\right)_T = V$$

The derivative of G with respect to T at constant p is another thermodynamic function called the **entropy,** S.

$$\left(\frac{\partial G}{\partial T}\right)_p = -S$$

Because the volume of a substance depends on temperature and pressure, the derivatives of volume with respect to these variables also have meaning. In fact, we've encountered both of them before.

$$\left(\frac{\partial V}{\partial T}\right)_p = \alpha V$$

$$\left(\frac{\partial V}{\partial p}\right)_T = -\beta V$$

In these equations, α is the coefficient of thermal expansion and β is the isothermal compressibility.

Because volume is the derivative of G with respect to p, the derivative of V with respect to T can be expressed as

$$\left(\frac{\partial V}{\partial T}\right) = \frac{\partial}{\partial T}\left(\frac{\partial G}{\partial p}\right)$$

The right side of the equation is called a second derivative of G and is usually written as $(\partial^2 G/\partial T \partial p)$. Therefore,

$$\left(\frac{\partial^2 G}{\partial T \partial p}\right) = \alpha V$$

Another second derivative of G is obtained from $\partial V/\partial p$.

$$\left(\frac{\partial V}{\partial p}\right) = \frac{\partial}{\partial p}\left(\frac{\partial G}{\partial p}\right) = \left(\frac{\partial^2 G}{\partial p^2}\right) = -\beta V$$

The remaining second derivative of G is

$$\left(\frac{\partial^2 G}{\partial T^2}\right) = \frac{\partial}{\partial T}\left(\frac{\partial G}{\partial T}\right) = -\left(\frac{\partial S}{\partial T}\right) = -\frac{C_p}{T}$$

where C_p is the constant-pressure **heat capacity,** which is the ratio of the heat energy absorbed by a system to its resulting temperature change.

Variations of the Gibbs function with temperature and pressure correspond to measurable properties of a system. Another derivative of particular significance is the change in the ratio G/T with 1/T.

$$\left(\frac{\partial(G/T)}{\partial(1/T)}\right)_p = -TS + G$$

The function $-TS + G$ is the enthalpy, which is symbolized by H. Enthalpy is central to any discussion of thermochemistry. The heat absorbed in a process that occurs at constant pressure is the change in the enthalpy of the system, ΔH.

Dependence of the Gibbs Function on the Amount of Substance

In addition to temperature and pressure, the amount of substance also affects the value of the Gibbs function (free energy) for an equilibrium system. For a system containing a mixture of substances, knowing the free energy of the system as a function, G, of temperature, pressure, and the amount of each component is to know all that thermodynamics can reveal about the system. Therefore, it is of significance to know the change in G when the amount of component "i" in a mixture is increased. This is the **chemical potential** of i, represented by $\mu_i = (\partial G/\partial n_i)$, where n_i is number of moles of component i. The nomenclature is well advised, because the chemical potential in a thermodynamic system behaves in the same manner as the gravitational potential in a mechanical system seeking equilibrium. We are all familiar with the fact that weights will tend to reduce their gravitational potential energy (i.e., go downhill) until reaching equilibrium at a stable minimum (floor of a valley). So it is with the amount of a component of a nonequilibrium mixture. If the value of the chemical potential of the component in one region of the mixture exceeds the value in another region, the component will flow from the first region to the second. For example, at $-5°C$ and 1 atm pressure, the chemical potential of liquid water exceeds that of ice, so under these conditions, all the liquid is converted to solid. If the temperature were 5°C, however, the order of the magnitudes of the chemical potentials would be reversed, and liquid water would be the stable form. At 0°C and 1 atm both chemical potentials have the same value; ice and water can coexist at equilibrium, and there is no driving force to convert one to the other.

In a mixture, the chemical potential of a species must be the same in all parts of the mixture at equilibrium. This requirement is a convenient starting point for deriving expressions for the colligative properties (vapor-pressure lowering, freezing-point depression, boiling-point elevation, osmotic pressure). These are derived by equating the chemical potentials of each component in all parts of the system. For example, to derive the expression for the freezing point of a solution, one begins by setting the chemical potential of the solvent in the solution equal to the chemical potential of the pure solid solvent.

The chemical potentials are also used to analyze an equilibrium involving a chemical reaction. Here we look at the Gibbs function of the products and compare it with that of the reactants. Because the products are different chemical species from the reactants, each substance has only one chemical potential in the reaction mixture (assuming the mixture is homogeneous). The Gibbs function of a mixture of x components is the sum of the products of the amount of each component, n_i, multiplied by its chemical potential, μ_i; that is,

$$G = n_1\mu_1 + n_2\mu_2 + n_3\mu_3 + \ldots + n_x\mu_x$$

The direction of a chemical reaction (with respect to the way it is written) is determined by the difference between $G_{products}$ and $G_{reactants}$. If the Gibbs function of the reactants exceeds that of the products, the reaction proceeds from reactants to products (left to right, as conventionally written), and vice versa. While the reaction proceeds, the chemical potentials of the individual components of the mixture change, because the chemical potential of a component in a mixture depends on the concentration of the component in the mixture. Generally, the chemical potential of a component decreases as its mole fraction decreases. Thus, if the reaction is going from left to right, the contribution from $G_{reactants}$ to the total G becomes smaller and that from $G_{products}$ becomes greater, until a point is reached where $G_{products} = G_{reactants}$. At that point no driving force exists for further chemical reaction. Equilibrium has been reached. At equilibrium a definite relation exists between the concentrations of the components, and this relation is expressed in the equilibrium constant. For the reaction $aA + bB \leftrightarrows dD + eE$, where a, b, d, and e are the stoichiometric coefficients in the chemical reaction (mole numbers), the equilibrium constant is

$$K_c = \frac{[D]^d[E]^e}{[A]^a[B]^b}$$

provided the mixture is ideal. In this expression the square brackets represent molar concentrations, and K_c is called the molar equilibrium constant.

Note that equilibrium thermodynamics does not furnish values for the chemical potentials; rather it furnishes relations between them. Experiments must be performed to discover the magnitude of the chemical potentials (and of derived quantities such as equilibrium constants). The direct calculation of the magnitude of thermodynamic properties from models of molecular interactions is the domain of **statistical mechanics.** Statistical mechanics begins with a theoretical model of the interactions between molecules and calculates bulk properties based on the molecular properties in the model. Even calculations of a relatively simple property, such as the density of liquid water as a function of temperature and pressure, are extremely demanding. The direct calculation of the chemical potential of water is even less successful. It is easier to do an experiment.

VAPOR PRESSURE AS A FUNCTION OF TEMPERATURE

When a volatile liquid substance is placed in a sealed container, some of the liquid evaporates, and then the substance exists in two **phases** (different homogeneous regions), liquid and vapor. Net evaporation continues as long as the chemical potential of the substance in the liquid phase, μ_L, is greater than that of the vapor phase, μ_V. As net evaporation proceeds, the partial pressure of the vapor increases, but when $\mu_L = \mu_V$, the partial pressure of the vapor reaches a fixed value, called the **vapor pressure** of the liquid.

For a system containing c components in p phases, we might expect to be able to vary the mole fractions of all but one (because the mole fractions must sum to 1) of the components in each phase. That is, for each phase we could vary c − 1 mole fractions, for a total of p(c − 1) variables. We could also vary the temperature and pressure. Thus, we might expect to have p(c − 1) + 2 independent intensive variables in the system. However, if the system is at equilibrium, the chemical potentials of each component are the same in all phases. Thus, the mole fractions of each component are *dependent* in all but one phase at equilibrium. This reduces the number of independent variables by c(p − 1), and the number of independent intensive variables (degrees of freedom) is f = p(c − 1) + 2 − c(p − 1) = c − p + 2. This relationship, f = c − p + 2, is called the **Gibbs phase rule.**

For a one-component liquid-vapor equilibrium, there are two phases and one component, so f = 1. Thus, there is only one independent variable among temperature and pressure (the mole fraction being fixed at 1 in all phases of a one-component system). Therefore, the vapor pressure, p, is dependent on (not independent of) T. For water, p = 18 torr at 20°C, and p = 760 torr at 100°C.

The relationship between temperature, T, and vapor pressure, p, can be found (in derivative form) from the requirement that at equilibrium the chemical potential of the substance is the same in both phases: $\mu_{\text{liquid}} = \mu_{\text{vapor}}$. Changes in the chemical potentials are given by $d\mu_{\text{liquid}} = V_L dp_L - S_L dT_L$ and $d\mu_{\text{vapor}} = V_V dp_V - S_V dT_V$. At equilibrium, $p_L = p_V$, $T_L = T_V$, and $d\mu_{\text{liquid}} = d\mu_{\text{vapor}}$; therefore, $(V_L - V_V)dp = (S_L - S_V)dT$. Thus, $dp/dT = \Delta S/\Delta V$. Because $\Delta S = \Delta H/T$, $dp/dT = \Delta H/T\Delta V$. This is the **Clapeyron equation.** For the special case of the liquid-vapor equilibrium with the assumptions $\Delta H_{\text{vap}} = $ constant and $\Delta V = V_{\text{gas}} = nRT/p$, integration of the Clapeyron equation yields

$$\ln\left(\frac{p_1}{p_2}\right) = \frac{\Delta H_{\text{vap}}}{R}\left(\frac{1}{T_2} - \frac{1}{T_1}\right)$$

which is a form of the **Clausius-Clapeyron equation.** The implication of this equation is that the logarithm of p is inversely proportional to T, and this is approximately true for most liquids.

CONCENTRATION SCALES

Various concentration scales are commonly used (and are featured in the demonstrations). Perhaps the most commonly used scale for describing the concentration of solutes in solution is **molarity:** M = moles solute/liters of solution. It is useful because moles of solute can be easily determined by measuring volume of solution, which is handy in stoichiometric volume relationships (e.g., titrations). Molarity has the disadvantage that its value is temperature and pressure dependent. Three scales which are temperature and pressure independent are weight fraction, wt. %, mole fraction, x, and molality, m.

The **weight fraction** is the mass of a component of a mixture divided by the total mass of the mixture. Compositions of tabulated system properties are often given in wt. % (100 times the weight fraction). The **mole fraction** (x = moles of the component divided by total moles) is appropriate when all components of a mixture are present in substantial amounts or with multicomponent solutions. The **molality** (m = mol solute/kg solvent) is used with two-component solutions when one component (the solute) is present in minor amount. The major component is the solvent.

LIQUID VAPOR EQUILIBRIUM WITH MULTIPLE COMPONENTS

Liquid-vapor equilibria between solutions and gas mixtures are very common. A set of rules for multiple components is easily produced, but its properties are most obvious for a two-component system—a binary system. For a binary system in liquid-vapor equilibrium, there are two components and two phases, and, therefore, there are $f = 2 - 2 + 2 = 2$ independent intensive variables from the collection T, p, x_1, y_1 (x_1 = mole fraction component 1 in the liquid, y_1 = mole fraction component 1 in the vapor). If p is fixed, then T depends on both x_1 and y_1. For an ideal solution, the value of T as a function of x_1 and y_1 can be calculated by applying **Dalton's law** of partial pressures to the vapor, **Raoult's law** to the vapor pressure of solution components, and the Clausius-Clapeyron equation to the vapor pressures of the pure components. Dalton's law states that the total pressure, p, of a mixture of gases is the sum of the individual pressures (partial pressures) of each gas. In the case of a mixture of two gases, $p = p_1 + p_2$. In an equilibrium between an ideal solution (methanol-water, for example) and an ideal gas mixture (most gas mixtures obey the ideal gas mixture equations quite closely), p_1 and p_2 are the partial pressures of the two components. According to Raoult's law, $p_1 = x_1 p_1^0$ and $p_2 = x_2 p_2^0 = (1 - x_1)p_2^0$, where p_1^0 and p_2^0 are the vapor pressures of the pure liquids. Therefore, the total vapor pressure of the solution is $p = x_1 p_1^0 + (1 - x_1)p_2^0$. Dalton's law gives the mole fractions in the vapor: $y_1 = p_1/(p_1 + p_2) = x_1 p_1^0/[x_1 p_1^0 + (1 - x_1)p_2^0]$. The vapor pressures of the pure liquids depend on temperature, and the temperature dependence can be found from the Clausius-Clapeyron equation. For example, calculation of the vapor pressure of pure methanol at 80.0°C requires the heat of vaporization of methanol, 38.9 kJ·mol^{-1}, and the vapor pressure at another temperature. At the normal boiling point of a liquid, its vapor pressure is 1 atm. The normal boiling point of methanol is 64.7°C. Then,

$$\ln\left(\frac{p_1^0}{1 \text{ atm}}\right) = \frac{38,900 \text{ J·mol}^{-1}}{8.314 \text{ J·mol}^{-1}\text{K}^{-1}}\left(\frac{1}{(273.15 + 64.7)\text{K}} - \frac{1}{(273.15 + 80.0)\text{K}}\right)$$

Solving this gives the vapor pressure of methanol at 80.0°C as 1.82 atm. A similar calculation using 44.3 kJ·mol^{-1}K^{-1} as the heat of vaporization of water gives the vapor pressure of water at 80.0°C as 0.445 atm. If the total pressure of methanol-water solution at 80.0°C is 1 atm, then its composition is given by 1 atm = x_1(1.82 atm) + $(1 - x_1)$(0.445 atm), the mole fraction of methanol in the liquid, x_1, is 0.403. The mole fraction of methanol in the vapor, y_1, is $x_1 p_1^0/p = (0.403)(1.82 \text{ atm})/(1 \text{ atm}) = 0.733$. These calculations can be performed for a range for temperatures giving the temperature dependence of the liquid and vapor compositions at constant pressure. When T versus x_1 and T versus y_1 are plotted on the same graph, the result is a temperature-composition diagram. Figure 2 is such a diagram for the methanol-water system.

Temperature-composition diagrams such as in Figure 2 are useful in understanding what happpens in **fractional distillation.** Figure 2 indicates that a 50-50 methanol-water liquid mixture has a vapor pressure of 1 atm at about 76°C. Therefore, it begins to boil at this temperature. At this temperature, the mole fraction of methanol in the vapor of this boiling solution is about 0.8. If this vapor were condensed to a liquid, the diagram indicates that this liquid containing 0.8 mole fraction of methanol would boil at about 68°C. The vapor of this liquid would contain methanol at a mole fraction of about 0.95. By repeating the boiling-condensing process through several stages, the mole fraction of methanol in the condensed liquid can be increased. In the ideal solu-

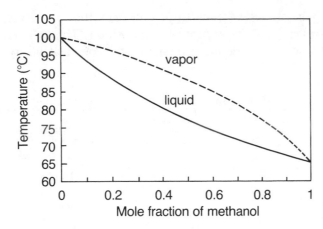

Figure 2. Temperature-composition diagram of a methanol-water mixture.

tion, the most volatile component is found at the top of a fractionating column, and the boiling liquid gradually becomes richer in the less volatile component.

The temperature-composition diagram for the ethanol-cyclohexane system is shown in Figure 3. This system does not follow the ideal solution equations. Rather the T-x and T-y curves have minima at the same point, indicating formation of a minimum boiling **azeotrope,** a mixture whose liquid and vapor have the same composition. Therefore, distillation of a mixture of the azeotrope composition is futile as a means of separation. Because it forms a minimum-boiling azeotrope, the ethanol-cyclohexane system produces its azeotrope at the top (the coolest part) of a column used for fractional distillation. The hydrogen chloride–water system shows a maximum-boiling azeotrope; data for this system are shown in Table 2. The composition of the boiling liquid approaches the azeotrope composition as material is boiled away.

The ideal solution expressions above indicate that the partial pressure, p_i, in the vapor phase should equal $x_i p_i^0$ (Raoult's law). Only when x_i for one of the components (the solvent) approaches unity does the equation hold for solutions of multiple volatile

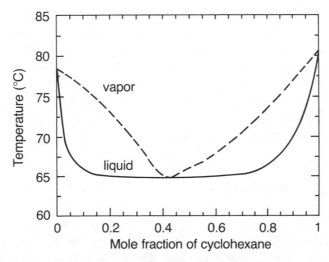

Figure 3. Temperature-composition diagram of an ethanol-cyclohexane mixture.

components in general. As x_i tends to 0, one finds that $p_i = K_i x_i$, where K_i, the **Henry's law** constant, is independent of composition. This relation for solutes and Raoult's law for the solvent define a model system of great utility, the ideal dilute solution.

Table 2. Temperature-Composition Data for HCl-Water System (p = 751.3 torr) [2]

t °C	wt. % HCl_L	wt. % HCl_G	x_{HCl}	y_{HCl}
107.3	21.79	31.00	0.121	0.182
107.5	21.36	27.54	0.1183	0.1581
107.6	20.79	22.97	0.1148	0.1284
108.27	20.242	—	—	—
107.8	20.09	18.49	0.1105	0.1008
107.2	18.09	7.55	0.0983	0.0388
105.5	15.97	2.80	0.0917	0.0140
103.1	12.45	0.75	0.0657	0.0037
101.5	8.67	0.21	0.0447	0.0010

SOLUBILITY

Another important equilibrium involves the amount of solute dissolved in a solvent in equilibrium with the pure solute. This **solubility** is often a strong function of temperature. For the (fairly uncommon) case of an **ideal solution,** one in which the solvent-solvent, solvent-solute, and solute-solute intermolecular forces are identical, the **mole fraction** of the solute in solution (x_2) as a function of temperature (T) is given by

$$\ln(x_2) = -\frac{\Delta H_{fus}}{R}\left(\frac{1}{T} - \frac{1}{T_0}\right)$$

where ΔH_{fus} is the **enthalpy of fusion** of the solute,
      T_0 is its melting point,
and   R is the gas constant: R = 8.314 $J \cdot mol^{-1}K^{-1}$.

This "ideal law of solubility" is analogous to the corresponding expression for the freezing-point depression of the solvent caused by a nonvolatile solute (a colligative property—see the following section). Because ΔH_{fus} is always positive, the ideal law of solubility predicts that x_2 (i.e., the solubility) increases with temperature *for all materials.* Counterexamples are readily found (see Demonstration 9.13), indicating the limited utility of the law. Interactions between solvent and solute molecules specific to the particular solvent and solute play a significant role in determining solubility, making universal "ideal" relationships quite rare.

COLLIGATIVE PROPERTIES

The colligative properties of solutions—vapor-pressure lowering, boiling-point elevation, freezing-point depression, and osmotic pressure—involve equilibria between phases. For example, freezing-point depression involves an equilibrium between the solvent of the solution and a pure solid (freezing) solvent. As mentioned earlier, in an equilibrium between phases, the chemical potential of a component is the same in each phase. For the equilibrium between solvent in solution and solid solvent we have $\mu_L(T,p) = \mu_S^0(T,p)$. Here, $\mu_L(T,p)$ is the chemical potential of the liquid solvent and $\mu_S^0(T,p)$ is the chemical potential of the pure solid solvent. In this notation, the superscript 0 indicates pure substance, and T and p indicate that μ depends on temperature and pressure. A model of the properties of a phase is often constructed by specifying the dependence of the chemical potentials on temperature, pressure, and composition. For example, for the solvent in dilute liquid solutions, it is approximately true that $\mu_L(T,p) = \mu_L^0(T,p) + RT\ln(x)$, where $\mu_L^0(T,p)$ is the chemical potential of the pure liquid solvent and x is its mole fraction in the solution. This means that $\mu_L^0(T,p) + RT\ln(x) = \mu_S^0(x)$. If this equation is solved for $\ln(x)$, differentiated with respect to T, rearranged, and integrated, we obtain $\ln(x) = (\Delta H_{fus}/RT_0^2)(T - T_0)$. The solvent mole fraction $(x_1 \equiv x)$ is related to the number of moles of solute (n_2) for a *dilute* solution $(x_1 \gg x_2)$ by

$$\ln(x_1) = \ln(1 - x_2) \approx -x_2 = -\frac{n_2}{n_1 + n_2} \approx -\frac{n_2}{n_1}$$

The ratio n_2/n_1 can be expressed in terms of the molality, m, of the solution: $n_2/n_1 = mM/1000$, where M is the molar mass of the solvent. Substituting for $\ln(x)$ and solving for $T_0 - T$ yields

$$T_0 - T = \left(\frac{MRT_0^2}{\Delta H_{fus}\,1000}\right)m = K_f m$$

The difference $T_0 - T$, which is represented by θ_f or ΔT_f, is the freezing-point depression, and the equation gives an explicit prescription for the freezing-point depression constant, K_f (which depends on the properties of the solvent but not of the solute).

If the solute in a very dilute solution dissociates or associates, the expression for the freezing-point lowering becomes $\theta_f = i\,K_f m$, where i is the number of species in solution per solute molecule dissolved. Thus, for a very dilute solution of sodium chloride in water, $i = 2$. Unfortunately, for only slightly concentrated ($m = 10^{-4}$ or larger) electrolyte solutions or moderately concentrated ($m = 10^{-2}$ or larger) nonelectrolyte solutions, departures from the assumed form of the chemical potential of the solvent $[\mu_L^0 + RT\ln(x)]$ are significant. The factor i can still be defined as the number which makes the equation $\theta_f = i\,K_f m$ correct. The freezing-point depression experiment determines i as a function of m. From this relationship, one can work backward to the chemical potential, and the concentration dependence of the chemical potential of the solvent can be determined. In principle, this yields, via the **Gibbs-Duhem equation,** the chemical potential of the solute and generates complete thermodynamic information. Thus, i is a measure of the system properties. Thermodynamics yields relations between macroscopic quantities; it does not yield numbers for isolated properties—this is the domain of statistical mechanics.

To obtain the boiling-point elevations, ΔH_{fus} is replaced by $-\Delta H_{vap}$, the **enthalpy of vaporization.** We find $\Delta T_b = T - T_b = K_b m$ and its extensions involving i.

Analysis of the osmotic effect proceeds differently. We have an equilibrium between pure solvent separated from a solution by a semipermeable membrane, which allows only solvent to pass through it. The pure solvent is at pressure p and the solution is at pressure $p + \pi$, where π is its osmotic pressure. Therefore,

$$\mu_L^0(T,p) = \mu_L^0(T,p + \pi) + RT\ln(x_1)$$

Substituting $-n_2/n_1$ for $\ln(x_1)$ yields $-n_2RT/n_1 = \mu_L^0(T,p) - \mu_L^0(T,p + \pi)$. As long as π is small, the difference $\mu_L^0(T,p) - \mu_L^0(T,p + \pi)$ is $-(\partial\mu_L^0/\partial p)\pi$ (this is the meaning of the derivative). The value of $\partial\mu_L^0/\partial p$ is $\overline{V}^0$, the molar volume of solvent. Therefore, $-n_2RT/n_1 = -\pi\overline{V}^0$. This rearranges to $\pi = cRT$, where $c = n_2/n_1\overline{V}^0 = n_2/V$ is the molarity of the solute. Again, the equation rests on the assumed form for the chemical potential of solvent in solution, namely $\mu_L^0 + RT\ln(x)$, which is true for ideal solutions.

For real solutions, the mole fraction x is replaced by the **activity,** a, of the solvent. Activity is defined by the equation $\mu_L(T,p) = \mu_L^0(T,p) + RT\ln(a)$. For the osmotic pressure, this definition yields (for a dilute solution) $\overline{V}^0\pi = -RT\ln(a)$. Similar modifications employing activity occur in the boiling-point elevation and freezing-point depression developments. For the boiling-point elevation, for example, the result for the equilibrium of chemical potentials of solution and pure vapor, after differentiation with respect to T, rearrangement, and integration, becomes:

$$\ln(a) = \frac{\Delta\overline{H}_{vap}^0}{R}\left(\frac{1}{T} - \frac{1}{T_b}\right)$$

The value of the activity is obtained by measuring the properties of the solution and using these to calculate its value. The utility of the activity is confirmed by the observation that the same value is obtained from different types of measurements (e.g., osmotic pressure and boiling-point elevation).

ADDITIONAL THERMODYNAMIC RELATIONSHIPS

Other properties of liquids, such as surface tension and **dielectric constant** can be related to pressure and temperature by thermodynamics. The **first law of thermodynamics** states that for every system there is a function of state, called the **internal energy** function and represented by U, whose value is determined by the values of the independent variables of the system. Furthermore, for thermally isolated processes (those in which the temperature of the surroundings does not change), this function has the property that $\Delta U = U_2 - U_1 = -w$, where w is the work done by the system. Work is defined in Newtonian mechanics as the product of the force on an object and the distance the object is moved by the force. Therefore, for a thermally isolated system, the value of ΔU can be determined by applying Newtonian mechanics. However, if a process involves a change in the temperature of the surroundings, that is, if it is not thermally isolated, then ΔU involves another factor, namely heat. The first law of thermodynamics is used to *define* heat. For a process that is not thermally isolated, $\Delta U = q - w$, where q is the heat absorbed by the system. There may be more than one path to get from state 1 to state 2. Because U is a function of state, ΔU will be independent of the path. However, both q and w will depend on how the process takes place. In a *cyclic process*, $U_2 = U_1$, so $\Delta U = 0$, and $q = w$. This can be interpreted as a statement of conservation of energy: for a process in which the initial and final states of the system are the same, the work (energy) done by the system is equal to the heat (energy) put into the system.

The first law is very general in that the nature of the work is not specified. Work can be in the form of pressure-volume work, where the volume of the system increases against a constant opposing pressure. In this case, $dw = p\,dV$, and $dU = dq - p\,dV$. To accommodate other variables, other forms of dw can be introduced. For Demonstrations 9.23 and 9.24, we use $dw = -\gamma\,dA$, where γ is the interfacial tension and A the area of the interface. In 9.25, $dw = -M\,g\,dh$, where M is the system mass, g the acceleration due to gravity, and h the location of the mass (center of mass) in the gravitational field. In 9.32, dw = −E dP, where E is the electric field strength and P the polarization. In all of these cases, because U depends on the independent variables, A, h, and P, dU depends on dA, dh, and dP.

The preceding narrative is mostly a short introduction to thermodynamics, focussing on the relations useful for interpreting the demonstrations in this chapter. These include expressions between pressure, temperature, volume, and amount of substance for pure materials, and relations deriving from the condition of equilibrium between phases or between reactants and products in a chemical reaction.

WATER

The most common and most unusual liquid is water. The oceans contain about 1.4×10^{21} kg of water, and there is a similar amount trapped in rocks as water of hydration. In spite of its being the most common liquid on the surface of the earth, virtually every thermodynamic and transport property of liquid water is unlike those of most other liquids.

Perhaps most noteworthy among water's anomalous properties is its existence as a liquid at room temperature. This is quite remarkable for a substance of so low a molar mass. The freezing point and the boiling point of water are much higher than one might expect from a comparison of these properties to those of similar substances. As the data in Table 3 and Figure 4 show, the freezing points and boiling points of isoelectronic sequences of hydride molecules decrease with decreasing molecular mass, except for water, where these properties show a jump.

Table 3. Melting and Boiling Points of Group IV and Group VI Hydrides [3, 4]

Formula	Molar mass (g/mol)	Freezing point (°C)	Boiling point (°C)
H_2Te	130	−49	−2
H_2Se	81	−60	−40
H_2S	34	−85	−61
H_2O	18	0	100
SnH_4	123	−150	−52
GeH_4	77	−165	−88
SiH_4	32	−185	−112
CH_4	16	−182	−164

From the sequence H_2Te, H_2Se, H_2S one might expect values of −100°C and −70°C for the freezing point and the boiling point of water; the actual values 0°C and 100°C are much higher.

Most other properties of water are anomalous as well. As it does for most liquids, the density of water increases when the temperature is lowered, but only down to 4°C. Below this temperature, the liquid density decreases. Even more unusual, the density of solid water (ice) is lower than that of the liquid. Water is one of only a few liquids in which the solid form floats, and this property has significant consequences for bodies of water such as lakes. The temperature at the bottom of a lake is never lower than 4°C,

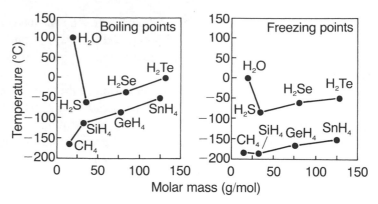

Figure 4. Boiling and freezing points of group IV and group VI hydrides.

unless the lake freezes solid from surface to bottom. This is an unlikely occurrence, because as the lake freezes, ice forms on the surface, creating an insulating layer that effectively inhibits heat transfer from the underlying water.

The changing density of water and the volume increase on freezing can be incorporated into a simple but informative experiment easily done by elementary school children [5]. A glass soda bottle is filled to the brim with water and sealed with a single-holed stopper fitted with a 20-cm glass tube. The tube is then half filled with water. When the bottle is chilled, the level of the liquid in the tube goes down as the water is cooled to 4°C and then begins to rise as the water is cooled further to 0°C, and a larger increase in volume occurs when ice forms in the bottle.

The relationships among the phases of water (solid, liquid, and gas) at various temperatures and pressures can be displayed by a phase diagram, which is a plot of pressure versus temperature. Curves on a phase diagram indicate the temperatures and pressures where two phases coexist at equilibrium. The phase diagram of water is shown in Figure 5. The curve on the lower right of the diagram indicates the pressures and temperatures at which liquid and gas are in equilibrium. It is a plot of the boiling point (temperature) of water at each pressure. At each pressure, this curve marks the temperature below which water is a liquid and above which it is a gas. Therefore, at p-T points on the high-temperature side of (i.e., below) the curve, water is a gas; at points above the curve, it is a liquid; and on the curve, liquid and gas are in equilibrium. The curve at the upper left is a plot of the freezing point at each pressure. On the low-temperature side of (i.e., left of) this curve, water is a solid, on the right it is a liquid, and on the curve, liquid and solid are in equilibrium. The curve at the lower left is a plot of the "sublimation point" versus pressure, and it divides p-T points representing solid above from gas below. Where these three curves meet is the triple point, the temperature and pressure at which all three phases exist in equilibrium. The triple point occurs at 611 Pa and 273.16 K. This temperature is exact by definition; it is used to define the size of the degree on the kelvin scale. At the end of the liquid-gas curve is the critical temperature, 374°C, and critical pressure, 2.23×10^7 Pa. It is not possible to have a liquid-gas equilibrium above this temperature, no matter what the pressure. This is the case because, above this temperature, liquid and gas are no longer distinguishable states. Moving to the right along the liquid-gas curve, the temperature increases and the density of the liquid decreases. At the same time, the pressure increases and the density of the gas increases. Eventually, a point (the critical temperature and pressure) is reached where the densities of the two phases become the same. At and beyond this point the phases are indistinguishable—there is but one phase, resembling a gas in that it completely fills its container. Of particular interest is the slope (dp/dT) of the plot for the liquid-ice equilibrium. The negative slope indicates that the freezing point rises as the pressure is decreased. According to the Clapeyron equation, $dp/dT = \Delta H_{fus}/T\Delta V_{fus}$. Because the process of melting (fusion) is endothermic—that is, $\Delta H_{fus} > 0$—the value of ΔV_{fus} must be negative when ice melts; that is, water expands on freezing. The high-pressure portion of the phase diagram reveals the existence of multiple phases of ice. This detail is not shown in Figure 5, but a portion of the diagram extending to high pressures (Figure 6) shows various phases of ice.

The isothermal compressibility of water decreases as the temperature is raised to 46°C, but above this it increases again. Other liquids exhibit monotone increases of this property with temperature. However, the anomalous behavior of water is suppressed when the water is compressed.

Generally, the viscosity of liquids increases with pressure. However, water is again an exception. Below 30°C the viscosity of water decreases as pressure is applied, but, under sufficient pressure, it increases as it does for other liquids.

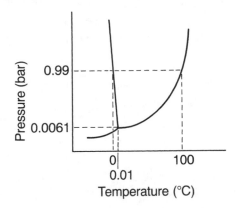

Figure 5. Phase diagram of water.

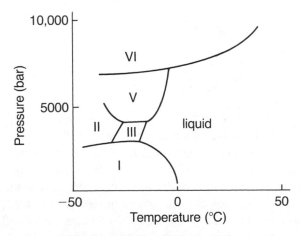

Figure 6. Phase diagram of ice.

The anomalous properties of water can be attributed to its molecular structure and the intermolecular forces this structure entails [6, 7]. Water molecules contain two hydrogen atoms bonded to an oxygen atom. The hydrogen-oxygen-hydrogen bond angle is about 105 degrees. The electrons in the hydrogen-oxygen bonds are attracted strongly by the oxygen atom, and the bonds are quite polar. Because the only electrons in the hydrogen atoms are the valence electrons, the shifting of these electrons toward the oxygen atom leaves a high concentration of positive charge in the vicinity of the hydrogen atoms. This concentration leads to strong attractive forces between a hydrogen atom of one water molecule and the electron-rich oxygen atom of another. These intermolecular forces are called **hydrogen bonds.** In addition to being quite strong as intermolecular forces go, hydrogen bonds between the water molecules are also directional. X-ray diffraction shows that the oxygen atoms in ice are surrounded by four hydrogen atoms (two covalently bonded and two hydrogen bonded). These four hydrogen atoms are situated tetrahedrally around the oxygen atom.

In ice the tetrahedral arrangement of hydrogen atoms around oxygen atoms is extensive. The structure of ice is shown in Figure 7. It consists of a network of water molecules, each held in place by hydrogen bonds to four other water molecules. This network of hydrogen bonds holds each molecule at a greater distance from its neighbors than it would be without the network. When ice melts to form liquid water, the network breaks down, and the molecules move more closely together. Hence, the density of ice, $0.913 \text{ g} \cdot \text{cm}^{-3}$, is less than the density of liquid water, $1.00 \text{ g} \cdot \text{cm}^{-3}$.

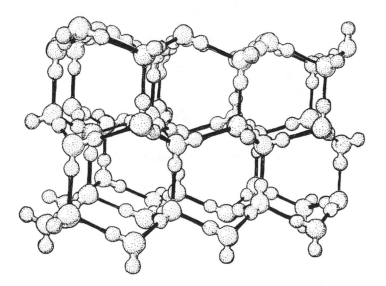

Figure 7. Structure of ice.

At 0°C ice melts to form liquid water, and the heat of fusion of ice is 6.02 kJ per mole of ice melted. This is substantially less than energy required to break 1 mole of hydrogen bonds, namely 20 kJ. Therefore, only a portion of the hydrogen-bonded network of ice is destroyed in going from the solid to liquid, and liquid water consists of clusters of hydrogen-bonded molecules. As the temperature is raised, the added energy breaks up these clusters. The high heat capacity of water reflects the strength of the hydrogen bonds and the amount of energy required to break them. Water has one of the highest heat capacities ($4.2 \text{ J} \cdot \text{g}^{-1}\,{}^{\circ}\text{C}^{-1}$) among liquids.

At normal atmospheric pressure, water boils at 100°C. This rather high boiling point for a substance with such a low molar mass reflects the strength of the hydrogen bonds. When water boils into the gas phase, virtually all of the hydrogen bonds are broken. Water vapor at lower temperature, however, can contain a significant number of dimers and trimers.

The unique temperature dependence of the density of liquid water (increasing with decreasing temperature to 4°C, then decreasing again as the temperature continues to decrease) is also a result of intermolecular hydrogen bonding. As water cools, its molecules lose kinetic energy, and their vibrations are less vigorous. Thus, in water as in most liquids, the molecules require less space as their vibrations diminish. However, in water, as the vibrations diminish, the molecules are more likely to assume the proper position for hydrogen bonding with a neighbor. When such positions are assumed and hydrogen bonds form, the water molecules become locked into a more open structure, and the volume increases. At 4°C, the number of hydrogen bonds in water increases sufficiently to offset the contraction due to less molecular vibration, and water expands with further decreases in temperature, leading to a decrease in its density.

The extensive hydrogen bonding in water also accounts for the anomalous pressure dependence of its viscosity. Extensive hydrogen bonding produces large clusters of water molecules which impede the flow of molecules past each other, leading to a rather high viscosity. As pressure is applied to water and it is compressed, the open structure required for hydrogen-bond formation is broken, so the size of the clusters decreases, decreasing the viscosity.

The decrease in water's isothermal compressibility with temperature is also attributable to changes in hydrogen bonding. As its temperature is increased, molecular motion increases, and the extent of hydrogen bonding in water decreases. The decrease in hydrogen bonding leads to less resistance to compression as temperature is increased.

Boiling water is part of our everyday experience. This is a nonequilibrium process (the bottom of the pot is hotter than the top), and how boiling takes place depends strongly on conditions. Three types of boiling in liquids can be distinguished: nuclear boiling, transition boiling, and film boiling [8, 9] (photographs are shown in reference 8). In nuclear boiling, the usual teakettle boiling, small bubbles of water vapor form at the same positions on the bottom of the pot, rise in the denser but cooler liquids, and collapse. This process continues until the liquid in the kettle becomes hot enough to allow the bubbles to break the surface. In transition boiling, the surface at which boiling is initiated is hotter, and large, irregularly shaped bubbles form. Finally, in film boiling (at a very hot surface) a film of vapor covers the whole surface of the hot body. This type of boiling occurs when a drop of water falls onto a hot surface. The drop darts about but takes a surprisingly long time to disappear. The reason is that the heat transfer is suppressed by the film of water vapor. Heat transfer becomes progressively less efficient as one goes from nuclear to film boiling. This sequence is accompanied by an increase in noise generated by the boiling. If a red-hot metal is quenched in water, the sequence of boiling takes place in reverse: first film boiling, then transition boiling, and finally nuclear boiling.

Freezing of water (to make ice cubes) is equally interesting [10]. Ice that forms slowly tends to be clear, whereas rapid freezing produces opaque ice. As ice forms, solutes (including air) are rejected and accumulate at the solid-liquid interface. If freezing is rapid, a large number of bubbles form (the solubility of air in water is exceeded at the interface) and are trapped by the advancing front of pure solid. At moderate freezing rates the bubbles grow in size in the direction of crystal growth, and one finds cylindrical bubbles known as ice worms. Finally, with very slow freezing the solutes have time to diffuse away from the interface and clear ice is formed. Similar removal of high local concentrations takes place when ice is grown in flowing water.

POLYMER SOLUTIONS

Solutions containing polymeric materials as solutes exhibit flow behavior quite different from that of nonpolymeric solutions. Not only their flow properties are unusual; their elastic properties are responsible for some remarkable behavior, too. For example, a polymeric liquid forced through a die will swell to a size considerably larger than the opening in the die. These elastic properties can cause a polymer solution to climb up the shaft of a rotating stirrer, rather than being thrown away from it.

The viscosity and elasticity of substances is the domain of **rheology,** the science of the flow and deformation of matter. Rheology deals with those properties of matter which determine how it deforms or flows when it is subjected to an external force. When subjected to stress (force per area), all materials deform to some extent. If the material is a fluid, the deformation is continuous, that is, the material flows. The relationship between an applied stress and the resulting deformation is a function that defines the rheological properties of the material.

For liquids composed of simple molecules in a single phase (e.g., water), the viscosity is independent of the stress, that is, the flow rate is proportional to the stress. Liquids for which this is the case are called **Newtonian liquids.** For liquids that contain large complex molecules such as polymers, the viscosity changes with the applied stress. Most polymer solutions (and molten polymers) decrease in viscosity as the deforming stress increases, commonly by factors of 10^3–10^4. The opposite behavior, dilatancy, is occasionally found (e.g., cornstarch in ethylene glycol–water mixtures). Reference 11 contains a large number of photographs with supporting discussion demonstrating the striking qualitative differences between the behavior of ordinary Newtonian liquids and polymeric liquids. One such difference is in bubble shapes, which in polymeric liquids are prolate spheroids ("footballs"), and in Newtonian liquids are oblate spheroids ("Frisbees").

The rheological properties of polymer solutions often exhibit extreme behavior of great practical significance, controlling the design of mixers, extruders, pumps, molds, and the formulation of paints, greases, and pastes. The solution's response to the non-linear mechanical forces that occur during stirring or extruding varies tremendously with the type of formulation. A few of the unusual rheological properties are highlighted in the demonstrations and are discussed in references 11–17.

Another property of polymeric solutions of great importance is their susceptibility to normal stresses. When stress is applied to a Newtonian liquid, the liquid deforms (flows) in a direction parallel to that in which the stress is applied. Non-Newtonian liquids, however, can flow in directions perpendicular to the applied stress. This is particularly significant with motor oil. The purpose of motor oil is to prevent contact between the pistons and the engine block. Piston motion produces a force on the oil perpendicular to the direction of piston movement. This force tends to reduce the viscosity of the non-Newtonian liquid oil. Furthermore, the temperature of the oil changes over the very large range from a cold start to final operating temperature, and the viscosity of the oil can be affected by this temperature change. Motor oils are rated on the SAE (Society of Automotive Engineers) scale. In these ratings, the numbers express the viscosity of the oil, with higher ratings for more viscous oil. Generally, a motor oil needs a viscosity of SAE 30–40 at operating temperatures. Because the viscosity of oil increases with decreasing temperature, the viscosity may be too high for the oil to be pumped through the engine when the engine is cold. Therefore, large numbers and amounts of additives are incorporated into motor oil to reduce the temperature dependence of the oil's viscosity. These additives cause the viscosity of the oil to increase with increasing temperature, just the opposite of the oil's behavior without them. Oils with these additives are called multigrade oils and are rated with two SAE numbers. Two common types are SAE 10W-40 and SAE 10W-30 motor oil. In these ratings, the first number expresses the low-temperature viscosity and the second number the high-temperature viscosity. (The "W" indicates that the oil is suitable for winter use in cold climates.) Generally, the greater the spread between the low- and high-temperature ratings, the more viscosity-modifying additives the oil contains. These additives can be degraded by prolonged use at high temperatures, forming deposits on engine parts. Although 10W-40 is often recommended, in fact it has a deleterious effect on engine parts when used over an extended period of time. Insist on 10W-30 motor oil.

Many plastic products are made by extruding a polymer melt through a die. Because the polymer melt is non-Newtonian, it is subject to "die swell," the expansion of the polymer stream as it exits the die. This swelling is not the same at the corners as at the center of an edge of the die, and to extrude a rectangular product, the die must be nonrectangular. Die swell may amount to several diameters of the original stream. Polymer additives in small amounts (10–500 ppm) do not significantly affect the viscosity, but they do greatly affect drag reduction (the pressure drop at a given flow rate during turbulent flow). Similar large changes in heat transfer and jet breakup (collapse of forced air bubbles) are found.

Polymer solutions are not restricted to the industrial sector. A growing field is biorheology. Examples in human biology include the control of air flow by mucous in the lungs and lubrication of the joints by synovial fluid. Procreation of the species is enhanced by the rheological properties of seminal fluid.

Polymer industries have been around for a long time, although the explosive growth started after World War II. The Mayan Indians made rubber balls from latex. Goodyear in 1839 added sulfur and heated the material to form a moldable product suitable for (carriage!) tires. Dunlop in 1888 created the pneumatic tire. Rubber is still used for tires today. Plastics have been in use for over a century. Nitrocellulose and camphor were molded into billiard balls around the time of the American Civil War. Plastics are replacing steel in construction—a prototype automobile was recently constructed from carbon-fiber impregnated plastic.

Polymer solutions exhibit nonideal behavior, for example, in their solvent vapor pressure and phase transitions. This nonideal behavior is a result of the extended structure imparted to the solution by the extensive intramolecular structure of the large solute molecules. The **Flory-Huggins theory** of polymer solution incorporates this effect [18, 19]. Polymer-solvent forces control to a large extent the molecular configuration of the solute in solution, also contributing to the nonideality of the solutions.

COLLOIDS

Closely related to solutions are the **colloids,** which are uniform dispersions of tiny particles of one substance (the dispersed phase) in another substance (the continuous phase). The size of the dispersed particles is much larger than single molecules, but still much too small to be individually visible. The designation of a colloid is somewhat arbitrary, but particles usually have diameters in the range of $10^{-7}-10^{-3}$ cm. Unlike solutions, colloids tend to degrade with time [20]. Colloids are very common, and the ingredients for demonstrating many of their properties can be purchased at the supermarket. The physical properties of many foodstuffs are a result of chemical agents which promote colloidal behavior.

Based on the nature of the continuous and dispersed phases, colloids can be classified into eight types:

Continuous phase	Dispersed phase	Name	Representative
gas	liquid	aerosol	fog
gas	solid	aerosol	fly ash
liquid	gas	foam	shaving cream
liquid	liquid	emulsion	hollandaise sauce
liquid	solid	sol	toothpaste
solid	gas	foam	bread
solid	liquid	gel	Jell-O
solid	solid	sol	metal-doped glass

The particles in colloidal systems are often charged, and repulsions between the similarly charged particles tend to stabilize the colloid. Irreversible changes of state can be induced by suitably large interactions with the colloid. For example, dissolving a salt in an aqueous colloid reduces the repulsions between the colloid particles by insulating them from each other. This insulation can allow the particles to approach each other and aggregate, forming particles too large to remain dispersed. The physical chemistry of colloids has a rich history, attracting the attention of Thomas Graham, Wilhelm Ostwald, and Michael Faraday among others.

The thermodynamic analysis of colloidal systems is somewhat treacherous because of the occurrence of relatively long-lived metastable states, where the system is in a quasi-equilibrium state. Colloidal systems do not succumb easily to rigorous analysis. The rheological properties are even more complex than those of concentrated polymer solutions. Time-dependent behavior of parameters of the constitutive relations occurs more often than not.

An example of a colloidal product that has captured the majority of the exterior house paint market is latex paint [21]. "Emulsion" (latex) paints are composed of a pigment dispersion (milled into water) and a latex formed by emulsion polymerization (really a sol, because the polymer particles, typically acrylic copolymers, are solids). These colloids are blended to produce the paint. However, the viscosity of the formulation is often low and the paint spreads too thinly. To remedy this a thickener is added (carboxy methyl cellulose or fumed silica, for example), producing a thixotropic gel, one that spreads smoothly but stiffens rapidly when not sheared, thus preventing drip. Latex paints have the added desirable feature that cleanup is possible with soap and water. Fungicides, defoamers, buffers, and freeze-thaw stabilizers are commonly added to the paint. The final product contains more than 50% solid by weight.

A colloidal product with a controversial history is shaving cream! Termed an aerosol foam (really a pressurized foam), shaving cream in the can is an emulsion with an aqueous-solution continuous phase and a dispersed phase (the propellant) consisting of a nontoxic liquid that boils below room temperature at atmospheric pressure. The propellant is kept liquid by the internal pressure in the can. Chlorofluorocarbons were a popular choice of propellant for this and other aerosol (out of the can) products. However, degradation products of these compounds react with the stratospheric ozone, resulting in an increase of ultraviolet radiation reaching the Earth's surface and, in turn, an increase in skin cancer [22]. Because pressurized cans are a convenience product, the aerosol industry was obliged to hasten the move to replace the propellant; hydrocarbons are a suitable choice but have their own safety hazard [23]. In the meantime, alternative means of dispensing have appeared; for example the "pump" that operates on air.

GLOSSARY

(Relevant demonstration numbers are shown in parentheses.)

activity: a quantity which relates the chemical potential of a pure substance, μ_i^0, to that of the substance in a mixture, μ_i; represented by a and defined by the equation $\mu_i = \mu_i^0 + RT\ln(a)$.

azeotrope: a solution in a liquid-vapor equilibrium where both phases have the same composition (9.11).

boiling-point elevation: the difference between the boiling point of a solution and that of the pure solvent; one of the colligative properties.

calculus: the branch of mathematics that treats continuously varying quantities and is characterized by the use of infinite processes involving the passage to a limit.

chemical potential: the derivative of the Gibbs function with respect to a mole number.

Clapeyron equation: in a liquid-vapor equilibrium the slope of a plot of pressure versus temperature equals the entropy difference divided by the difference in volume of the two phases.

Clausius-Clapeyron equation: an integrated form of the Clapeyron equation (9.5).

coefficient of thermal expansion: a relation between volume and temperature at fixed pressure of the form $(1/V)(\partial V/\partial T)_p$ (9.3).

colligative properties: a property numerically the same for a group of substances independent of their chemical natures (9.19, 9.20, 9.21, 9.22).

colloid: a phase dispersed to such a degree that the surface interactions become an important factor in determining its properties (9.40, 9.41, 9.42).

condense: to change from the gaseous state to the liquid or solid state.

condensed state: a state of relatively high density.

Dalton's law: the pressure of a gas mixture equals the sum of the partial pressures of its components.

density: mass divided by volume (9.2).

derivative: an expression of the sort dy/dx which equals the slope of a plot of y versus x. If other variables are fixed, d is replaced by ∂ and the variables which are constrained are shown as subscripts. A rigorous definition of the derivative is provided by the calculus.

dielectric constant: the property of a substance equal to the ratio of the capacitance of a capacitor filled with the substance to the capacitance of the capacitor when empty (9.32).

diffusion: the transport property involving the flow of matter (9.28, 9.29, 9.30).

enthalpy of fusion: the heat absorbed during melting (9.21).

enthalpy of vaporization: the heat absorbed during boiling (9.6, 9.22).

entropy: a thermodynamic function, S, defined by the second law of thermodynamics. S has the property that its change is zero when the system is carried around a path whose starting and ending points coincide.

equation of state: a relationship among pressure, volume, temperature, and amount of substance.

equilibrium: a condition in which the system properties do not change with time.

first law of thermodynamics: a statement that there exists a thermodynamic function of state, the internal energy, U, with the property that $\Delta U = -w$ for thermally isolated processes, where w is the work done by the system.

Flory-Huggins theory: a theory of the thermodynamic properties of polymer solutions that allows for the extended nature of polymer molecules.

fluid states: states that flow freely under stress.

fractional distillation: method of separating several volatile components of differing boiling points by boiling the mixture and condensing the vapor in several stages (9.10).

freezing-point depression: the difference between the freezing point of a solution and the freezing point of the pure solvent; one of the colligative properties.

fusion: the change from solid to liquid phase (9.21).

Gibbs-Duhem equation: an equation (in differential form) relating changes among the intensive variables which describe the system.

Gibbs function: the fundamental thermodynamic function for systems whose independent variables are chosen to be temperature, pressure, and the mole numbers.

Gibbs phase rule: the relation $f = c - p + 2$, where f is the number of intensive variables which can be varied independently, c is the number of components, and p the number of phases.

heat capacity: the heat absorbed divided by the temperature change.

Henry's law: the partial pressure of a slightly soluble gas equals a constant multiplied by its mole fraction in solution (9.18).

hydrogen bond: the interaction between a hydrogen atom in one molecule and an electronegative atom in another molecule which leads to a moderately strong but thermally labile intermolecular bond.

ideal gas law: the relation $pV = nRT$ (R a constant).

ideal solution: a solution for which the chemical potential obeys the equation $\mu_i = \mu_i^0 + RT\ln(x_i)$, where x_i is the mole fraction of a component.

integration: the process of continuously summing infinitely small changes; the inverse of differentiation; corresponds to finding the area under a curve.

intensive property: a property of a substance that is independent of the amount of substance.

internal energy: internal energy, U, is defined by the first law of thermodynamics as $\Delta U = -w$ in thermally isolated systems, where w is the work done by the system.

isothermal compressibility: a relation between volume and pressure at fixed temperature of the form $-(1/V)(\partial V/\partial p)_T$.

molality: moles of solute per kilogram of solvent.

molarity: moles of solute per liter of solution.

mole fraction: moles of a species divided by total moles.

Newtonian liquid: a liquid which flows immediately upon the application of a force, and whose rate of flow is directly proportional to the magnitude of the force applied.

osmotic pressure: the pressure exerted on a solution containing nonvolatile solute in equilibrium with pure solvent; one of the colligative properties.

partial derivative: a derivative taken treating all but one of the independent variables as constants.

phase: a homogeneous region.

Raoult's law: the partial pressure in the gas phase equals the product of the mole fraction in solution multiplied by the vapor pressure for a chosen species.

rheology: the science treating the deformation and flow of matter.

solubility: the amount of solute in solution at equilibrium for a given amount of solvent when excess solute is present (9.13).

states of matter: gas, liquid, or solid.

statistical mechanics: the formalism that allows averaging over many of the molecular coordinates to compute macroscopic properties.

sublimation: the change directly from solid to gas state.

surface tension: surface energy (J/m^2) resulting from imbalanced intermolecular forces at interfaces (9.23, 9.24, 9.25).

thermodynamics: the study of the laws governing the conversion of energy from one form to another, the direction of heat transfer, and the availability of energy to do work.

transport properties: properties relating the flow of matter (or momentum, heat, etc.) under stress (concentration gradient, shearing force, temperature gradient, etc.) (9.27, 9.28, 9.31).

vaporization: the change from liquid to gas phase (9.7).

vapor pressure: the pressure of a pure material in a liquid-vapor equilibrium (9.7, 9.9).

vapor-pressure lowering: the difference between the vapor pressure of the solvent in a solution and the vapor pressure of the pure solvent; one of the colligative properties (9.5).

viscosity: the (negative) constant of proportionality between a shearing force and the velocity gradient perpendicular to it (9.27).

weight fraction: grams of solute divided by kilograms of solution.

REFERENCES

1. G. W. Castellan, *Physical Chemistry,* 3d ed., Addison-Wesley Publishing Co.: Reading, Massachusetts (1983).
2. *International Critical Tables,* Vol. 3, 1st ed., McGraw-Hill: New York (1928).
3. R. C. Weast, Ed., *CRC Handbook of Chemistry and Physics,* 66th ed., CRC Press: Boca Raton, Florida (1985).
4. J. A. Dean, Ed., *Lange's Handbook of Chemistry,* 13th ed., McGraw-Hill: New York (1985).
5. B. E. McBryde and F. W. Brown, *Sci. and Children* (Jan.): 23 (1985).
6. L. Pauling, *The Nature of the Chemical Bond,* 3d ed., Cornell University Press: Ithaca, New York (1960).
7. D. Eisenberg and W. Kauzmann, *The Structure and Properties of Water,* Oxford University Press: Oxford (1969).
8. J. W. Westwater, *Sci. Am.* 190 (June) (1954).
9. J. C. May, *SciQuest* (Oct.) (1979).
10. S. B. Chalmer, *Sci. Am.* 200 (Feb.) (1959).

11. R. B. Bird, T. C. Armstrong, and O. H. Hassager, *Dynamics of Polymeric Liquids, Fluid Mechanics,* Vol. 1, 2d ed., John Wiley and Sons: New York (1987).

12. A. S. Lodge, *Elastic Liquids,* Academic Press: New York (1964).

13. J. Walker, *Sci. Am.* 186 (Nov.) (1978).

14. A. A. Collyer, *Phys. Educ.* 8:11 (1973).

15. A. A. Collyer, *Phys. Educ.* 8:333 (1973).

16. A. A. Collyer, *Phys. Educ.* 9:38 (1974).

17. A. A. Collyer, *Phys. Educ.* 9:313 (1974).

18. T. L. Hill, *An Introduction to Statistical Thermodynamics,* Addison-Wesley Publishing Co.: Reading, Massachusetts (1960).

19. P. J. Flory, *Principles of Polymer Chemistry,* Cornell University Press: Ithaca, New York (1953).

20. H. R. Kruyt, Ed., "Irreversible Systems," in *Colloid Science,* Vol. 1, Elsevier: Amsterdam (1952).

21. R. R. Meyers and J. S. Long, "Formulations, Part I," in *Treatise on Coatings,* Vol. 4, Marcel Dekker: New York (1975).

22. R. S. Stolarski, *Sci. Am.* 258 (Jan.):30 (1987).

23. P. L. Layman, *Chem. Eng. News* 64 (Apr. 28):29 (1986).

9.1

Volume Changes upon Mixing

When two liquids are mixed, the volume of their solution is less than the sum of their individual volumes (Procedure A). When a different pair of liquids is mixed, the volume of their solution is greater than the sum of their individual volumes (Procedure B).

MATERIALS FOR PROCEDURE A

4 drops blue food coloring

4 drops yellow food coloring

50 mL distilled water

50 mL absolute (or 95%) ethanol, C_2H_5OH, or absolute methanol, CH_3OH

10 cm black electrical tape

2 50-mL volumetric flasks, with stoppers

100-mL volumetric flask, with stopper

long stemmed funnel to fit 100-mL volumetric flask

MATERIALS FOR PROCEDURE B

50 mL ethyl acetate, $C_2H_5OCOCH_3$

50 mL carbon disulfide, CS_2 (**Use only in a fume hood.**)

10 cm black electrical tape

2 50-mL volumetric flasks, with stoppers

100-mL volumetric flask, with stopper

long-stemmed funnel to fit 100-mL volumetric flask

PROCEDURE A

Preparation

Wrap a piece of black electrical tape around the neck of each of the three volumetric flasks so the bottom edge of the tape is just above their calibration marks. This will make the marks easier to see.

Place 4 drops of blue food coloring in one of the 50-mL volumetric flasks and 4 drops of yellow food coloring in the other. Fill one of the flasks to the mark with distilled water and the other with absolute ethanol. Stopper and agitate the flasks to color the liquids uniformly.

Presentation

Pour the liquid from one of the 50-mL volumetric flasks through the funnel into the 100-mL volumetric flask. Then pour the liquid from the other 50-mL volumetric flask into the 100-mL flask. Stopper and agitate the 100-mL volumetric flask to mix the liquids. When the liquids are uniformly mixed, the color of the mixture will be green, and its volume will be less than 100 mL.

PROCEDURE B

Preparation

Wrap a piece of black electrical tape around the neck of each of the three volumetric flasks so the top edge of the tape is just below the calibration marks. This will make the marks easier to see. In a fume hood, fill one of the flasks to the mark with ethyl acetate and the other with carbon disulfide. Stopper the flasks.

Presentation

This demonstration should be presented in a fume hood. Pour the liquid from one of the 50-mL volumetric flasks through the funnel into the 100-mL volumetric flask. Then pour the liquid from the other 50-mL volumetric flask into the 100-mL flask. Stopper and agitate the 100-mL volumetric flask to mix the liquids. When the liquids are uniformly mixed, the volume of the mixture will be more than 100 mL.

HAZARDS

Ethanol, methanol, and ethyl acetate are flammable.

Carbon disulfide is extremely flammable and toxic; it should be handled only in a fume hood. The explosive range is 1–50% (v/v) in air, and its flash point is −33°C. The vapor is irritating to the eyes and is malodorous.

DISPOSAL

The alcohol-water mixture should be flushed down the drain with water.

The carbon disulfide–ethyl acetate mixture should be discarded in a receptacle for flammable organic wastes.

DISCUSSION

Procedure A shows that the volume of a 50-50 mixture of ethanol and water is less than the sum of the volumes of the separate components. Procedure B shows that the volume of a 50-50 mixture of ethyl acetate and carbon disulfide is greater than the sum of the volumes of the individual liquids.

One of the properties of the ideal solution is that the volume of mixing (the difference between the volume of the solution and the total volume of its components) is 0. This demonstration shows both negative (Procedure A) and positive (Procedure B) deviations from ideality. The demonstration can be used to illustrate molecular packing in liquids, or in connection with partial molal quantities [1].

The mixture of 50 mL of absolute ethanol with 50 mL of water has a volume of about 96.5 mL at 20°C. (The error in the final volume for the experiment is ±0.2 mL.) The effect of temperature is negligible. The anticipated volume change upon mixing, if any, is easily calculated using solution densities from the literature [2]. The pertinent data are presented in the table. At 20°C the density of water is 0.9982 g/mL, that of ethanol is 0.7893 g/mL. Therefore, the mass of 50.00 mL of water is 49.91 g, and that of 50 mL of ethanol is 39.47 g. The mass of the mixture is 89.38 g, and this mixture is 44.16% ethanol by weight. The density of a 44.16% mixture of ethanol and water is 0.9265 g/mL. Therefore, the volume of 89.38 g of the mixture is 96.47 mL. A similar calculation shows that the effect is nearly the same when methanol is used. In place of absolute ethanol, 95% ethanol could also be used. Nominal 95% ethanol is a constant-boiling azeotrope of ethanol and water, and it is 95.57% ethanol by weight. If 95% ethanol is used in Procedure A, the volume of the mixture will be 96.8 mL.

Densities of Liquids at 20°C [2]

Components	Wt. %		Density (g/mL)
	Component A	Component B	
(A) Ethanol	100.00	0.00	0.7893
(B) Water	0.00	100.00	0.9982
Mixture of A and B	44.16	55.84	0.9265
(A) Methanol	100.00	0.00	0.7917
(B) Water	0.00	100.00	0.9982
Mixture of A and B	44.23	55.77	0.9267
(A) Carbon disulfide	100.00	0.00	1.2559
(B) Ethyl acetate	0.00	100.00	0.8890
Mixture of A and B	58.55	41.45	1.0580

A volume increase occurs in the carbon disulfide–ethyl acetate system. The final volume of the mixture in Procedure B is 101.4 mL, so the effect is only a third as large as in the alcohol-water mixture in Procedure A. The effect would be more visible if the demonstration were done on a larger scale, but both of these liquids are rather expensive and more difficult to dispose of than alcohol and water.

The demonstration can be made to produce quantitative results. The distance from the meniscus to the calibration mark on the neck of the volumetric flask can be measured. This distance and the diameter of the neck can be used to compute the volume change. Alternatively, the neck of the large volumetric flask can be marked in 0.1 mL units before it is used. Because the volume of the ethanol-water mixture in Procedure A is about 96 mL, the calibration mark on the neck of the 100-mL volumetric flask must be at least 4 cm above the bottom of the neck to keep the meniscus of the mixed liquids in the neck.

An interpretation of the direction of the volume change in molecular terms is somewhat treacherous considering the complex nature of the intermolecular forces, especially in hydrogen-bonded liquids. As described in the introduction to this chapter, the structure of liquid water is open, because hydrogen bonds between the molecules hold them in specific orientations to each other. Rather crudely we could say that the open (relatively low-density) ice-like structure of water is broken down by introducing ethanol, resulting in a compacting of the solution relative to the pure liquids. The structures of the ethyl acetate and carbon disulfide molecules are quite different from each other, and therefore the attractive forces between ethyl acetate molecules are different from those between carbon disulfide molecules. The attractive forces between carbon disulfide and ethyl acetate molecules in a mixture of the liquids are smaller than those in the neat liquids. Thus this system expands upon mixing. This picture of the relative strength of the attractive forces is supported by the fact that this system forms a minimum-boiling azeotrope [3], that is, an azeotrope in which the boiling point of the mixture is lower than the boiling points of the components. Azeotropy is the subject of Demonstration 9.11.

Two like liquids are ethanol and methanol, and we expect only a small volume of mixing. The densities of the pure liquids are nearly the same and the solution densities can be represented by a linear function of the composition (wt. %) [2]. This reflects the fact that the volume of mixing is significantly less than the error in its determination and thus it is unobservable in our relatively insensitive demonstration and even in experiments in which the accuracy is two orders of magnitude better. We are able to conclude that the volume of mixing for the ethanol-methanol system is, for all practical purposes, 0.

REFERENCES

1. F. Daniels, R. A. Alberty, J. W. Williams, C. D. Cornwell, P. Bender, and J. E. Harriman, *Experimental Physical Chemistry,* 7th ed., McGraw-Hill: New York (1970).
2. *International Critical Tables,* Vol. 3, 1st ed., McGraw-Hill: New York (1928).
3. L. H. Horsley, *Azeotropic Data,* Vol. 3, American Chemical Society: Washington, D.C. (1973).

9.2

Density and Miscibility
of Liquids

A tube containing four layers of liquid is shaken and allowed to rest. After several minutes, the liquids have separated into the original four layers (Procedure A). Another tube contains four layers of liquid, and on the surface of each liquid floats a solid object (Procedure B).

MATERIALS FOR PROCEDURE A

2–10 mg oil-soluble dye (e.g., sudan yellow)

100 mL bis(2-chloroethyl)ether, $(ClCH_2CH_2)_2O$

100 mL perfluoro-1,3-dimethylcyclohexane, $C_6F_{10}(CF_3)_2$†

100 mL distilled water

100 mL mineral oil, with density 0.7–0.8 g/mL

500-mL glass cylinder, with stopper

heat-shrink tubing to fit over cylinder and its stopper, or Parafilm

MATERIALS FOR PROCEDURE B

100 mL mercury

100 mL carbon tetrachloride, CCl_4

100 mL distilled water

100 mL mineral oil, with density 0.7–0.8 g/mL

500-mL glass cylinder, with stopper

1–3-cm pieces of each of the following to fit inside cylinder:

    brass and/or copper (e.g., a U.S. cent)

    Teflon and/or Masonite

    nylon and/or cork

heat-shrink tubing to fit over cylinder and its stopper, or Parafilm

---

† Eighty percent perfluoro-1,3-dimethylcyclohexane is available from the Aldrich Chemical Company, 940 W. Saint Paul Avenue, Milwaukee, Wisconsin 53233; catalog no. 28,231-6. Ninety-nine percent purity can also be used and is also available but 80% purity is much less expensive and works well.

PROCEDURE A [1]

Preparation

Dissolve a small amount of dye (start with a volume about equal to the head of a pin) in 100 mL of bis(2-chloroethyl)ether to give an intense but distinct color. Pour 100 mL each of perfluoro-1,3-dimethylcyclohexane, the colored ether, distilled water, and mineral oil into the 500-mL cylinder. Seal the cylinder with its stopper and seal the junction with Parafilm or heat-shrink tubing. Shake the cylinder gently for about 15 seconds. Some of the dye dissolved in the ether will dissolve in the mineral oil, coloring it. Allow the cylinder to rest for several minutes. The liquids will separate into distinct layers.

Presentation

Display the cylinder containing four distinct liquid layers: colorless fluorocarbon on the bottom, colored ether above that, colorless water above the ether, and colored mineral oil floating at the top. Shake the cylinder gently for several seconds to mix the liquids. Allow the cylinder to rest for a while, and the liquids will again separate into the four distinct layers. Repeat the shaking and resting procedure. The more vigorous the shaking, the longer the time required for the liquids to separate into distinct layers, but the liquids always separate eventually.

PROCEDURE B

Preparation

Pour 100 mL of mercury into the 500-mL cylinder. Slowly pour 100 mL of carbon tetrachloride onto the mercury, 100 mL of distilled water onto the carbon tetrachloride, and 100 mL of mineral oil onto the water. Drop the small pieces of brass and/or copper into the cylinder; these will sink through the top three liquids but float on the surface of the mercury. Carefully drop the pieces of Teflon and/or Masonite into the cylinder; these will fall through the oil and water and rest on the surface of the carbon tetrachloride. Drop the pieces of nylon and/or cork into the cylinder. The nylon will sink through the mineral oil and float at the junction of the oil with the water. If cork is also dropped in, it will remain at the top.

Close the cylinder with its stopper and seal the junction with Parafilm or heat-shrink tubing. Because the mineral oil and carbon tetrachloride are miscible, the cylinder should not be agitated!

Presentation

Place the assembled cylinder on display where the pieces of solid can be seen floating at the interfaces between the liquids.

HAZARDS

Bis(2-chloroethyl)ether is toxic and irritating to the skin and mucous membranes. Prolonged exposure should be avoided.

The toxicological properties of perfluoro-1,3-dimethylcyclohexane are unknown. Prolonged exposure to the liquid or vapor should be avoided.

Mercury is extremely toxic and should be handled with care to avoid prolonged or repeated exposure to the liquid or vapor. Continued exposure to the vapor can result in severe nervous disturbance, insomnia, and depression. Continued skin contact also can cause these effects, as well as dermatitis and kidney damage. Mercury should be handled only in well-ventilated areas. Mercury spills should be cleaned up immediately by using a capillary attached to a trap and an aspirator (see figure). Small amounts of mercury in inaccessible places should be treated with zinc dust to form a nonvolatile amalgam.

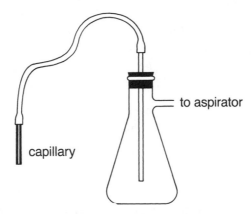

to aspirator

capillary

Device for recovering spilled mercury.

Carbon tetrachloride is toxic and carcinogenic. Exposure to the liquid and vapor should be minimized.

The apparatus used in these demonstrations is sealed to prevent the escape of any toxic fumes.

DISPOSAL

The apparatus can be stored sealed for repeated performances of this demonstration. The liquids in the cylinder from Procedure A should be discarded in a container for halogen-containing waste solvents. The liquids above the mercury in the cylinder from Procedure B should be decanted from the mercury and discarded in a container for halogen-containing waste solvents. The mercury should be purified by pouring it through a funnel lined with a filter paper having a pinhole at the bottom of the cone. The purified mercury should be returned to its container.

DISCUSSION

In Procedure A four mutually immiscible liquids of differing densities are sealed in a tube. These liquids settle into four separate layers with the densest liquid at the bottom and the least dense on top. Because these liquids are mutually immiscible—that is, no liquid will mix with any of the others—the tube containing them can be shaken and they will again settle into separate layers. In Procedure B, four liquids of differing densities are layered in a tube with the densest liquid on the bottom and the least dense on the top. Solid objects of intermediate densities float between the liquids. The densities of the materials used in this demonstration are given in the table. Because the liquids are not mutually immiscible, this tube cannot be shaken without causing some mixing of them—in particular, carbon tetrachloride and mineral oil.

Density of Materials at 25°C [2]

Solid	Liquid	Density (g/cm^3)
Cork		0.2
	Mineral oil	0.9
	Water	0.998
Teflon		1.2
	Bis(2-chloroethyl)ether	1.22
Masonite		1.3
	Carbon tetrachloride	1.595
	Perfluoro-1,3-dimethylcyclohexane	1.8
Brass		8.5
Copper		8.92
	Mercury	13.456

Density columns can be made with household products. A six-layer column can be made by layering, in turn from bottom to top, Aunt Jemima syrup (brown), Prestone antifreeze (yellow-green), Dawn dishwashing detergent (blue), Revlon Flex Balsam shampoo (gold), water (colored as desired with food coloring), and Oops paint remover (colorless) [3]. Solids can be added to see which level they reach, as in Procedure B.

A set of seven mutually immiscible liquids is known [4]. From top to bottom, these liquids are: heptane, aniline, water, fluorocarbon oil, phosphorus, gallium, and mercury. The melting point of phosphorus is 44°C and that of gallium is 38°C, so they are liquids only at elevated temperatures. However, they can be supercooled to room temperature fairly easily to produce a column containing seven liquids. If this column is shaken too vigorously, the supercooled liquids will solidify. Furthermore, because the densities of heptane-saturated aniline and water are quite similar, these two require a rather long time to separate after the column is mixed. A five-layer column of immiscible liquids can be produced from the set described in Procedure A by adding mercury to the system. However, mercury reacts with most dyes, forming a solid product. If the column containing dye and mercury is shaken too often, the inside of the container will soon be coated with unsightly solids.

At least two questions are raised by this demonstration: Why do liquids (and solids) vary in density? And what determines whether liquids dissolve in each other? Atomic volumes do not increase as rapidly with increasing atomic number as do atomic masses. Thus, there is a very loose correlation between increasing density and increasing atomic number. Replacing atoms of lower atomic number with atoms of higher atomic number tends to increase the density of the substance. For example, CH_2Cl_2 is less dense than $CHCl_3$, which is less dense than CCl_4. The question of miscibility can be discussed in terms of the similarity of attractive intermolecular forces between components of the mixture. A simple rule is "like dissolves like." By "like" we mean that the molecular sizes, shapes, and functional groups are sufficiently similar to produce similar intermolecular forces. Even great differences in molecular structure can be tolerated before the species become "unlike." Clearly the materials for Procedure A are unlike, and we are not surprised to find that they are immiscible. The rule should not be pushed too hard, for in some instances we can observe the opposite of what is expected; for example, in Demonstration 9.1 ethyl acetate and carbon disulfide are found to mix in all proportions.

REFERENCES

1. D. A. Ucko, R. Schreiner, and B. Z. Shakhashiri, *J. Chem. Educ.* 63:1081 (1986).
2. R. C. Weast, Ed., *CRC Handbook of Chemistry and Physics,* 66th ed., CRC Press: Boca Raton, Florida (1985).
3. C. DiSapio, *South-western Connecticut Chemmunicator* 5:22 (1987).
4. J. H. Hildebrand and R. L. Scott, *The Solubility of Nonelectrolytes,* 3d ed., Reinhold Publishing Co.: New York (1950).

9.3

The Dependence of Volume on Temperature: Coefficients of Thermal Expansion

The heights of mercury, ethanol, and water columns in identical thermometers are measured at two different temperatures. From these data, the relative coefficients of thermal expansion of mercury, water, and ethanol are obtained.

MATERIALS

25 mL absolute ethanol, intensely colored with several milligrams of methyl violet **(See Hazards section before handling methyl violet.)**

25 mL distilled water, intensely colored with food coloring or a water-soluble pH indicator (e.g., methylene blue)

3 identical mercury-in-glass thermometers, −10°C to +110°C

diamond saw

vacuum pump, with trap

heat gun

pinch clamp

2 500-mL beakers

hot plate

2 rubber dropping bulbs, or 2 pinch clamps and 2 pieces of rubber tubing to fit thermometers

PROCEDURE

Preparation

Construct an ethanol-filled thermometer as follows: Use a diamond saw to remove the very tip of one of the −10°C to +110°C mercury-in-glass thermometers, leaving most of the expansion space at the top (see Figure 1). Attach the open end of the thermometer to a trap attached to a vacuum pump (see Figure 2). Pump the mercury from the thermometer into the trap; heating the thermometer bulb with a heat gun and tapping the thermometer will hasten the process. Use a pinch clamp to seal the rubber hose from the thermometer to the trap, and remove the evacuated thermometer and the hose from the trap. Immerse the free end of the hose in a solution of methyl violet in ethanol and open the pinch clamp. The colored ethanol solution will fill the thermometer.

Repeat the procedure described in the preceding paragraph with another ther-

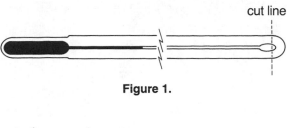

Figure 1.

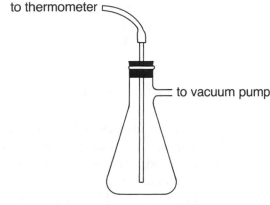

Figure 2.

mometer, but use distilled water colored with food coloring or a pH indicator in place of the colored ethanol.

Momentarily place the bulbs of both thermometers in a beaker of boiling water to force out a small amount of the liquid. After the thermometers have cooled to room temperature, check that the column of liquid in the thermometers ends somewhere between −10°C and +20°C. If the column ends above 20°C, return the bulb to the boiling water to force out more liquid. If the column ends below −10°C, the thermometer must be pumped out and refilled as described in the first paragraph. Close the open end of each thermometer with a rubber dropping bulb or a piece of rubber tubing sealed with a pinch clamp.

Reheat the beaker of water to about 35°C and have another beaker of water at room temperature.

Presentation

Place the bulbs of all three thermometers in the beaker of water at room temperature. Record the position of the top of the liquid in the thermometers (in °C). Move the thermometers to the beaker of water at 35°C and record the position of the liquid in each.

These data can be used to determine the relative coefficients of thermal expansion of the three liquids, as described in the discussion.

HAZARDS

Ethanol is flammable and should be kept away from open flames.

Methyl violet is toxic when ingested; its oral LD50† in mice is 105 mg/kg. It also causes persistent staining of the skin.

† See the note on page 28.

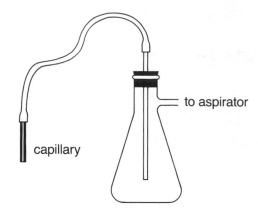

Figure 3. Device for recovering spilled mercury.

Mercury is extremely toxic and should be handled with care to avoid prolonged or repeated exposure to the liquid or vapor. Continued exposure to the vapor can result in severe nervous disturbance, insomnia, and depression. Continued skin contact also can cause these effects, as well as dermatitis and kidney damage. Mercury should be handled only in well-ventilated areas. Mercury spills should be cleaned up immediately by using a capillary attached to a trap and an aspirator (see Figure 3). Small amounts of mercury in inaccessible places should be treated with zinc dust to form a nonvolatile amalgam.

DISPOSAL

The thermometers can be stored for repeated performances of this demonstration, or the modified thermometers can be discarded in a solid-waste receptacle.

Transfer the mercury from the trap to a container for waste mercury.

DISCUSSION

Most materials expand when they are heated. The extent of the expansion for a given temperature change varies from one liquid to another, as shown in this demonstration. The magnitude of the volume change is expressed by the coefficient of thermal expansion, α, for a material. The coefficient of thermal expansion is the fractional change in volume with temperature at constant pressure, and is expressed mathematically as $\alpha = (1/V)(\partial V/\partial T)_p$, where V is the volume of the material, T its temperature, and p the pressure.

The coefficient of thermal expansion of Pyrex is small compared with that of the liquids used in this demonstration. Therefore, the volume of the thermometer is virtually constant over the operating temperature range of the thermometer, and the operation of the thermometer is determined by the volume of the bulb and column, V, the diameter of the capillary, d, and the coefficient of thermal expansion of the liquid. For a typical thermometer, $V = 0.2$ cm^3 and $d = 0.02$ cm. The coefficients of thermal expansion for the liquids used in this demonstration are given in the table. In a typical mercury thermometer, a change of 1°C causes a fractional change of $(1.818 \times 10^{-4}\,\mathrm{K}^{-1})$ $(1\,\mathrm{K}) = 1.8 \times 10^{-4}$ in volume. This corresponds to an actual volume change of $(1.8 \times 10^{-4})(0.2\ \mathrm{cm}^3) = 3.6 \times 10^{-5}$ cm^3. Because the capillary has a diameter of 0.02 cm, it has a cross-sectional area of 3.1×10^{-4} cm². The volume change causes a change in the height of the mercury of $(3.6 \times 10^{-5}$ cm³$)/(3.1 \times 10^{-4}$ cm²$) = 0.12$ cm. The change in the height of the column of other liquids is proportional to their coefficients of thermal expansion, provided the dimensions of the thermometers are the same. The actual readings of the thermometers filled with ethanol and with water depend on the amount of liquid left in each thermometer.

Coefficients of Thermal Expansion and Isothermal
Compressibilities for Materials at 20°C [1]

Substance	α (K^{-1})	β (bar^{-1})
Mercury	1.818×10^{-4}	3.85×10^{-6}
Water	2.07×10^{-4}	4.53×10^{-5}
Ethanol	1.12×10^{-3}	1.10×10^{-4}
Pyrex	3.3×10^{-6}	—

It is instructive to consider what happens if the thermometer is heated above the temperature where the interior is completely filled with liquid (about 125°C for a $-10°C$ to $+110°C$ thermometer). In particular, the pressure within the thermometer is significant. To the extent that V is constant, the change in pressure with temperature is $(\partial p/\partial T)_V$, which is equal to α/β, the ratio of the coefficient of thermal expansion to the isothermal compressibility (isothermal compressibility, β, is equal to $-(1/V)(\partial V/\partial p)_T$). This ratio is 45 bar/K for mercury, 10 bar/K for ethanol, and 4.6 bar/K for water. Considering that normal atmospheric pressure is about 1 bar, it is clear that overheating a thermometer is an unsound practice! The enormous pressures produced when the temperature is raised above the temperature where the thermometer is completely filled with liquid will certainly cause the thermometer to explode. For this reason, the tops of the thermometers used in this demonstration are not sealed after the thermometers are filled. Another reason for not sealing them is that sealing can introduce strains in the glass and make the thermometer susceptible to cracking.

Both mercury and alcohol capillary thermometers are in common use. Mercury thermometers can be used over a range of temperatures from above the melting point of mercury ($-39°C$) up to fairly elevated temperatures (0–360°C in a standard type). Smaller temperature ranges are achieved by reducing the diameter of the capillary. Ethanol has a coefficient of thermal expansion about six times that of mercury. This allows a greater sensitivity (column height versus temperature) for the same thermometer dimensions.

Liquid-in-glass thermometers are subject to a wide number of systematic errors, which include distillation of the liquid, irregular motion of the meniscus, irreproducible immersion (stem correction), nonrigidity of the bulb, flow with time of the glass (from manufacture and from temperature cycling), nonuniform bore, and temperature dependence of the coefficient of thermal expansion of the liquid. These are effects, except for the distillation of the liquid, on the order of 0.01°C or less. Although the mercury-in-glass thermometer is not suited for a standard of temperature, it is suitable for routine and convenient use where moderate accuracy (typically a few hundredths of a degree) is required.

Gases also expand when heated. For an ideal gas the coefficient of thermal expansion, α, is equal to $1/T$. Demonstration 5.6 in Volume 2 of this series illustrates the thermal expansion of gases in the following manner: A shelled, hard-boiled egg is seated in the mouth of a flask that has been warmed. When the flask is chilled in an ice bath, the egg pops into the flask. The flask is then inverted so the egg lodges in the neck of the flask. When the flask is warmed, the egg pops out. Another example of the same effect consists of cooling a fixed amount of gas at constant pressure. The gas can be contained in a balloon, and when liquid nitrogen is poured over it, the balloon shrinks.

REFERENCE

1. J. A. Hall, *The Measurement of Temperature*, Chapman and Hall, Ltd.: London (1966).

9.4

Boiling Water in a Paper Cup:
Heat Capacity of Water

A paper cup filled with water does not burn in the flame of a Bunsen burner (Procedure A). A balloon bursts when it is held over a match, but another that contains a small amount of water does not (Procedure B).

MATERIALS FOR PROCEDURE A

ca. 1 liter tap water

1-liter beaker

2 paper cups

crucible tongs

Bunsen burner

MATERIALS FOR PROCEDURE B

ca. 25 mL tap water

2 rubber balloons

matches

PROCEDURE A

Preparation

Fill a 1-liter beaker with water.

Presentation

Grasp an empty paper cup with the tongs and hold it in the flame of the Bunsen burner. The cup will char and begin to burn. Remove the cup from the burner flame and dunk it in the 1-liter beaker of water to extinguish the fire.

Pour about 60 mL of water into another paper cup. Grasp the cup with tongs and hold it over the flame of the Bunsen burner. The paper of the cup may darken, but it will not burn. After about a minute, the water in the cup will begin to boil.

PROCEDURE B

Preparation

Inflate one of the balloons and tie a knot in its neck to seal it. Pour about 25 mL of tap water into the other balloon. Inflate and seal the second balloon.

Presentation

Light another match and grasp the balloon containing water. Bring the burning match up to the bottom of the balloon, taking care to hold the match directly below the water in the balloon. The balloon will not burst.

DISCUSSION

This demonstration illustrates the large heat capacity of water by contrasting the effect of a flame on a paper cup (Procedure A) or on a balloon (Procedure B) in the presence of water with the effect in its absence. In both cases, the presence of water imparts a surprising resistance to the usual effect of the flame on these objects. An empty paper cup begins to burn almost as soon as it is placed in a flame. However, when the cup contains water, the cup does not burn; in fact, water can be boiled in the paper cup. Similarly, a balloon bursts immediately upon being placed in a flame, but if the balloon contains a small amount of water, it does not burst.

Flammable substances, such as paper and rubber, need only to be heated to their kindling temperature to cause them to burn in air. Both paper and rubber have rather low heat capacities; that is, it requires relatively little heat to raise their temperatures. Water, on the other hand, has quite a large heat capacity. It requires 4.2 joules of heat energy to raise the temperature of 1 g of water by 1°C. When the water-filled paper cup is placed in the flame, the water absorbs most of the heat energy of the flame, and it slows the increase in temperature. Furthermore, as long as there is liquid water in the cup, the temperature at the bottom of the cup cannot exceed 100°C, the boiling point of water. Because the kindling temperature of paper is considerably above 100°C, the cup will not begin to burn until all of the liquid water has evaporated.

A similar process prevents the bursting of the balloon that contains water. When the rubber of the balloon is heated it softens. In the absence of water, the temperature of the rubber rises to a point where the rubber becomes too soft to withstand the pressure of the air in the balloon, and the balloon bursts. If there is liquid water at the spot where the balloon is being heated, the water absorbs the heat and prevents the temperature from rising above 100°C as well as the rubber from becoming too soft to withstand the internal pressure of the balloon.

The high heat capacity of water can be attributed to its extensive network of hydrogen bonding. When the temperature of a substance increases, the motion of its constituent molecules increases. In order for molecular motion to increase, intermolecular forces must be overcome. The hydrogen bonds between water molecules are quite strong as intermolecular forces go, and overcoming them requires quite a bit of thermal energy. Therefore, to increase the temperature of water by a certain amount requires more energy than it does to increase the temperature of most other liquids by the same amount. This is what makes water so effective as a fire extinguisher for many types of fires: the heat that would otherwise raise the temperature of nearby flammables to their kindling temperatures is absorbed by the water instead, preventing the spread of the fire.

9.5

Vapor Pressure
of Pure Liquids and Solutions

Volatile liquids are injected into the space above the mercury in several mercury barometers, and their vapor pressures are determined from the changes in the heights of the mercury columns (Procedure A). The same method is used with pure solvent and a solution containing a non-volatile solute to illustrate Raoult's law (Procedure B). Mixtures of volatile liquids are used to illustrate ideal and nonideal solution behavior (Procedure C).

MATERIALS FOR PROCEDURE A

1 mL of each of four of the following liquids: water, methanol, absolute ethanol, diethyl ether, n-hexane, n-heptane, or acetone

special barometer apparatus (See Procedure A for assembly instructions.)

4 1-mL syringes with needles

meter stick

MATERIALS FOR PROCEDURE B

11 g decane, $C_{10}H_{22}$

40 mL diethyl ether, $(C_2H_5)_2O$

special barometer apparatus (See Procedure A for assembly instructions.)

balance accurate to 0.1 g

4 1-mL syringes with needles

meter stick

MATERIALS FOR PROCEDURE C

10 mL n-hexane, $CH_3(CH_2)_4CH_3$

20 mL n-heptane, $CH_3(CH_2)_5CH_3$

5 mL methanol, CH_3OH

10 mL absolute ethanol, C_2H_5OH

7 mL acetone, CH_3COCH_3

10 mL chloroform, $CHCl_3$

special barometer apparatus (See Procedure A for assembly instructions.)

balance accurate to 0.1 g

4 1-mL syringes with needles

meter stick

PROCEDURE A

Preparation

Construct the apparatus illustrated in Figure 1. The apparatus consists of four mercury barometers backed by a steel plate painted white and holding magnetic pointers to indicate the height of the mercury column in each barometer. Each barometer tube is a glass tube 90 cm long with an inside diameter of 11 mm; each is sealed at one end. Set up each barometer as follows: Wearing rubber gloves, fill the tube completely with mercury and cover the open end with the tip of a finger so all air is excluded from the tube. Invert the sealed tube over a dish containing mercury to a depth of about 3 cm. With the open end of the tube beneath the surface of the mercury, uncover the opening. Some of the mercury will drain from the barometer tube, but the height of the mercury in the tube will stabilize at about 75 cm. Clamp the inverted tube to the stand.

Carefully bend the needles of four 1-mL syringes into a semicircular shape. Fill each syringe with about 1 mL of a different one of the following liquids: water, methanol, ethanol, diethyl ether, n-hexane, n-heptane, or acetone.

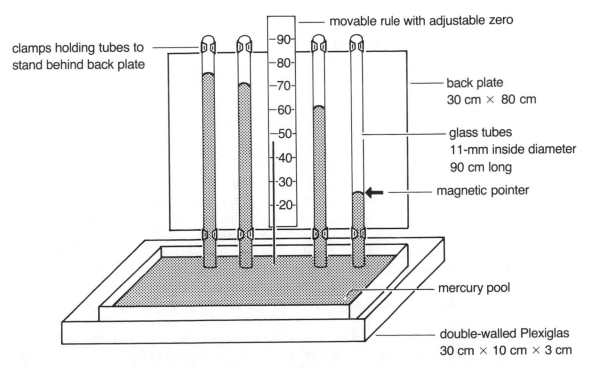

Figure 1. Vapor pressure apparatus. Diethyl ether, ethanol, and water were injected into the barometer tubes, from right to left. No liquid was injected into the barometer tube on the far left.

Presentation

Measure the height of the column of mercury in the barometer tubes. Insert the tip of one of the syringe needles into the open end of one of the barometers, which is submerged in the mercury. Holding the tip of the needle adjacent to the inner wall of the barometer tube, slowly inject the contents of the syringe into the tube. The liquid will rise to the top of the tube and float on the mercury, and some of the liquid will vaporize, depressing the mercury column. (If the liquid is injected quickly and into the center of the tube, some of it may vaporize before reaching the top of the mercury column, pushing some of the mercury ahead of it to the top of the tube. This could occur with enough force to fracture the barometer tube.) Measure the height of the mercury column after injecting the liquid. The difference in the heights of the mercury column before and after the injection of the liquid is equal to the vapor pressure of the liquid.

Repeat the procedure described in the preceding paragraph with the liquids in the other three syringes, using a different barometer for each.

PROCEDURE B

Preparation

Construct the barometer apparatus as described in the first paragraph of Procedure A.

Carefully bend the needles of four 1-mL syringes into a semicircular shape.

Prepare a solution with a mole fraction of 0.1 of decane in diethyl ether by dissolving 1.5 g of decane in 8 g of ether. Draw 1 mL of this solution into one of the syringes. Similarly, fill two other syringes with 1 mL of 0.2 and 0.3 mole fraction solutions, prepared by dissolving 3.4 g and 5.9 g of decane, respectively, in 8-g aliquots of ether. Fill the fourth syringe with 1 mL of pure diethyl ether.

Presentation

Measure the height of the column of mercury in the barometer tubes. Insert the needle tip of the syringe containing pure ether into the open end of one of the barometers, which is submerged in the mercury. Inject the contents of the syringe into the tube. The liquid will rise to the top of the tube and float on the mercury, and some of the liquid will vaporize, depressing the mercury column. (If the liquid is injected quickly and into the center of the tube, some of it may vaporize before reaching the top of the mercury column, pushing some of the mercury ahead of it to the top of the tube. This could occur with enough force to fracture the barometer tube.) Measure the height of the mercury column after injecting the liquid. The difference in the heights of the mercury column before and after the injection of the liquid is equal to the vapor pressure of the liquid.

Repeat the procedure described in the preceding paragraph with the solutions in the other three syringes. The solutions will have a vapor pressure proportional to the mole fraction of the solvent in the solution.

PROCEDURE C

Preparation

Construct the barometer apparatus as described in the first paragraph of Procedure A.

Prepare the following four solutions by mixing the indicated amounts of each liquid.

    Solution 1: 8.6 g *n*-hexane and 10.0 g *n*-heptane
    Solution 2: 3.2 g methanol and 4.6 g absolute ethanol
    Solution 3: 4.6 g absolute ethanol and 10.0 g *n*-heptane
    Solution 4: 5.8 g acetone and 12 g chloroform

Each of these solutions contains a 0.50 mole fraction of each component.

Carefully bend the needles of four 1-mL syringes into a semicircular shape. Fill each syringe with 1 mL of a different one of the four solutions.

Presentation

Measure the height of the column of mercury in the barometer tubes. Insert the tip of needle of the syringe containing solution 1 into the open end of one of the barometers, which is submerged in the mercury. Inject the contents of the syringe into the barometer tube. The liquid will rise to the top of the tube and float on the mercury, and some of the liquid will vaporize, depressing the mercury column. (If the liquid is injected quickly and into the center of the tube, some of it may vaporize before reaching the top of the mercury column, pushing some of the mercury ahead of it to the top of the tube. This could occur with enough force to fracture the barometer tube.) Measure the height of the mercury column after injecting the liquid. The difference in the heights of the mercury column before and after the injection of the liquid is equal to the vapor pressure of the solution. Record the vapor pressure.

Repeat the procedure described in the preceding paragraph with the solutions in the other three syringes. Record the vapor pressures of these solutions. The vapor pressures of solutions 1 and 2 are consistent with Raoult's law, whereas that of solution 3 shows a positive deviation, and solution 4 a negative deviation.

HAZARDS

Mercury is extremely toxic and should be handled with care to avoid prolonged or repeated exposure to the liquid or vapor. Continued exposure to the vapor can result in severe nervous disturbance, insomnia, and depression. Continued skin contact also can cause these effects, as well as dermatitis and kidney damage. Mercury should be handled only in well-ventilated areas. Mercury spills should be cleaned up immediately by using a capillary attached to a trap and an aspirator (see Figure 2). Small amounts of mercury in inaccessible places should be treated with zinc dust to form a nonvolatile amalgam.

Methanol, ethanol, diethyl ether, *n*-hexane, *n*-heptane, and acetone are extremely flammable.

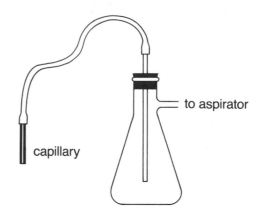

Figure 2. Device for recovering spilled mercury.

Chloroform is carcinogenic and poisonous when inhaled.

Methanol is poisonous when ingested or inhaled.

Diethyl ether forms explosive peroxides when stored for extended periods of time. Surplus diethyl ether should not be stored.

DISPOSAL

The barometers should be disassembled by a process the reverse of the way they were assembled. It is not practicable to collect the small amounts of volatile liquids used in this demonstration; they should be allowed to evaporate in a well-ventilated area away from open flames or sparks. The mercury and glass components of the apparatus must be cleaned before they can be used again. The mercury can be cleaned by passing it through a pinhole in the center of a piece of filter paper suspended in a funnel. The mercury should be returned to a sealed container for storage. The glass tubes should be rinsed with acetone and allowed to dry in a well-ventilated area.

DISCUSSION

In Procedure A the vapor pressures of several pure liquids are determined in a barometer [1]. Procedure B uses a similar apparatus to show the effect of a nonvolatile solute on the vapor pressure of a liquid [2, 3]. The vapor pressures of mixtures with two volatile components are demonstrated in Procedure C [4, 5].

When each barometer tube is set up, the space above the mercury contains only mercury vapor at an extremely low pressure, and the height of the column of mercury indicates the atmospheric pressure. When a liquid is added to the space above the mercury, the vapor of the liquid joins that of the mercury, increasing the pressure of the gas above the mercury. This forces the column of mercury down, and the height of this column is now equal to the difference between atmospheric pressure and the pressure of the liquid's vapor. Once an equilibrium is established between the liquid and its vapor, the pressure of the vapor is the "vapor pressure" of the liquid. Therefore, the vapor pressure of the liquid can be determined by measuring the height of the mercury column and subtracting it from its height before the liquid was added. Diethyl ether, with a vapor pressure of 442 torr at 20°C, shows a large change in the height of the mercury column. Even water, with a vapor pressure of 18 torr at 20°C shows a clearly visible effect.

Raoult's law gives the pressure above a solution as a function of the solution composition (x is the mole fraction) and the vapor pressure of the pure components. For a mixture containing two components, Raoult's law is expressed in the equation

$$p = x_1 p_1^0 + (1 - x_1) p_2^0$$

where x_1 is the mole fraction of one component,
$\quad\quad$ p_1^0 the vapor pressure of that component,
and $\quad$ p_2^0 the vapor pressure of the other component.

The predictions of Raoult's law can be compared with the observed pressure using the data in the table. The change in solution composition caused by formation of the vapor should be considered a possible source of error. With 1 mL at 20°C of a 50-mole-percent methanol-ethanol solution, only 0.01 of the total moles in the system is found in the vapor phase; the rest is excess liquid whose composition is consequently affected. In this demonstration, at most 0.03 of the moles of the n-heptane–n-hexane mixture is vaporized.

Vapor Pressure of Liquids at 20°C [7]

Substance	p^0 (torr)
Water	17.535
Ethanol	43.9
Methanol	88.7
Diethyl ether	442.2

The vapor pressure of a liquid is the pressure of the vapor in equilibrium with the liquid in a closed container. The vapor pressure of a liquid is temperature dependent, and the change in the vapor pressure of a pure liquid is related to the change in temperature of the liquid by the Clausius-Clapeyron equation.

$$\ln\left(\frac{p_2}{p_1}\right) = \frac{\Delta \bar{H}_{vap}}{R}\left(\frac{1}{T_1} - \frac{1}{T_2}\right)$$

In this equation, p_2 is the vapor pressure at temperature T_2, p_1 is the vapor pressure at T_1, $\Delta\bar{H}_{vap}$ is the molar heat of vaporization of the liquid, and R is the gas constant. By measuring the vapor pressure of a liquid at various temperatures, its heat of vaporization can be determined with this equation. The heat of vaporization is always positive (vaporization is an endothermic process), so the vapor pressure increases with temperature.

Two demonstrations which feature vapor pressure are the collapsing can (Demonstration 5.1 in Volume 2 of this series) and the iodine solid-vapor equilibrium (Demonstration 5.20, also in Volume 2). To collapse a can, water in the bottom of a can is boiled until virtually all of the air has been driven out. The can is then sealed. It collapses as it cools to room temperature. The can can be returned to near its original shape by reheating it. Alternatively, when the can is connected to a vacuum pump, it collapses as the air is pumped from it. The iodine solid-vapor equilibrium is shown by sealed tubes containing iodine solid and vapor placed in baths at different temperatures. The violet color of the iodine vapor is most intense in the warmest tube and least intense in the coldest tube.

REFERENCES

1. N. Egan, P. C. Ford, and A. R. Burkett, *J. Chem. Educ.* 53:303 (1976).
2. E. Koubek and M. J. Elert, *J. Chem. Educ.* 59:357 (1982).
3. E. Koubek and D. R. Paulson, *J. Chem. Educ.* 60:1069 (1983).
4. A. R. Gordon and W. J. Hornibrook, *Can. J. Res.* 24B:263 (1946).
5. J. B. Ferguson, M. Freed, and A. C. Morris, *J. Phys. Chem.* 37:87 (1933).
6. L. H. Horsley, *Azeotropic Data,* Advances in Chemistry Series, American Chemical Society: Washington, D.C. (1963).
7. *International Critical Tables,* 1st ed., Vol. 3, McGraw-Hill: New York (1928).

9.6

Evaporation as
an Endothermic Process

An aspirator reduces the pressure in a boiling flask containing acetone. The acetone then boils, and its temperature decreases (Procedure A). A "drinking duck" toy sips water from a glass until the level of the water in the glass falls below the reach of its beak (Procedure B).

MATERIALS FOR PROCEDURE A

See Materials for Demonstration 1.1 in Volume 1 of this series.

MATERIALS FOR PROCEDURE B

tap water, to fill drinking glass

drinking glass, ca. 10 cm deep (e.g., "old-fashioned" glass or jelly tumbler)

"drinking duck" toy (see figure in Procedure B)†

PROCEDURE A

See the Procedure of Demonstration 1.1 in Volume 1 of this series.

PROCEDURE B

Preparation and Presentation

Fill the drinking glass with tap water. Set the "drinking duck" next to the glass and tilt the body of the duck until its beak dips in the water (see figure). Hold the beak of the duck in the water until the duck's head becomes damp, then release the duck so its body returns to an upright position. After several seconds the liquid in the lower bulb of the duck's body will begin to rise in the central tube. The liquid will flow through the tube into the head of the duck. At the same time, the duck will gradually tip, until its beak dips in the water. When the body of the duck becomes horizontal, the liquid will flow from the duck's head, through the tube, and back into the lower bulb. When the liquid returns to the lower bulb, the duck returns to a vertical position. The duck will

---

† Such toys are available under a variety of names from toy stores and novelty shops.

repeat this "drinking" cycle until the level of the water in the glass falls below the reach of its beak.

HAZARDS

The drinking duck toy is made of very thin glass and breaks easily. Breaking the toy will release the volatile liquid it contains, which may be flammable.

DISPOSAL

The drinking duck toy can be retained for repeated use. It will continue to function indefinitely. If the toy breaks, the liquid contents should be allowed to evaporate in a well-ventilated area. The remaining materials can be discarded in a solid-waste receptacle.

DISCUSSION

This demonstration shows the spontaneous transfer of matter from the liquid to the vapor phase, evaporation, when the system pressure is less than the vapor pressure. It also shows that vaporization is an endothermic process. Procedure A shows that the rate of evaporation of acetone is increased when the pressure of the gas over the liquid is lowered: the acetone begins to boil. The temperature of the acetone also decreases, indicating that evaporation is an endothermic process. Procedure B uses the evaporation process to produce motion in the drinking duck toy.

The endothermic nature of the evaporation process is responsible for the operation of the drinking duck [1]. The duck consists of a central glass tube and two glass bulbs (see figure). The upper end of the glass tube opens directly into the upper bulb that forms the head of the duck. This upper bulb has a small protrusion on the side, representing the duck's beak, and the entire bulb is coated with an absorbant material that functions as a wick. The lower end of the glass tube extends almost to the bottom of the lower glass bulb. The lower bulb is a reservoir of volatile liquid (often dichloromethane colored with a dye). The duck is supported on a pivot near the center of the tube.

When the liquid is in the lower bulb, the mass of the bottom of the duck is greater than that of the top, and the duck assumes an upright position. When the head of the duck is dry, the vapor pressure of the liquid in the bottom is the same as it is in the head (ignoring the small hydrostatic pressure of the liquid in the tube), and the liquid stays in the lower bulb. When the head of the duck is moistened with water, the water evaporates from the wick, and the endothermic evaporation process cools the head. Because the head is cooler than the lower bulb, the vapor pressure of the liquid is lower in the head than in the lower bulb. The pressure differential between the two bulbs causes the liquid to rise through the tube into the head. As liquid flows into the head, the head becomes heavier, and the duck begins to tip at the pivot. When the duck tips far enough, the lower end of the tube rises out of the liquid, allowing the higher pressure vapor in the lower bulb to enter the upper bulb. As the vapor pressures equalize between the bulbs, the liquid returns to the lower bulb, and the duck returns to an upright position. If the duck is positioned in such a way that its beak dips in water when it tips, its head will remain damp, the evaporation of water will continue, the head will remain cooler than the lower bulb, and the tipping cycle will repeat.

REFERENCE

1. C. F. Bohren, *Clouds in a Glass of Beer,* John Wiley and Sons: New York (1987).

9.7

Liquid-Vapor Equilibrium

Sealed tubes containing bromine liquid and vapor are placed in baths at different temperatures. The reddish brown color of the bromine vapor is most intense in the warmest tube and least intense in the coldest tube. This is described in Demonstration 5.19 in Volume 2 of this series. This demonstration shows that vapor pressure increases with temperature.

9.8

Boiling Liquids
at Reduced Pressure

Water is made to boil at 50–60°C under reduced pressure created by an aspirator or by the condensation of water vapor. This is described in Demonstration 5.21 in Volume 2 of this series. This demonstration shows that vapor pressure decreases with decreasing temperature.

9.9

Vapor Pressure of a Solution
(A Corridor Display)

Water and sulfuric acid are sealed in a vacuum desiccator. Over a period of several weeks, the volume of liquid in the beaker containing water decreases and that in the beaker containing acid increases (Procedure A). Crystals are placed next to a small puddle of liquid in a covered dish on an overhead projector. After several minutes, the crystals have turned to drops of liquid (Procedure B).

MATERIALS FOR PROCEDURE A

40 mL distilled water

40 mL concentrated (18M) sulfuric acid, H_2SO_4

2 100-mL beakers

2 labels for beakers

gloves, plastic or rubber

vacuum desiccator (without desiccant) large enough to hold 2 100-mL beakers

MATERIALS FOR PROCEDURE B

overhead projector

10–20 grains of table salt (sodium chloride, NaCl)

distilled water in a wash bottle

petri dish, with cover

glass microscope slide to fit inside petri dish

PROCEDURE A

Preparation

Label one of the 100-mL beakers "water" and the other "sulfuric acid." Wearing gloves, pour 40 mL of distilled water into the beaker labelled "water" and 40 mL of concentrated sulfuric acid into the other beaker. Place the beakers side by side in the desiccator and seal it. Connect the vacuum hose from a water aspirator to the desiccator and evacuate it. (Do not use a vacuum pump to evacuate the desiccator—the pressure inside may be lowered to the point where the water will boil.)

Presentation

Place the desiccator where it can be viewed undisturbed for several weeks. The level of liquid in the water beaker will decrease and that in the sulfuric acid beaker will increase. The volume of pure water will decrease to 30 mL in about 2 days and to 20 mL in about 10 days. It requires about 2 months for essentially all of the water to evaporate and condense into the acid solution.

PROCEDURE B [1]

Preparation and Presentation

Place the petri dish on the overhead projector, and set a glass microscope slide on the bottom of the petri dish. Sprinkle 10–20 grains of table salt onto the glass slide. Carefully add distilled water from the wash bottle to the petri dish, wetting only the bottom of the dish and not the salt crystals or the top of the slide. Place the cover on the petri dish. Over a span of about 20 minutes, the salt crystals will grow into hemispherical drops as they absorb water vapor from the air inside the covered dish and dissolve in the condensed water.

Uncover the petri dish. After several hours, all of the water will have evaporated from the condensed droplets, leaving patches of salt-crystal residue where the drops had been.

HAZARDS

Because concentrated sulfuric acid is both a strong acid and a powerful oxidizing agent, it must be handled with great care. Spills should be neutralized with an appropriate agent, such as sodium bicarbonate ($NaHCO_3$), and then wiped up.

DISPOSAL

Pour the liquid from the acid beaker into a 1-liter beaker containing water. Slowly add sodium bicarbonate to the water until fizzing stops. Then flush the solution down the drain with water.

DISCUSSION

In this demonstration, liquid water evaporates at one location and condenses at another inside a sealed container. In Procedure A, pure water evaporates inside an evacuated desiccator and condenses into a sulfuric acid solution. The process requires about a day for a noticeable amount of water to move from one place to the other, and therefore, is suited to long-term display. In Procedure B, water evaporates in a covered petri dish and condenses into a sodium chloride solution. This process is displayed on an overhead projector and requires about 20 minutes for a significant change to occur.

The system inside the desiccator contains water in three phases: pure liquid, a solution of water in sulfuric acid, and water vapor. That these three phases are not in equilibrium is revealed by the fact that a change, albeit slow, occurs in the system: the volume of pure water decreases and the volume of the solution increases. The system moves toward an equilibrium between water in three phases, an equilibrium it can never reach. In thermodynamic terms, at equilibrium the chemical potentials of water in these three phases are equal. The chemical potential of the water in the beaker containing pure water, μ_1, is equal to μ_L^0, where μ_L^0 is the chemical potential of pure water. The chemical potential of water in the beaker containing the solution, μ_2, is equal to $\mu_L^0 + RTln(a)$, where a is the activity of the water in the solution. The system in the desiccator will be at equilibrium when $\mu_1 = \mu_2$, that is, when $\mu_L^0 = \mu_L^0 + RTln(a)$. Clearly, this can be the case only when a = 1. In the solution of water in sulfuric acid, a < 1 and a increases as the fraction of water in the solution increases. The system approaches equilibrium by transferring water from the beaker of pure water to the beaker of solution. However, because the beaker of solution contains sulfuric acid, the activity of water in the solution will never reach 1, and equilibrium will never be attained as long as pure liquid water remains in the desiccator.

The water is transferred from one beaker to the other through the vapor phase. The net transfer occurs as the two containers of water in the liquid phase attempt to establish their equilibrium vapor pressures inside the desiccator. The partial pressure of pure water is higher than that of the water in the solution. Therefore, as the pure water evaporates to establish its equilibrium vapor pressure, water vapor condenses in the solution to lower the partial pressure to the equilibrium value for the solution.

The volume of water decreases more rapidly than the volume of sulfuric acid increases. This occurs because the density of sulfuric acid solutions is greater than that of pure water. For example, the density of pure H_2SO_4 is 1.8 g/mL and that of a 1:1 (v/v) mixture of H_2SO_4 and H_2O is 1.5 g/mL, whereas the density of pure water is 1.0 g/mL [2]. This information indicates that after all of the water has condensed into the sulfuric acid, the volume of acid solution in the beaker is

$$\frac{(40 \text{ mL } H_2SO_4)(1.8 \text{ g/mL}) + (40 \text{ mL } H_2O)(1.0 \text{ g/mL})}{1.5 \text{ g/mL } 1:1 \text{ solution}} = 75 \text{ mL } 1:1 \text{ solution}$$

This is less than the sum of the volumes of the separate liquids, namely 80 mL.

The transfer of water is very slow at atmospheric pressure, because molecules of water vapor are continually colliding with molecules of air as they make their way from one beaker to the other. Therefore, the desiccator is partly evacuated to speed up the process. However, even at a pressure of about 30 torr, the process is still slow, and it requires several days for a 20% change in the volume of the liquids. Further evacuation is impossible, because the water establishes its vapor pressure of about 25 torr at room temperature, and the water would boil if the pressure were reduced below this value. Demonstration 5.16 in Volume 2 of this series shows the effect of pressure and intermolecular collisions on the rate of gas phase diffusion.

The other component of the solution, sulfuric acid, is in the same situation as the water, and one might expect the sulfuric acid to transfer to the other beaker. However, because the vapor pressure of sulfuric acid is very low, this transfer is extremely slow, and before any appreciable amount can move from the acid beaker to the water beaker, all of the water has moved to the acid beaker.

In Procedure B, salt grains are used instead of sulfuric acid. Molecules of water transfer through the gas phase from the pure water to the salt particles. Once sufficient water has condensed on the salt to form a saturated solution of sodium chloride in water, the operation and the explanation of this procedure are the same as those in Procedure A. The initial transfer of water molecules to the surface of the salt occurs because the water molecules are more strongly attracted to the salt crystal than they are to each other. Therefore, once they adhere to the salt, they evaporate from it more slowly than they do from each other, that is, from pure water. As more and more water condenses onto the salt, eventually the water molecules begin to surround the ions in the salt crystal, and the salt begins to dissolve, forming a drop. (A discussion of drop formation and its relationship to surface energy is presented in Demonstration 9.25.)

REFERENCES

1. C. F. Bohren, *Clouds in a Glass of Beer,* John Wiley and Sons: New York (1987).
2. *International Critical Tables,* Vol. 3, 1st ed., McGraw-Hill: New York (1928).

9.10

Separating Liquids: Fractional Distillation

Samples of pot liquid and distillate are collected from fractional distillation of a water-methanol solution. The pot liquid is not flammable, whereas the distillate is flammable.

MATERIALS

100 mL distilled water

50 mL methanol, CH_3OH

2 250-mL round-bottomed flasks, with standard taper joints

ring stand, with 3 clamps

water-cooled condenser, with standard taper joints to fit flask

distillation head to fit condenser

rubber tubing to connect condenser to cold-water tap and sink

heating mantle to fit 250-mL flask

variable transformer (e.g., Variac)

thermometer, $-10°C$ to $+110°C$, with suspension ring at top

string or heavy thread, ca. 1 m long

barometer

2 petri dishes

matches

watch glass, with diameter larger than that of petri dish

PROCEDURE

Preparation

At the site of the presentation, or on a cart, assemble the distillation apparatus as illustrated in Figure 1. Clamp one of the 250-mL round-bottomed flasks to the stand and insert the condenser in the neck of the flask. Clamp the condenser to the stand. Mount the distillation head to the top of the condenser. Attach the other 250-mL flask to the distillation head and clamp the flask to the stand. With a length of rubber tubing, attach the lower cooling-water connector of the condenser to the cold-water tap. Attach

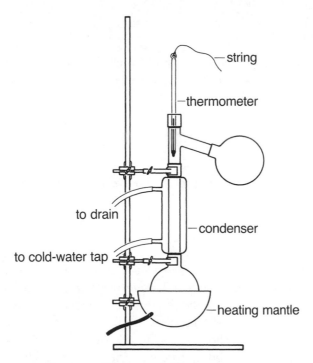

Figure 1. Fractional distillation apparatus.

another length of rubber tubing to the upper connector of the condenser and run the tubing to the drain in a sink. Mount the heating mantle on the lower flask and connect its power cord to the variable transformer. Tie the string securely to the suspension ring at the top of the thermometer; the string will allow the thermometer to be lowered through the distillation head and condenser into the flask.

Place 100 mL of distilled water and 50 mL of methanol in the lower 250-mL flask. Turn on the cooling water to the condenser. Turn on the variable transformer and gradually increase its setting until the condensing liquid just enters the distillation head. Do not allow any of the liquid to enter the upper flask. Allow the reflux to reach steady state as indicated by a constant temperature of the vapor at the top of the condenser.

Presentation

Increase the setting of the variable transformer until liquid flows into the upper flask. Insert the thermometer in the distillation head so its bulb is just below the arm leading to the upper flask. Record the temperature of the vapor at the top of the column. Lower the thermometer until its bulb is immersed in the boiling liquid. Record the temperature of the boiling liquid. Also record the ambient pressure. After 20–30 mL of liquid has collected in the upper flask, remove the heating mantle from the apparatus and turn off the variable transformer. Remove the upper flask from the apparatus and fill one of the petri dishes with liquid from the flask. Strike a match and ignite the liquid in the dish. Because the flames are difficult to see, demonstrate that the liquid is indeed burning by holding a fresh match over the dish and allowing the burning liquid to ignite it. Extinguish the flames by covering the petri dish with the watch glass. Remove the lower flask from the apparatus and fill the other petri dish with liquid from this flask. Try to ignite its contents with a match; it will not burn.

HAZARDS

Methanol burns with a pale blue flame which is difficult to see. Therefore, care must be taken to avoid burns. Because methanol vapors are toxic, the distillation should be carried out in a well-ventilated area.

DISPOSAL

The waste liquids should be flushed down the drain with water.

DISCUSSION

Fractional distillation is a process in which a mixture of liquids is partly vaporized, the vapor condensed, and the condensate partly revaporized and condensed in order to obtain a liquid that is enriched in one component to a desired degree. These successive vaporizations and condensations are usually not carried out as individual steps, but instead are accomplished by passing the vapor through a tube (column) with a large surface area above the boiling liquid. The surface inside the tube allows the vapor to establish equilibrium with its liquid throughout the tube. Because the temperature inside the tube decreases the further it is from the boiling liquid, the equilibrium composition of the liquid in the column varies along the column. Elaborate columns such as that illustrated in Figure 2 are used in industry to obtain several distillates of differing compositions.

If a mixture behaves as an ideal solution, it is possible to obtain a component of the mixture in a pure form. Essentially pure methanol can be obtained by fractional distillation from a methanol-water mixture. An analysis of this system based on the ideal-gas and ideal-solution models illustrates the character of the temperature-composition

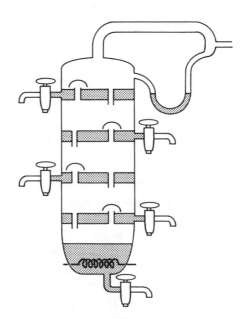

Figure 2. Commercial fractional distillation column.

diagram [*1–3*]. The vapor pressures of water and methanol as functions of temperature are given reasonably accurately by [*4*]

$$\log_{10}p^0 = A - B/T$$

where $A_{methanol} = 8.8017$,
$\quad B_{methanol} = 2002$,
$\quad A_{water} = 8.7384$, and
$\quad B_{water} = 2185.8$,
$\quad$ p is in torr,
and $\quad$ T is in kelvin.

From the observed temperature of the boiling liquid, this equation allows calculation of the vapor pressures p_1^0 and p_2^0 of the two pure components. The total vapor pressure of the mixture is the sum of the vapor pressure of each component multiplied by its mole fraction in the mixture:

$$p = x_1p_1^0 + x_2p_2^0 = x_1p_1^0 + (1 - x_1)p_2^0$$

When the mixture is boiling at atmospheric pressure, the total vapor pressure of the boiling liquid is equal to atmospheric pressure. Therefore the mole fraction of methanol in the mixture, x_1, can be calculated. The mixture used in this demonstration is about 30% methanol by weight, and the value of x_1 should be close to 0.194, if the amount of the mixture in the vapor phase in the column is not too large. The composition of the distillate is almost pure methanol, and the temperature at the top of the column is about that of the boiling point of methanol at ambient pressure. Only a small surface area on which the vapor can condense and return to the boiling liquid is needed to produce a distillate very concentrated in methanol, because nearly pure methanol results after only a few successive liquid-to-vapor equilibria. Starting from x_1 at 0.1940 and p at 740 torr, the equilibrium temperature, T, and the mole fraction in the vapor, y_1, can be calculated. (Note: $y_1p = x_1p_1^0$.) The phase rule (discussed in the introduction to the chapter) shows that only two variables are independent. After the calculation for the first liquid-vapor equilibrium, x_1 is replaced by y_1 and the calculation of the next stage on the temperature-composition diagram can be made. The calculation can be repeated for successive stages, and the results of such a calculation are shown in the table. The boiling point of pure methanol at 740 torr as calculated from the equation above is 64.315°C.

Temperature-Composition Data for
Methanol-Water (p = 740 torr)

x_1	y_1	t (°C)
0.1940	0.4734	87.897
0.4734	0.7770	77.013
0.7770	0.9328	68.893
0.9328	0.9825	65.600
0.9825	0.9956	64.644
0.9956	0.9989	64.397
0.9989	0.9997	64.335

Fractional distillation is the major technique employed by the multibillion-dollar petroleum industry to separate crude oil into commercial products (e.g., light solvents, gasoline, kerosene, motor oil, tar). Enormous plants have been constructed for the purpose. The process is complicated by the large number of components in crude oil and by the formation of azeotropes (see Demonstration 9.11).

REFERENCES

1. H. C. Van Ness and M. M. Abbott, *Classical Thermodynamics of Nonelectrolyte Solutions—with Applications to Phase Equilibria,* McGraw-Hill: New York (1983).
2. G. W. Castellan, *Physical Chemistry,* 3d ed., Addison-Wesley Publishing Co.: Reading, Massachusetts (1983).
3. F. Daniels, R. A. Alberty, J. W. Williams, C. D. Cornwell, P. Bender, and J. E. Harriman, *Experimental Physical Chemistry,* 7th ed., McGraw-Hill: New York (1970).
4. R. C. Weast, Ed., *CRC Handbook of Chemistry and Physics,* 63d ed., CRC Press: Boca Raton, Florida (1982).

9.11

Failing to Separate Liquids: Azeotropy

A solution of hydrogen chloride in water is distilled in a simple still. After some has distilled, the temperature of the boiling liquid becomes constant. The pH of the constant-boiling solution is measured.

MATERIALS

600 mL concentrated (12M) hydrochloric acid, HCl

400 mL distilled water

simple laboratory still, with a capacity of at least 1 liter

thermometer, readable to the nearest 0.01°C around 106°C

barometer

150-mL beaker

pH meter and electrode reading to at least −1 (e.g., Corning model 145: pH = −2 to +14)

PROCEDURE

Preparation

Dilute 600 mL of concentrated (12M) hydrochloric acid to 1 liter with distilled water. Fill the distillation flask of the still with the diluted acid. Start the distillation 2–3 hours before the demonstration is to be presented. Adjust the distillation rate to 3 mL/minute.

Presentation

Measure the temperature of the boiling liquid periodically during the distillation, and note the ambient pressure, until three-fourths of the solution has been distilled. Stop the distillation, and allow the still to cool to near room temperature. Pour some of the undistilled liquid remaining in the distillation flask into the 150-mL beaker, and measure the pH of the solution. (If a pH meter that reads to −1 is unavailable, the solution can be diluted 1:10 to get into the 0–14 range of most pH meters.)

HAZARDS

Concentrated hydrochloric acid can burn the skin and damage clothing. Hydrochloric acid vapors are extremely irritating to the eyes and respiratory system.

DISPOSAL

The remaining liquid and distillate should be combined, neutralized with sodium bicarbonate ($NaHCO_3$) until fizzing stops, and flushed down the drain with water.

DISCUSSION

Some liquid mixtures can be separated by the process of distillation (see Demonstration 9.10). However, there are solutions that are impossible to alter by distillation. Solutions that deviate sufficiently from ideal behavior form mixtures in which the mole fraction of the components in the vapor is the same as that in the liquid. Therefore, condensing the vapor will form a liquid identical with the boiling liquid. Such a mixture is called an azeotrope. The compositions of many azeotropic mixtures have been studied and are tabulated in reference works [1]. Hydrochloric acid (a solution of hydrogen chloride in water) forms an azeotrope, and a solution of accurately known composition can be prepared by the procedure in this demonstration.

When the forces between the different molecules in a solution are significantly different from those between the molecules of the pure components, there is often a minimum or maximum in the pressure-composition plot for the solution, and a corresponding maximum or minimum in the temperature-composition plot. (The figure is a temperature-composition plot for the ethanol-cyclohexane system, which shows a minimum. A figure showing the plot for the solution used in this demonstration is not included here because the curve is so nearly flat, it is difficult to see the maximum.) At the extremum (relative minimum or maximum), the compositions of the vapor and liquid

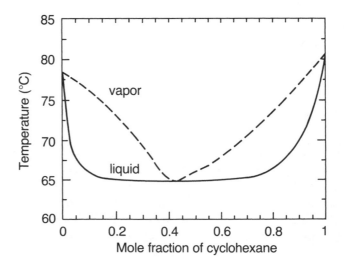

are the same and distillation is useless as a means of separating the components. Azeotrope formation is very common; a survey of 18,000 binary systems showed 6000 azeotropes [2–3].

The strong solvation of the hydrogen (and chloride) ion produces a minimum in the pressure-composition relationship and a corresponding maximum in the temperature-composition relationship. The azeotrope composition is a weak function of pressure, which means this system can be used to form acid of accurately known concentration ("constant-boiling acid") without the need for standardization.

p(torr)	wt. % HCl	T_b °C
500	20.916	97.578
700	20.360	106.424
760	20.222	108.584
800	20.155	110.007

The preparation of the azeotrope is time consuming and should be started several hours ahead of the determination of the pH. At 20°C the density of the solution (760 torr) is 1.0991 g/cm^3, yielding a molarity of 6.094 and a pH of -0.78 [1]. If the pH meter is calibrated and functioning properly, the calculated value agrees with the reading of the pH meter. The boiling point corresponds to the ambient pressure. The azeotrope can be approached from the other side by starting with a solution composition less than 6M.

Minimum-boiling azeotropes are also common. Such an azeotrope appears at the top of a fractionating column. The determination of the azeotrope composition as a function of pressure for such systems is an experiment suitable for the physical chemistry laboratory [4].

A system long studied and still of commercial significance is ethanol-water. Fractional distillation of fermentation products yields 190-proof ethanol, that is, ethanol containing 5% water by volume (proof = twice the volume percent of ethanol). To remove the final amounts of water, benzene can be added to 95% ethanol, and a ternary (three-component) azeotrope containing 74.1% benzene, 18.5% ethanol, and 7.4% water is distilled off at 64.9°C and 1 atm. This leaves ethanol free of water (absolute ethanol), but the ethanol may contain traces of benzene if a slight excess of benzene is used to assure complete removal of water. The benzene makes absolute ethanol unsuitable for human consumption.

REFERENCES

1. *International Critical Tables,* Vol. 3, 1st ed., McGraw-Hill: New York (1928).
2. W. Swietoslawski, *Azeotropy and Polyazeotropy,* Macmillan Co.: New York (1963).
3. G. W. Castellan, *Physical Chemistry,* 3d ed., Addison-Wesley Publishing Co.: Reading, Massachusetts (1983).
4. F. Daniels, R. A. Alberty, J. W. Williams, C. D. Cornwell, P. Bender, and J. E. Harriman, *Experimental Physical Chemistry,* 7th ed., McGraw-Hill: New York (1970).

9.12

Salting Out:
Making Liquids Immiscible

Methanol and water are mixed and form a homogeneous solution. Solid potassium carbonate is added to the solution, and it dissolves. More is added, and the solution separates into two liquid phases. When more potassium carbonate is added, a three-phase system of one solid and two liquid phases forms.

MATERIALS

150 g distilled water

150 g methanol, CH_3OH

175 g potassium carbonate, K_2CO_3

500-mL beaker

magnetic stirrer, with stir bar

PROCEDURE

Preparation and Presentation

Mix 150 g of distilled water and 150 g of absolute methanol in a 500-mL beaker. The two liquids are miscible and form a homogeneous solution.

Place the stir bar in the beaker and set the beaker on the stirrer. Add 25 g of potassium carbonate to the solution and stir the mixture until all of the solid dissolves. Repeat the addition of potassium carbonate in 25-g aliquots. After 75 g have been added and all of the solid dissolved, there will be two layers of liquid in the beaker, and the lower layer will be thinner than the upper layer. After 100 g have dissolved, the lower layer will still be the thinner of the two, but it will be thicker than it had been. The layers will be about the same thickness after 125 g have dissolved. After 150 g have been added, the mixture will remain cloudy, because not all of the solid dissolves; that is, the mixture is now composed of three phases—two liquids and one solid. After 175 g have been added the mixture will be even cloudier, because no more solid dissolves.

HAZARDS

Methanol is flammable, as well as being poisonous when ingested.

DISPOSAL

The waste mixture should be flushed down the drain with water.

DISCUSSION

Although methanol and water are miscible (i.e., soluble in all proportions), when potassium carbonate is added gradually to a 50-50 (w/w) mixture of these liquids, a three-component solution forms. This solution is a homogeneous mixture of water, methanol, and potassium carbonate. When more potassium carbonate is added, a point is reached where the mixture separates into two liquid phases, one with a methanol-rich solvent, and the other water rich. As more potassium carbonate is added, the two phases become more dissimilar. The water-rich phase loses methanol to the methanol-rich phase. Eventually, no more potassium carbonate will dissolve. At this point the mixture becomes a three-phase mixture, comprising two liquid phases and a solid phase, and the compositions of all three phases become fixed. The addition of more potassium carbonate changes only the amount of solid phase in the mixture.

The separation of an organic phase from an aqueous phase by the addition of a salt is called *salting out*. This phenomenon of salting out is common when salts are added to aqueous solutions of nonelectrolytes [1]. From a molecular standpoint, the strong hydration of the electrolyte ties up the water and makes it unavailable for the relatively weak hydrogen bonding with the nonelectrolyte [2]. Because it is the hydrogen bonding between water and the nonelectrolyte that keeps it in solution, the solubility of the nonelectrolyte decreases when the hydrogen bonding is disturbed.

Salting out of proteins, which are highly charged species, has been used as a method of purification for more than 90 years. By control of salt concentration, pH, or temperature, protein mixtures can be separated. Although organic phases can be forced out of aqueous solution by the addition of electrolytes, other separation methods, notably distillation, are preferred.

According to the Gibbs phase rule, $f = c - p + 2$, where p is the number of phases in the system, c is the number of components, and f the number of degrees of freedom, that is, the number of factors (temperature, pressure, mole fractions) that can vary. If the temperature and pressure are fixed, the number of remaining degrees of freedom in the water–methanol–potassium carbonate system is $3 - p$. When there is one phase the mole fractions of water and methanol can be varied independently (the mole fraction of the third component is determined by the other two). When there are two phases, specifying the mole fraction of one component in a chosen phase determines the remaining mole fractions in that phase and the composition of the other phase as well. When there are three phases, there are no degrees of freedom, and the composition of all three phases is fixed. The number of phases present in a mixture of three components can be depicted in a triangular phase diagram, such as in the figure [3]. Each point in and on the triangle represents a particular composition of the mixture, expressed as a weight fraction. The fraction of a particular component represented at a point is determined by the ratio of the distance of the point from the side of the equilateral triangle opposite the corner labelled with the name of the component to the height of the triangle. The area labelled "solution" contains all compositions for which a single solution is found. The central, quasi-semicircular region contains the composi-

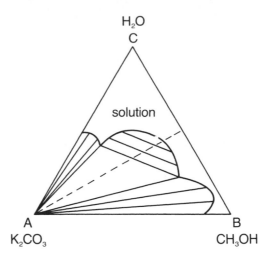

Phase diagram for water−methanol−potassium carbonate mixture.

tions which produce two liquid phases in equilibrium. The arcs touching the AC and AB sides of the triangle define the compositions for which solid K_2CO_3 and a single solution are in equilibrium. Finally, the triangular region is the three-phase region, where solid K_2CO_3 and two solutions of fixed composition are in equilibrium.

REFERENCES

1. C. J. O. R. Morris and P. Morris, *Separation Methods in Biochemistry,* John Wiley and Sons: New York (1976).
2. *Thermodynamic Behavior of Electrolytes in Mixed Solvents,* Vol. 2, American Chemical Society, Advances in Chemistry Series: Washington, D.C. (1976).
3. G. W. Castellan, *Physical Chemistry,* 3d ed., Addison-Wesley Publishing Co.: Reading, Massachusetts (1983).

9.13

Effect of Temperature on the Solubility of Manganese(II) Sulfate Hydrates

When a sealed test tube containing large pink crystals and clear solution is heated in a water bath, the crystals dissolve forming a clear pink solution, and when the temperature increases further, a white precipitate forms in the solution [1, 2].

MATERIALS

50 g anhydrous manganese(II) sulfate, $MnSO_4$, or 56 g manganese(II) sulfate monohydrate, $MnSO_4 \cdot H_2O$

50 mL distilled water

600 mL tap water

several drops of 6M hydrochloric acid, HCl (To prepare 1 liter of stock solution, slowly pour 500 mL of concentrated [12M] HCl into 400 mL of distilled water, and dilute the resulting solution to 1.0 liter.)

250-mL Erlenmeyer flask, with stopper

test tube, 2 cm × 15 cm, with stopper

3-cm length of 1-inch heat-shrink tubing (optional)

heat gun (optional)

refrigerator

600-mL beaker

hot plate

PROCEDURE [1]

Preparation

To prepare a saturated solution of manganese(II) sulfate, combine 50 g of anhydrous $MnSO_4$ and 50 mL of distilled water in a 250-mL Erlenmeyer flask and stopper the flask. The formation of a saturated solution will take about a week. Swirl the flask periodically during this period.

Decant the pale pink, clear solution from the Erlenmeyer flask into the test tube, and seal the tube with the rubber stopper. The stopper can be secured to the tube by encasing the stopper and the lip of the tube in heat-shrink tubing and shrinking the tubing to a snug fit with a heat gun.

269

Cool the test tube in a refrigerator until a readily visible amount of pink crystals has formed in the solution. This may take several days. Formation of crystals in the cooled solution can be speeded up by freezing a small amount of the solution and then shaking the test tube to disperse the fine powder that forms. The powder provides nuclei for crystal growth.

Prepare a hot bath by filling a 600-mL beaker with tap water, adding a few drops of 6M HCl to prevent scale formation, and heating the water to boiling on the hot plate.

Presentation

Display the crystals in the test tube. Place the tube in the hot-water bath. As the solution in the tube warms, the crystals will dissolve. Leave the tube in the water bath, and a white precipitate will form. It may take up to an hour for a large amount of the fine white precipitate to form.

HAZARDS

When the tube is heated in boiling water, a pressure differential of up to 1 atm may develop. This is likely to cause the stopper to be ejected if it is not seated securely, so that end of the tube should be aimed away from the audience.

Concentrated hydrochloric acid can cause burns to the skin, and the vapors are irritating to the eyes and respiratory system.

DISPOSAL

The sealed test tube can be reused in repeat performances of this demonstration. After several days at room temperature, the white precipitate that formed in the hot solution will have dissolved. Then, the pink crystals can be regenerated by chilling the test tube. Otherwise, the contents of the tube can be flushed down the drain with water.

DISCUSSION

In this demonstration a single solution is used to show that the solubility of a substance can decrease with either decreasing or increasing temperature. When an aqueous solution of manganese sulfate is cooled, a precipitate forms. When the same solution is heated, another precipitate forms. Both of these precipitates are hydrates of manganese(II) sulfate. The temperature dependence of the solubility of the various hydrates of manganese(II) sulfate is given in the table. Manganese(II) sulfate is most soluble at about 27°C, where the tetra- and pentahydrates have equal limiting solubility. Below 10°C large crystals of pink heptahydrate form, and its solubility decreases markedly with decreasing temperature. Above 27°C the monohydrate is the least soluble form, and its solubility decreases with increasing temperature.

A large amount of crystals form both when the tube is heated and when it is cooled. The table indicates that at 27°C, 66 g of $MnSO_4$ dissolve in 100 g of water, and the composition of a saturated solution is $66/166 = 39.8\%$ $MnSO_4$. Therefore, 100 g

Solubility of Manganese (II) Sulfate Hydrates as a Function of Temperature (in g $MnSO_4$ per 100 g water) [2, 3]

t (°C)	$MnSO_4 \cdot H_2O$	$MnSO_4 \cdot 4H_2O$	$MnSO_4 \cdot 5H_2O$	$MnSO_4 \cdot 7H_2O$
0	—	—	—	53.2
10	—	—	59.5	60.0
20	—	64.5	62.9	—
30	65.1	66.4	67.8	—
40	61.9	68.8	—	—
50	58.2	72.6	—	—
60	55.0	—	—	—
70	52.0	—	—	—
80	48.0	—	—	—
90	42.5	—	—	—
100	34.0	—	—	—

of saturated solution at 27°C contains 39.8 g of $MnSO_4$. At 0°C, 53 g of $MnSO_4$ are soluble in 100 g of water, for a composition of 34.6% $MnSO_4$. When 100 g of solution saturated at 27°C is cooled to 0°C, 39.8 g − 34.6 g = 5.2 g of $MnSO_4$ must leave the solution, and it leaves in the form of $MnSO_4 \cdot 7H_2O$. The molar mass of $MnSO_4$ is 151 g/mol, and that of $MnSO_4 \cdot 7H_2O$ is 277 g/mol. Therefore, it takes (5.2 g)(277/151) = 9.5 g of $MnSO_4 \cdot 7H_2O$ to remove 5.2 g of $MnSO_4$ from the solution. When the solution is heated to 100°C, only 34.0 g of $MnSO_4$ are soluble in 100 g of water, for a composition of 34.0/134.0 = 25.4% $MnSO_4$. When 100 g of solution saturated at 27°C is heated to 100°C, 39.8 g − 25.4 g = 14.4 g of $MnSO_4$ must leave the solution in the form of $MnSO_4 \cdot H_2O$. This corresponds to 16.1 g of $MnSO_4 \cdot H_2O$. The data in the table show that the solubility of the monohydrate decreases with increasing temperature, whereas the heptahydrate exhibits the opposite trend. Both types of behavior are common, but this system is unusual in that both cases can be observed in a single system.

The scale that forms in teakettles and hot-water pipes is a result of the temperature-dependence of the solubility of calcium carbonate in water. As temperature increases, the solubility of $CaCO_3$ in water decreases. When water, especially hard water containing relatively high amounts of dissolved calcium salts, is heated in a teakettle or a boiler, the solubility of the dissolved $CaCO_3$ decreases and the solid is deposited inside the kettle or the boiler.

The temperature dependence of solubility can be understood with the aid of the relation

$$\left[\left(\frac{\partial \ln \gamma}{\partial \ln m}\right)_{T,p} + 1\right]\left(\frac{\partial \ln m}{\partial T}\right)_{e,p} = \frac{\Delta H_{DS}}{RT^2}$$

where m is the molality of the solute in the saturated solution,

   γ its activity coefficient (equal to unity in infinitely dilute solution),

and ΔH_{DS} the differential heat of solution, that is, the heat absorbed when 1 mole of solute dissolves in (an infinite amount of) saturated solution at temperature T.

The subscript e refers to the change of molality of the saturated solution with temperature at equilibrium. The term in square brackets is always positive, so the change in solubility, expressed as dm/dT ≡ m d(ln m)/dT, will have the same sign as ΔH_{DS}. This is in accord with a loose interpretation of the principle of LeChatelier. If the solution process (the differential heat of solution into the saturated solution) is endothermic ($\Delta H_{DS} > 0$), the solubility increases with temperature and vice versa. The system reacts to a temperature increase by shifting the solid-solution equilibrium in a direction to absorb heat. Note that the system does not "undo" the change (the temperature increase). This is a *constraint* imposed on the system.

For ammonium nitrate the differential heat of solution is large (see Demonstration 9.14) and positive, and we find the solubility of ammonium nitrate increases strongly with temperature. The differential heat of solution of lithium chloride in water is hard to determine accurately because it is close to 0, but it is positive (see Demonstration 9.15). Lithium chloride shows only a small increase in solubility with temperature. Experiments on heat of solution determine the integral heat of solution, the heat absorbed when 1 mole of solute dissolves in sufficient solvent to produce a solution of a specified molality. The dependence of the integral heat of solution on molality can be used to calculate the differential heat of solution (at a chosen molality). The integral heat of solution of lithium chloride is negative (exothermic process), like that of copper sulfate.

The solution of ammonium nitrate in water is strongly endothermic, and this system is used for instant cold packs. Calcium chloride is commonly used in chemical hot packs.

REFERENCES

1. L. A. Bateman and W. C. Fernelius, *J. Chem. Educ.* 14:315 (1937).
2. K. A. Kobe, *J. Chem. Educ.* 16:183 (1939).
3. R. C. Weast, Ed., *CRC Handbook of Chemistry and Physics,* 66th ed., CRC Press: Boca Raton, Florida (1985).

9.14

Chemical Cold Pack: Dissolution as an Endothermic Process

First-aid cold packs demonstrate spontaneous endothermic reactions. This is described in Demonstration 1.2 in Volume 1 of this series. That the cold pack reaction is endothermic is not unusual. Of greater interest is the fact that it takes place spontaneously at constant temperature and pressure. This means that the value of the Gibbs function (the Gibbs free energy) decreases (see the introduction to this chapter and Demonstration 9.13).

9.15

Heat of Solution
of Lithium Chloride

The dissolution of lithium chloride in water is accompanied by the release of heat. This is described in Demonstration 1.8 in Volume 1 of this series. This system is of interest in that the solubility of lithium chloride in water *increases* with temperature, despite the fact that the integral heat of solution is negative. It is the *differential* heat of solution that determines the temperature dependence of solubility (see Demonstration 9.13).

9.16

Heat of Hydration
of Copper(II) Sulfate

The addition of a small quantity of water to copper sulfate powder in a test tube results in the evolution of heat. This is described in Demonstration 1.9 in Volume 1 of this series. This demonstration shows that ions draw water molecules into clusters by forming weak bonds, with the consequent evolution of heat. The process tends to decrease the system volume, and its reverse is responsible for the volume increase upon neutralization (see Demonstration 9.17).

9.17

Volume Increase
upon Neutralization†

A 500-mL sample of 2M hydrochloric acid and a 500-mL sample of 2M sodium hydroxide are combined in a 1-liter volumetric flask. The volume of the mixture is about 20 mL more than 1 liter.

MATERIALS

500 mL 2M sodium hydroxide, NaOH, in a 500-mL volumetric flask (To prepare this solution, dissolve 40 g of NaOH in 250 mL of distilled water in a beaker, allow the solution to cool, pour it into a 500-mL volumetric flask, rinse the beaker and pour the rinse into the flask, and dilute the solution in the flask with distilled water to the 500-mL mark.)

500 mL 2M hydrochloric acid, HCl, in a 500-mL volumetric flask (To prepare this solution, pour 83 mL of concentrated [12M] HCl into 250 mL of distilled water in a 500-mL volumetric flask, swirl the flask to mix the solution, allow the solution to cool, and dilute it to the 500-mL mark on the volumetric flask.)

1-liter volumetric flask with its calibration mark at least 8 cm below the top of its neck

funnel to fit 1-liter flask

ca. 10 cm electrical tape

PROCEDURE

Preparation

Wrap a piece of electrical tape around the neck of the 1-liter volumetric flask so the top edge of the tape is just below the calibration mark. This will make the mark easier to see.

Presentation

Place the funnel in the neck of the 1-liter volumetric flask. Pour the 2M NaOH solution from the 500-mL volumetric flask into the 1-liter flask. Then pour about half of the 2M HCl solution from its volumetric flask into the 1-liter flask. Swirl the large flask

† We wish to thank Alfred A. Rottino of Half Hollow Hills High School, Dix Hills, New York, for calling this demonstration to our attention.

to mix the solutions. The flask will become warm. Pour the remainder of the acid into the large volumetric flask. Swirl it to mix the solutions. Stopper the large flask and invert it several times to mix its contents. Set the flask down and remove the stopper to relieve pressure built up from the exothermic reaction of the mixture. The level of the liquid in the flask is well above its calibration mark. Allow the solution to cool to room temperature. The volume will decrease by a small amount as the solution cools, but even when the solution is at room temperature, its volume is about 20 mL more than 1 liter.

HAZARDS

Solid sodium hydroxide and its concentrated solutions can cause burns to the skin, eyes, and mucous membranes, and dust from the solid is irritating to the eyes and respiratory system.

Concentrated hydrochloric acid can cause burns to the skin, and the vapors are irritating to the eyes and respiratory system.

DISPOSAL

The waste solution (1M NaCl) should be flushed down the drain with water.

DISCUSSION

When 500 mL of an aqueous solution are combined with 500 mL of another aqueous solution, the volume of the resulting solution is usually 1 liter. In this demonstration, the volume is greater than 1 liter because a reaction occurs when these two solutions are mixed. The hydrogen ions in the hydrochloric acid solution combine with the hydroxide ions in the sodium hydroxide solution to form water molecules (solvent) in the mixture. Therefore, the mixture does not contain as many particles as were present in the two separate solutions, and one might anticipate a volume *decrease* upon reaction. However, this is not what is observed.

When 500 mL of 2M hydrochloric acid and 500 mL of 2M sodium hydroxide are mixed, the resulting solution is 1M sodium chloride. Because the neutralization reaction is exothermic, the solution is warm immediately after the two solutions are combined. The maximum temperature the solution reaches (T_{final}) can be estimated from the relation

$$Q = (T_{final} - T_{initial})Cm$$

where $T_{initial}$ is the starting temperature of the solutions,
   Q is the heat liberated in the reaction,
   C is the specific heat of the solution,
and   m the mass of the solution.

The value of Q is determined by the heat of neutralization for the reaction of hydrogen ions with hydroxide ions [$H^+(aq) + OH^-(aq) \longrightarrow H_2O(l)$], namely -55.9 kJ/mol. The reaction in this demonstration produces 1 mole of water, and therefore, releases 55.9 kJ of heat. Because the product solution is mostly water, the specific heat of the solution can be approximated with the specific heat of water, 4.2 J/g. Similarly the mass of the solution can be approximated by the mass of a liter of water, 1000 g. Then,

$$55{,}900 \text{ J} = (T_{final} - 21°C)(4.2 \text{ J/g})(1000 \text{ g})$$
$$T_{final} = 34°C$$

The mixture does not get quite this warm because some of the heat is absorbed by the container, and some is lost to the surroundings. However, even a temperature difference of 13°C would increase the volume of 1 liter of water by only 3 mL. The volume increase observed is close to 20 mL, and this volume increase persists even after the solution has returned to its initial temperature. Therefore, there must be some other factor responsible for the increased volume.

The volume change in this reaction can be calculated from data on the partial molar volumes of the solutes in the solutions [1]. The partial molar volume of a solute is the change in the system volume when 1 mole of solute is added to a solution whose volume is sufficiently large to prevent the concentration of the solute from changing appreciably. These partial molar volumes are determined from precise measurements of the density of solutions as a function of concentration. The tabulated values of partial molar volumes at infinite dilution are not much different from those for the finite concentrations used here, so we can calculate the volume change for the reaction using the tabulated values. The volume of a mixture is given by $V = \Sigma n_i \bar{V}_i$, where n_i is the number of moles of component i, and $\bar{V}_i$ its partial molar volume. For the reaction

$$HCl(aq) + NaOH(aq) \longrightarrow NaCl(aq) + H_2O(l)$$

the volume change is

$$\Delta V = V_{final} - V_{initial} = \bar{V}^0_{NaCl} + \bar{V}^0_{H_2O} - \bar{V}^0_{HCl} - \bar{V}^0_{NaOH} =$$
$$16.6 + 18.0 - 18.0 - (-5.2) = 21.9 \text{ cm}^3$$

For this reaction, at sufficient dilution, the Na^+ and Cl^- are spectator ions; that is, they have no effect on the observed volume change, because they are present in the separate solutions and in the mixture. The process occurring in the reaction could just as easily be written as

$$H^+(aq) + OH^-(aq) \longrightarrow H_2O(l)$$

This point of view is supported by the fact that (within experimental error) the difference of partial molar volumes of strong electrolytes with a common ion is independent of the common ion. In addition, the heat of the neutralization reaction (-55.90 kJ/mol) is independent of the spectator ions. (The exothermic nature of this acid-base neutralization reaction is shown in Demonstration 1.5 in Volume 1 of this series.) If we were to know the partial molar volumes of the ions (H^+ and OH^-) separately, we could determine ΔV from that information. These quantities are not available from thermodynamic measurements, because in aqueous solutions these ions do not exist independently of other ions. However, values can be calculated from a statistical mechanical model of the molecular structure of a solution of these ions. Such a procedure yields $\bar{V}^0_{H^+} = -5.4 \text{ cm}^3\text{mol}^{-1}$, and hence $\bar{V}^0_{OH^-} = \bar{V}^0_{H_2O} - \bar{V}^0_{H^+} - \Delta V = 1.5 \text{ cm}^3$. The reason we want the partial molar volumes of the ions is that it is easier to interpret the volume increase upon neutralization if the effect of the spectator ions is ignored.

The value of a particular partial molar volume of an ion can be broken down into a number of parts. The most important in the present case is electrostriction, the drawing in of matter in a nonuniform electric field (see Demonstration 9.32). The small hydrogen ion (especially) and the hydroxide ion break down the relatively open structure of pure water and form solvation spheres with a compact geometry. The insertion of a proton in water actually causes a decrease in system volume (*negative* partial molar volume). The hydroxide ion does have a positive partial molar volume, but it is not as large as it would be if electrostriction did not occur. The water produced by neutralization, on the other hand, has the same open structure as the solvent.

REFERENCE

1. R. A. Horne, Ed., *Water and Aqueous Solutions*, John Wiley and Sons: New York (1971).

9.18

Effect of Temperature and Pressure on the Solubility of Gases in Liquids

When a glass of cold tap water is allowed to warm to room temperature, bubbles form on the sides of the glass. When a bottle of seltzer is opened and poured into a glass, bubbles form and the liquid effervesces.

MATERIALS

cold-water tap at site of presentation

bottle of seltzer

2 drinking glasses

PROCEDURE

Preparation and Presentation

Draw a drinking glass full of cold tap water and allow it to warm to room temperature. As the water warms, bubbles of gas will form on the sides of the glass.

Open a bottle of seltzer and pour it into a drinking glass. Bubbles of gas will form throughout the liquid.

DISCUSSION

This demonstration agrees with the usual observations that gases become less soluble in a liquid when the temperature of the liquid increases and when the pressure of the gas over the solution decreases. The first behavior is illustrated when bubbles of gas form on the sides of a glass of cold water as the water warms to room temperature. The second is shown by the bubbles of gas that form when a bottle of seltzer (carbonated water) under pressure is opened to the atmosphere.

At low concentrations of gas in liquid, the solubility of the gas follows Henry's law,

$$p = K x$$

where p is the pressure of the gas over the solution,

x is the mole fraction of the gas in the solution,

and   K is the Henry's law constant.

The chart lists values of the Henry's law constant for several common gases [1, 2]:

Gas	K at 25°C (torr)	K at 10°C (torr)
N_2	6.51×10^7	5.02×10^7
O_2	3.30×10^7	2.46×10^7
CO_2	1.26×10^6	7.9×10^5

A larger Henry's law constant means a lower mole fraction in solution at a given pressure (i.e., a lower solubility). Gases with relatively low boiling points (e.g., N_2 and O_2) tend to be less soluble than those with higher boiling points (e.g., CO_2). The low boiling point is a signature of low intermolecular forces. As the data indicate, a glass of cold water warmed to room temperature will release bubbles of gas as the dissolved gases become less soluble.

The data indicate that the solubility of carbon dioxide at 1 atm (760 torr) of pressure is 3.35×10^{-2}M at 25°C. When carbon dioxide dissolves in water, it can react to a small extent with the water to form carbonic acid (H_2CO_3), and the carbonic acid can dissociate to form H^+ and HCO_3^- ions.

$$CO_2(aq) + H_2O(l) \leftrightharpoons H_2CO_3(aq) \qquad K_{eq} = 2.7 \times 10^{-3}$$

$$H_2CO_3(aq) \leftrightharpoons H^+(aq) + HCO_3^-(aq) \qquad K_a = 1.7 \times 10^{-4}$$

The equilibrium constant for the overall reaction is

$$CO_2(aq) + H_2O(l) \leftrightharpoons H^+(aq) + HCO_3^-(aq) \qquad K = K_{eq} K_a = 4.6 \times 10^{-7}$$

This constant indicates that the hydrogen-ion concentration in a saturated solution of CO_2 is $[4.6 \times 10^{-7}][3.35 \times 10^{-2}]^{1/2} = 1.2 \times 10^{-4}$, and the pH of the solution is 3.90. The pH of gastric juices ranges from 1 to 3. Therefore, introducing seltzer (or other carbonated beverages) into the stomach forces the equilibrium between ions and aqueous CO_2 toward the $CO_2(aq)$, resulting in a supersaturated solution and the departure of CO_2 from the solution (burp!).

In carbonated beverage containers, the pressure of CO_2 is several atmospheres. This pressure in excess of atmospheric pressure is responsible for the firmness of otherwise flexible containers. When the container is opened, the pressure drops rapidly and foaming results. The rapid frothing that occurs when a carbonated beverage is poured over ice is a result of the rough surface of the ice. The rough surface provides many nucleation sites at which bubbles can form, so many tiny bubbles are produced. The surface of glass is much smoother, so there are fewer sites at which bubbles can form, and the bubbles that do form grow relatively large.

When a bottle of beer is opened, the concentration of carbon dioxide in solution does not drop instantaneously (or even rapidly) to its equilibrium value given by Henry's law. Rather, bubbles form on nucleation sites and rise to the surface, growing in size as they ascend. If one wishes to hasten the process (making flat beer), salt grains can be added to the beer providing additional nucleation sites. The bubbles in beer can be termed clouds in a glass of beer, because the suspension of carbon dioxide bubbles in the liquid is similar to the suspension of water drops in a gas (a cloud) [3]. In fact, a cloud of this type, water drops in a gas, forms in the neck of a beer bottle when it is opened. There are no solid particles suspended in the space above the liquid in a beer bottle, because any that were there when the bottle was capped have long ago settled out. The lack of solid particles means that the nucleation sites for drop formation must be clusters of water molecules that coalesce by collisions with each other in a process called homogeneous nucleation. Cluster formation is enhanced at low temperatures—in fact, the gas in the neck of the bottle (mostly carbon dioxide, but also significant proportions of water vapor) is cooled by about 30°C by the expansion of the gas when the bottle is opened.

It takes time for the carbon dioxide in solution to escape to the vapor, so carbonated beverages retain their "fizz" by virtue of staying away from equilibrium. Local high concentrations of the heavier-than-air carbon dioxide in the vapor phase above the drink tends to keep the carbon dioxide in solution. Eventually, the excess gas escapes, and the drink becomes "flat" with a very low concentration of CO_2 in solution. Other demonstrations dealing with the solubility of carbon dioxide are 4.10 (Reactions Between Carbon Dioxide and Limewater) in Volume 1 of this series, and 6.2 (Reactions of Carbon Dioxide in Aqueous Solution), 6.4 (Carbon Dioxide Equilibria and Reaction Rates), and 6.29 (Facilitated Transport of Carbon Dioxide Through a Soap Film) in Volume 2.

The pressure dependence of the solubility of nitrogen gas is a concern for deep sea divers. Deep under the sea, both the divers and the air they breathe are at high pressure because of the weight of the water. Because the air is at high pressure, it is more soluble in the divers' blood than it is at the surface of the water. If the divers ascend quickly and the pressure is lowered suddenly, air will leave the blood too quickly and bubbles of gas will form in the blood stream with painful and potentially fatal consequences (the bends). Slow decompression (lowering of the pressure) to allow a gradual shift in the solubility equilibrium is advised. Sometimes, helium is used in place of nitrogen in the gas mixture breathed by divers, because helium is less soluble than nitrogen. Therefore, less helium dissolves at high pressure, so decompression requires less time.

Fish respond to the temperature dependence of the solubility of oxygen. As water warms, it holds less dissolved oxygen. If the water becomes too warm, the oxygen in solution is depleted to the point where fish may suffocate. Each species of fish is adapted to a particular temperature range, but no fish can exist in hot water where the concentration of dissolved oxygen is negligible.

REFERENCES

1. R. A. Alberty and F. Daniels, *Physical Chemistry,* 5th ed., John Wiley and Sons: New York (1980).
2. R. C. Weast, Ed., *CRC Handbook of Chemistry and Physics,* 66th ed., CRC Press: Boca Raton, Florida (1985).
3. C. F. Bohren, *Clouds in a Glass of Beer,* John Wiley and Sons: New York (1987).

9.19

Osmosis Through the Membrane of an Egg

Raw eggs with their shells dissolved become smaller and lighter when they are soaked in sugar syrup, and they become larger and heavier when they are soaked in water [1–3].

MATERIALS

3 raw chicken eggs

ca. 500 mL vinegar

1.2 liters distilled water

400 mL 20% sugar syrup (To prepare this solution, dissolve 85 g of table sugar in 350 mL of water.)

10 mg methylene blue or methylene green pH indicator

several paper towels

4 600-mL beakers

balance

glass stirring rod

PROCEDURE

Preparation

Soak the three eggs in vinegar for 2–3 days, until the shell has dissolved and the translucent membrane remains as the only covering. (Using a stronger acid may hydrolyze the egg membrane, making it too fragile to handle.) Carefully remove the eggs from the vinegar, taking care not to rupture the membrane. Blot the eggs dry with paper towels.

Pour 400 mL distilled water into three of the 600-mL beakers and pour 400 mL sugar syrup into the remaining beaker.

Presentation

Weigh the eggs and record their masses. Keeping track of each egg, place one of the eggs in the beaker of sugar syrup and another in one of the beakers of water. Add about 10 mg methylene blue to one of the remaining beakers of water and stir the mixture. Place the remaining egg in the colored water.

After the eggs have been immersed in the liquids for about 30 minutes, carefully remove them. Blot them dry. Reweigh the eggs and record their masses. The eggs that were soaked in water will be heavier, and the one soaked in the sugar syrup will be lighter. Furthermore, the interior of the egg soaked in the colored water will be colored. Place the colored egg in the last beaker of clear water. Soon, the water will become colored.

The eggs can be returned to their respective beakers to soak overnight to produce an enhanced effect. The egg soaked in sugar solution will become quite limp.

DISPOSAL

Waste solutions should be flushed down the drain with water. The eggs should be discarded in a waste receptacle.

DISCUSSION

A bird's egg has a hard outer shell made substantially of calcium carbonate ($CaCO_3$), and this shell is lined with a membrane that is permeable to water. When the egg is soaked in vinegar, the acetic acid in the vinegar dissolves the shell:

$$CaCO_3(s) + 2 HC_2H_3O_2(aq) \longrightarrow Ca(C_2H_3O_2)_2(aq) + H_2O(l) + CO_2(g)$$

In the decalcified egg, its liquid contents are held only by the membrane. The liquid contents of the egg are a complex solution containing proteins, lipids, carbohydrates, and salts.

In the demonstration the egg and the solution in which it is immersed are initially at the same temperature and pressure. Therefore, by the process of osmosis, water will flow through the egg's membrane from the region of high concentration to the region of low concentration. When the decalcified egg is placed in water, the water flows into the egg. When the decalcified egg is placed in the sugar solution, water flows out of the egg. The sugar solution is more concentrated than the contents of the egg, and the sugar molecules are too large to pass through the membrane. The change in mass will be about 1% in 30 minutes. The egg membrane is durable and will withstand gentle handling.

Osmosis can be used to refresh tired (limp) celery. The crispness of celery results from the pressure of water in the cells in the celery stalk. When this water is lost through transpiration from the leaves and evaporation at the surface of the stalk, the celery becomes limp. The water lost by transpiration and evaporation can be returned to the celery by soaking it in water. The cells contain an aqueous solution whose solutes cannot cross the cell walls. However, water can move through the walls, and it moves from the region where it is more concentrated (the pure water soak) to the region where it is less concentrated (the cell interiors). As the pressure of the water in the cells increases, the crispness of the celery returns. A similar technique is used and works for the same reasons in making radish "roses" for garnish. Several cuts are made in the radish around its perimeter; then the radish is soaked in water. As water enters the radish cells, they swell and cause the cuts to open, leading to the appearance of petals around the radish. Reversible changes such as these can be observed with dandelion stems. If the end of a dandelion stem is slit parallel to its length and the stem is placed in water, the end will flare and curl up. If the stem is then placed in salt water, the end will uncurl and return to its original shape. Vegetables sold in supermarkets are frequently coated with wax to prevent water from escaping and the produce from softening. The wax is a solvent-impermeable barrier.

The theory of the osmotic effect is developed in the introduction to this chapter.

REFERENCES

1. B. Cocanour and A. S. Bruce, *J. College Sci. Teach.* 15:127 (1985).
2. V. L. Mullin, *Chemistry Experiments for Children,* Sterling Publishers: New York (1961).
3. L. A. Ford, *Chemical Magic,* T. S. Denison and Co.: Minneapolis, Minnesota (1959).

9.20

Osmotic Pressure of a Sugar Solution

A membrane bag is filled with sugar solution and attached to a glass tube. When the bag is immersed in water, the level of liquid rises in the glass tube.

MATERIALS

2-liters distilled water

ca. 400 g sucrose (table sugar), $C_{12}H_{22}O_{11}$

ca. 10 mg iodine, I_2 (or other colored substance soluble in hexane)

2 mL hexane, C_6H_{14}

hot plate

1-liter beaker

25 cm of cellulose dialysis tubing or synthetic sausage casing (available at meat markets)

2-m glass tube, with inside diameter of 8 mm

glass-working torch

1-holed rubber stopper, no. 4–6

single-edged razor blade

2-holed rubber stopper, no. 8–12 (hole sizes compatible with glass tubing)

10 cm glass tube, with inside diameter of 8 mm

20-cm length of rubber tubing to fit 10-cm glass tube

pinch clamp

ring stand, at least 50-cm tall

clamp for ring stand

30 cm string

white poster board, strip 10 cm wide with a total length of 2 m

thermometer, $-10°C$ to $+110°C$

gloves, plastic or rubber

test tube, 10 mm $\times$ 75 mm

600-mL beaker

PROCEDURE

Preparation

Boil the dialysis tubing or sausage casing for several minutes in water. Allow the tubing to cool in the water.

Construct the apparatus illustrated in the figure. Flare one end of the 2-m glass tube to allow liquid to be poured easily into the tube. Insert the other end of the tube through the 1-holed stopper and position the stopper about 35 cm from the unflared end of the tube. Cut a groove around the side of the 2-holed rubber stopper. Insert the 10-cm glass tube through one of the holes of the stopper, and insert the 2-m glass tube through the other hole. Attach the 20-cm rubber tubing to the 10-cm glass tube in the rubber stopper and close the free end of the rubber tubing with a pinch clamp. Mount the 2-m glass tube to the ring stand by clamping it at the 1-holed stopper. Tie a knot in one end of the boiled dialysis tubing. Insert the 2-holed stopper in the open end of the dialysis tubing and secure it with a string tied snugly in the groove in the stopper. Fasten a 10-cm wide strip of white poster board along the height of the 2-m glass tube. This can be accomplished by punching small holes in the poster board and looping string through the holes and around the tube.

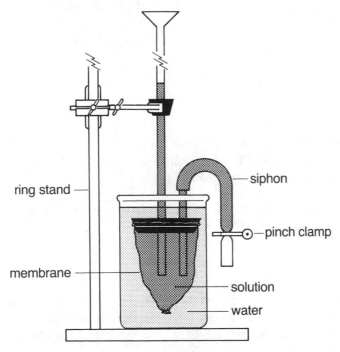

Osmotic pressure apparatus.

Place the 1-liter beaker on the stand and lower the apparatus so the cellulose "bag" is in the beaker. Fill the beaker with distilled water to cover the 2-holed stopper.

Prepare a sucrose solution by adding table sugar to 100 mL of boiling distilled water until the temperature of the boiling solution rises to 108°C. This will require about 400 g of sugar. Allow the sugar solution to cool. The concentration of this solution is 10.1 molal [*1*], which corresponds to a concentration of 77.6% (w/w) sucrose [*2*].

Empty the 1-liter beaker of water and place the empty beaker under the apparatus. Open the pinch clamp on the rubber tubing and pour sugar solution into the 2-m glass tube until the solution fills the cellulose bag. Close the pinch clamp and pour more sugar solution into the 2-m tube until the level of the solution in the tube is about 15 cm above the 2-holed stopper. Tighten the strings if any of the solution leaks from the cellulose bag. Wearing gloves, dissolve a crystal of iodine in 2 mL of hexane in the test tube and pour the solution into the 2-m tube. Put the clamped end of the rubber tubing in the 600-mL beaker.

Presentation

Fill the 1-liter beaker with distilled water to cover the 2-holed stopper. The level of liquid in the 2-m tube, as indicated by the colored hexane floating on its surface, will begin to rise at an initial rate of about 3 cm/minute. After the solution column rises above the top of the siphon, the pinch clamp should be opened to allow any air remaining in the siphon to be expelled. The rate at which the column rises will decrease as the sugar solution inside the membrane is diluted by the water diffusing into it. After about an hour, the liquid will approach the top of the 2-m tube. To prevent it from overflowing, drain off some of the liquid into the 600-mL beaker by opening the pinch clamp. The rise of solution and draining of the 2-m tube can be repeated many times. Replenish the water in the 1-liter beaker as necessary to keep its level above the 2-holed stopper.

HAZARDS

Wear gloves while handling the iodine container. Iodine is a very strong oxidizing agent, and contact with the skin can result in severe burns. Because iodine vaporizes readily at room temperature to yield toxic fumes, adequate ventilation must be provided when it is handled.

Hexane is flammable and should be kept away from open flames.

DISPOSAL

Use a dropping pipette to remove the hexane solution of iodine from the apparatus and discard the solution in a container for waste organic solvents. Flush the remaining aqueous solutions down the drain with water. Discard the membrane in a standard waste receptacle.

DISCUSSION

In this demonstration, pure water and a concentrated sucrose solution are separated by a membrane that is permeable to water molecules but not to sucrose molecules. Water molecules flow through the membrane from a region of high water concentration (pure water) to one of low water concentration (the sucrose solution). The pressure that must be applied to the sucrose solution to stop the net flow of water through the membrane is the osmotic pressure of the solution.

No attempt is made to stop the flow of water through the membrane in this demonstration. Rather, the effect of this flow is observed. As the water flows into the bag formed by the membrane the volume of the solution is increased, causing its level in the attached tube to rise. The solution in the tube will rise until the downward pressure of the liquid is equal to the osmotic pressure of the solution.

The osmotic pressure of the original sucrose solution in the bag can be calculated using the following equations [3]:

$$\bar{V}^0 \pi = -RT \ln(a) \qquad \ln(a) = \frac{\Delta H^0_{vap}}{R} \left(\frac{1}{T'} - \frac{1}{T_0} \right)$$

where π is the osmotic pressure,

$\bar{V}^0$ is the molar volume of the solvent at the temperature, T, of the osmotic pressure experiment,

ΔH^0_{vap} is the enthalpy of vaporization of the solvent at its boiling point, T_0,

and T' is the boiling point of the solution at the same pressure.

These equations indicate that the osmotic pressure of the original solution is over 400 bar (1 atm = 1.01325 bar). The familiar equation for osmotic pressure, $\pi = cRT$, where c is the molarity of the solution, is an approximation that applies only to dilute solutions. The solution used in this demonstration is most certainly not dilute, and an application of this equation will give a result far from the actual value of the osmotic pressure.

The pressure of 400 bar corresponds to a height of a solution having a density of 1.39 g/mL [2] of 3000 meters, indicating the impracticality of observing the equilibrium height. The actual equilibrium height will be less, because the solution is diluted when water crosses the membrane into the solution. Dilution of the solution reduces the osmotic pressure and makes accurate calculations on this system difficult. Even if a tall tube were used to contain the rising column, the membrane would rupture once the pressure differential reaches a few bar.

The (equilibrium) osmotic effect can be understood as a balance between the tendency of water molecules to flow from a region of high concentration to one of lower concentration (the solution) and the effect of increasing pressure that tends to drive solvent molecules out of the solution phase.

REFERENCES

1. *International Critical Tables*, Vol. 3, 1st ed., McGraw-Hill: New York (1928).
2. R. C. Weast, Ed., *CRC Handbook of Chemistry and Physics*, 66th ed., CRC Press: Boca Raton, Florida (1985).
3. G. W. Castellan, *Physical Chemistry*, 3d ed., Addison-Wesley Publishing Co.: Reading, Massachusetts (1983).

9.21

Getting Colder:
Freezing-Point Depression

When salt is added to a mixture of ice and water, the temperature of the mixture decreases. The temperature change is shown with a large-display thermometer (Procedure A) or with an air thermometer (Procedure B). After salt is sprinkled onto a string lying on an ice cube, the string freezes to the cube and is used to lift it (Procedure C).

MATERIALS FOR PROCEDURE A

250 mL distilled water

250 g ice

50 g sodium chloride, NaCl

600-mL beaker

large-display thermometer (e.g., digital thermometer)

stirring rod

MATERIALS FOR PROCEDURE B

15 g methanol, CH_3OH

1 liter distilled water

ca. 5 mg methylene blue

500 g ice

50 g sodium chloride, NaCl

freezing-point depression apparatus (See Procedure B for materials and assembly instructions.)

mercury-in-glass thermometer, $-10°C$ to $+110°C$

marker

stirring rod

meter stick

MATERIALS FOR PROCEDURE C

ice cube

drinking glass filled with tap water

string, ca. 50 cm long

shaker of table salt

PROCEDURE A

Preparation and Presentation

Pour about 250 mL of distilled water into the 600-mL beaker. Insert the probe of the large display thermometer in the water. Note the temperature. Add 250 g of ice to the beaker. Note how the temperature falls to the freezing point of water. Add 50 g of sodium chloride to the ice water and stir the mixture. Note that the temperature falls to below the freezing point of water.

PROCEDURE B

Preparation

Assemble a freezing-point depression apparatus as illustrated in Figure 1. The apparatus uses an air thermometer, which can be constructed from glass tubing, as shown in Figure 1. (A similar, but less durable, version can also be assembled as shown in Figure 2.) Flare one end of a 2-m long glass tube with an outside diameter of 8 mm to facilitate pouring liquid into the tube. The other end of the 2-m tube is sealed in a reservoir made from glass tubing with an outside diameter of about 4 cm. This reservoir should be about 15 cm deep, and the 2-m tube should extend about 13 cm into the

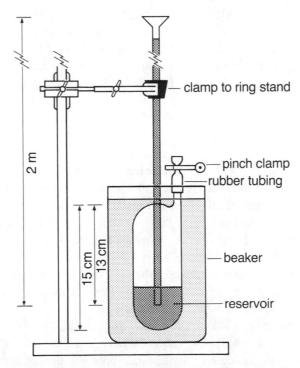

Figure 1. Freezing-point depression apparatus using an air thermometer.

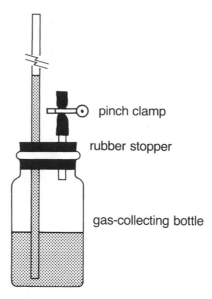

Figure 2. Alternate air thermometer.

reservoir. The reservoir should also have near its top a short side arm made from glass tubing with an outside diameter of 8 mm. (The air thermometer shown in Figure 2 uses a gas-collection bottle as a reservoir. The 2-m tube is inserted through one of the holes of a 2-holed rubber stopper, and the side arm is inserted through the other hole.) A 10-cm piece of rubber tubing is attached to the side arm of the thermometer, and the tubing is sealed with a pinch clamp. Set the reservoir inside a 1-liter beaker and clamp the thermometer to a ring stand. Using adhesive tape, fasten a 10-cm-wide strip of white poster board to the stand immediately behind the thermometer.

Mix 15 g of methanol with 85 g of distilled water. (The methanol serves as an antifreeze.) To increase the visibility of the liquid, dissolve several milligrams of methylene blue in the solution. Open the pinch clamp, then pour the methanol-water solution through the flared opening in the 2-m glass tube until the level of the liquid is about 2 cm above the lower end of the tube. Fill the beaker with water and allow it to reach ambient temperature.

Presentation

Seal the rubber tubing with the pinch clamp, trapping air at ambient temperature and pressure in the reservoir. Pour the methanol-water solution into the flared top of the apparatus to bring the liquid level to about 1 m above that in the reservoir. Measure the temperature of the water with a mercury thermometer. Mark the height of the liquid on the white background, labelling it with the measured temperature.

Add ice to the beaker to lower the temperature of its contents to 0°C. The level of the methanol-water in the tube will drop. Measure the temperature of the liquid in the beaker with the mercury thermometer. Mark the level of the liquid on the white background and label this level with the new temperature.

Add 50 g of sodium chloride to the beaker and stir the mixture. The level of methanol-water in the apparatus will have fallen farther. Mark the level.

With the meter stick, measure the distance between the top two marks. This distance corresponds to the difference in the first two recorded temperatures. Measure the distance between the middle and lower marks. Convert this distance into a temperature difference by multiplying it by the ratio of the first temperature difference to the distance between the top two marks. Subtract this temperature difference from the middle temperature to obtain an extrapolated value of the temperature of the ice-salt-water mixture. Measure the temperature with the mercury thermometer and compare this measured temperature with the extrapolated value.

PROCEDURE C [1]

Preparation and Presentation

Drop the ice cube into the glass of water. Dip one end of the string into the water to dampen it. Lay the damp end of the string on the ice cube and drape the rest of the string over the rim of the glass and onto the table. Sprinkle table salt on the portion of the string on top of the ice cube. After about 1 minute, gently pull up on the string. The ice cube will be frozen onto the string and can be lifted out of the glass.

HAZARDS

Methanol is poisonous when ingested or inhaled, as well as being flammable.

DISPOSAL

The salt water should be flushed down the drain with water.

The liquid in the air thermometer can be stored in the reservoir, or it can be flushed down the drain with water.

DISCUSSION

This demonstration shows that the temperature of a mixture of ice and water decreases when salt is added to it. In Procedure A, a large-display digital thermometer is used to show the effect; Procedure B uses an empirical thermometer with gas as the working substance to reveal the temperature change. In Procedure C, salt causes a damp string to freeze onto an ice cube.

The decrease in the temperature of ice water when salt is added is a result of the depression of the freezing point of the water caused by the salt dissolved in it. For an ideal solution, the magnitude of the depression, ΔT_f, is related to the concentration of the solution by

$$\Delta T_f = i \, K_f m$$

where m is the molality of the solution,

K_f is the freezing-point depression constant of the solvent,

and i is the mole number of the nonvolatile solute.

The mole number is the effective number of moles of solute particles per mole of dissolved species. For sodium chloride, $i = 2$ (2 moles of ions are produced for each mole of

NaCl that dissolves). For water, the freezing-point depression constant, K_f, is $1.86°C/m$. In Procedure A, the concentration of the NaCl in the water, assuming none of the ice melts (which is not quite correct), is 3.4m. Therefore, if the solution in the demonstration is ideal, the freezing point of the water falls by

$$\Delta T_f = (2)(1.86°C/m)(3.4m) = 12°C$$

A temperature of $-12°C$ is not reached in this demonstration because of significant deviations from ideality. The table indicates the freezing-point depression for aqueous solutions of various concentrations. In order to preserve the equality of the freezing-point depression expression, the mole number is often interpreted as an effective mole number. The data in the first two columns of the table can be used in $\Delta T_f = i K_f m$ to calculate the effective mole number as a function of solution concentration. As the table shows, the effective mole number of NaCl in aqueous solution decreases as the concentration of the solution increases, until the concentration becomes very high. One can perform the same calculation with the data for methanol, but because methanol is a nonelectrolyte, the calculated values of i do not lend themselves to a simple physical interpretation.

Freezing-Point Depression of Aqueous Solutions [2]

Solute	Molality (m)	Magnitude of depression (ΔT_f)	Effective number of moles (i)
Sodium chloride	0.00100	0.003676	1.977
	0.0100	0.03606	1.939
	0.100	0.3470	1.866
	1.00	3.387	1.821
	2.00	6.927	1.863
Methanol	0.157	0.278	0.952
	1.000	1.814	0.975
	2.000	3.724	1.001
	5.150	10.53	1.028

The melting of ice is an endothermic process; that is, ice absorbs heat as it melts. When salt is added to ice, it lowers the melting point of ice. However, in order for the ice to melt it must absorb heat from the surroundings. When salt is sprinkled on the damp string and the ice cube, some of the ice melts. It absorbs heat from the surroundings, which include the string. As heat is removed from the damp string, the water in the string freezes, causing it to adhere to the ice cube.

Street ice can be melted by spreading salt on the surface. As the solution concentration of sodium chloride increases, the freezing point decreases until the limit of solubility of sodium chloride in water is reached. This means that ice cannot be melted by use of salt at temperatures below $-21°C$ ($-6°F$). Salting the roads when the temperature is near freezing ($0°C$, $32°F$) and roads tend to be slick has the advantage of keeping water in liquid form so it can drain away. The disadvantage is that the runoff kills trees (maples are particularly susceptible) and pollutes lakes. An alternative is to use sand to improve traction, but the solids ultimately clog storm sewers. Salting of streets also contributes to accelerated rusting of iron automobile parts and deterioration

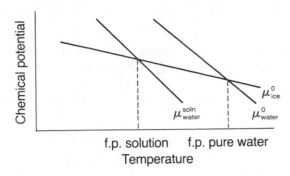

Figure 3. Change in chemical potential of ice, water, and salt-water with temperature.

of concrete. The current trend in some areas is to be sparing in the use of salt and sand at the expense of some increase in automobile accidents.

To keep the radiator fluid in the engine of an automobile from freezing, ethylene glycol can be added. A 50-50 (v/v) mixture freezes at about −22°F. However, ethylene glycol is not really a nonvolatile solute, and the above equations do not apply.

Freezing-point depression is one of the colligative properties, and the relevant theory is presented in the introduction to this chapter. A qualitative explanation of the thermodynamics of freezing-point lowering is easy to grasp. At the freezing point of the pure solvent, the chemical potentials of water and ice are equal. The chemical potentials of water and ice both decrease with increasing temperature, but that of ice decreases more slowly (see Figure 3). Adding solute to the solution decreases the chemical potential of the liquid water and leaves that of ice unchanged. This requires that the chemical potential–temperature curves intersect at a lower equilibrium temperature.

The empirical thermometer used in Procedure B employs a liquid to show the change in the volume of the gas with changes in temperature. In order to prevent the liquid in the thermometer from freezing in the water-ice-salt mixture, the liquid must have a lower freezing point than the temperature of this mixture. Such a liquid is the mixture of water and methanol. The solute methanol depresses the freezing point of the water inside the thermometer, just as the salt depresses the freezing point of the water outside the thermometer.

Calibration of the gas thermometer on the Celsius scale can be done theoretically from a model of the apparatus. The required information includes the volume of the gas within the thermometer at known temperature and pressure, the inner diameter of the reservoir, the inner and outer diameters of the tube, the composition and density of the reservoir fluid, and the partial pressures of methanol and water over the reservoir as a function of temperature. This last information can be generated from the vapor pressures of methanol and water taken from the literature [2], treating the solution as ideal. The analysis involves the solution of five simultaneous equations in the unknowns of moles in the vapor phase, pressure of the vapor phase, volume of the vapor phase, height of the liquid in the reservoir, and height of the liquid in the tube. The equations can be found in a physical chemistry text [3]. Rather than using this theoretical analysis to calculate the height of the column of liquid in the tube from accurate measurements of the apparatus's dimensions, the analysis can be used to show that the height of the column is proportional to the temperature, and therefore, the thermometer can be calibrated empirically.

REFERENCES

1. D. Herbert, *Mr. Wizard's Supermarket Science,* Random House: New York (1980).
2. R. C. Weast, Ed., *CRC Handbook of Chemistry and Physics,* 66th ed., CRC Press: Boca Raton, Florida (1985).
3. G. W. Castellan, *Physical Chemistry,* 3d ed., Addison-Wesley Publishing Co.: Reading, Massachusetts (1983).

9.22

Getting Hotter: Boiling-Point Elevation by Nonvolatile Solutes

The boiling point of a solution is higher than the boiling point of the pure solvent (Procedure A). A quantitative relationship between the boiling point and the concentration of the solution can be presented (Procedure B).

MATERIALS FOR PROCEDURE A

300 mL distilled water

100 g calcium chloride dihydrate, $CaCl_2 \cdot 2H_2O$

600-mL beaker

stirring hot plate, with stir bar

large-display thermometer (e.g., digital thermometer)

MATERIALS FOR PROCEDURE B

60 mL cyclohexane

60 mL 1.0 molal naphthalene in cyclohexane (To prepare about 100 mL of solution, dissolve 10 g of naphthalene in 100 mL of cyclohexane.)

refluxing apparatus (See Procedure B for materials and assembly instructions.)

large-display thermometer with 0.1°C resolution (e.g., digital thermometer [such as Omega 5800: −30°C to +100°C, and with reduced accuracy, over 100°C])

PROCEDURE A

Preparation and Presentation

Pour about 300 mL of distilled water into the 600-mL beaker and put the stir bar in the beaker. Set the beaker on the stirring hot plate and insert the probe of the large-display thermometer in the water. Turn on the stirrer and heater and note that the temperature of the water rises until it starts to boil, at which point the temperature stops increasing. While the water is boiling, add 100 g of $CaCl_2 \cdot 2H_2O$. The water will stop boiling and the temperature will fall. Gradually the temperature of the solution will increase until the solution starts to boil. The temperature of the boiling solution will be 2−3° higher than that of boiling pure water.

PROCEDURE B

Preparation

Assemble the refluxing apparatus as shown in the figure. The boiling flask is a three-necked, 100-mL, round-bottomed flask having $^{14}\!/_{20}$ standard-taper joints. The joints must be used with Teflon sleeves rather than grease to prevent contamination of the solutions. The heating element is either a Nichrome-wire coil mounted in one of the side necks of the flask or a heating mantle on the bottom of the flask. Either heat source requires a variable power supply to control the intensity of the heating. The reflux condenser is water cooled and seated in the center neck of the flask. (The reflux condenser may or may not be offset to the side as shown in the figure; use whatever type is available.) The flask's remaining side neck contains a temperature probe, either a thermistor or thermocouple, inserted through a temperature adapter. The temperature probe is attached to the large temperature display.

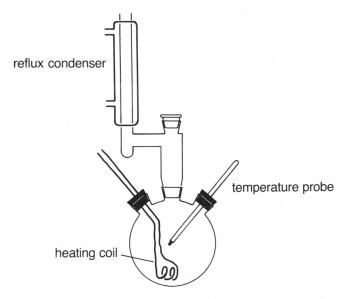

Refluxing apparatus.

Presentation

Remove the condenser from the flask and pour 60 mL of cyclohexane into the flask. Seat the condenser in the neck of the flask. Adjust the temperature probe so that its end is covered by the liquid in the flask. With the cooling water running through the condenser, turn on the heater. Once a steady reflux is obtained, the boiling point of the solvent at ambient pressure is displayed on the thermometer. Record this boiling point. Turn off the heater and allow the apparatus to cool.

Once the flask has cooled, remove it from the apparatus and empty it. Pour 60 mL of 1.0 molal naphthalene in cyclohexane into the flask. Seat the condenser in the neck of the flask. Adjust the temperature probe so that its end is covered by the liquid in the flask. With the cooling water running through the condenser, turn on the heater. Once a steady reflux is obtained, the boiling point of the solution at ambient pressure is displayed on the thermometer. Record this boiling point.

HAZARDS

Cyclohexane is flammable and should be kept away from open flames.

Naphthalene is toxic when ingested or absorbed through the skin during prolonged direct contact.

DISPOSAL

The aqueous solution should be flushed down the drain with water. The cyclohexane solution and cyclohexane should be discarded in a receptacle for combustible organic wastes.

DISCUSSION

This demonstration shows a rise in the boiling point of a liquid with the addition of a nonvolatile solute. In Procedure A, calcium chloride is dissolved in boiling water, and the boiling point of the water increases, providing qualitative agreement with the predictions of an idealized model. In Procedure B, a solution of naphthalene in cyclohexane has a higher boiling point than the pure solvent. Because the concentration of the solution is known, the quantitative relationship between the boiling-point elevation and the concentration predicted by the ideal model can be tested.

The increase in the temperature of a boiling liquid when a soluble substance is added to it is an indication of the elevation of the boiling point of the liquid caused by the solute. For an ideal solution and a nonvolatile solute, the magnitude of the elevation, ΔT_b (or θ_b), is related to the concentration of the solution by

$$\theta_b = \Delta T_b = iK_b m,$$

where m is the molality of the solute,

K_b is the boiling-point elevation constant of the solvent,

and   i is the mole number of the solute.

The mole number is the effective number of moles of solute particles per mole of dissolved species. For calcium chloride, $i = 3$ (3 moles of ions are produced for each mole of $CaCl_2$ that dissolves) [1]. For water, the value of the boiling-point elevation constant is $0.512°C/m$, and for cyclohexane it is $2.79°C/m$ [2]. In Procedure A, the concentration of the $CaCl_2$ in the water is 2.0 m. Therefore, if the solution in the demonstration is ideal, the boiling point of the water rises by

$$\Delta T_b = (3)(0.512°C/m)(2.0 \text{ m}) = 3.1°C$$

A temperature of 103°C is not reached in Procedure A because aqueous solutions of electrolytes deviate from ideality at high concentrations. The deviations from ideality exceed experimental error if the solution molality exceeds 1×10^{-4}. In solutions above this concentration, the effective mole numbers of electrolytes are significantly smaller than predicted from the number of ions per formula unit. In Procedure B, the concentration of the naphthalene in cyclohexane is 1.0 molal. Therefore, if the solution is ideal, its boiling-point elevation will be

$$\Delta T_b = (1)(2.79°C/m)(1.0 \text{ m}) = 2.79°C$$

The observed boiling-point elevation is quite close to this value, indicating that the solution is nearly ideal.

The value of the boiling-point elevation constant depends only on physical properties of the solvent, and is related to these properties by the equation

$$K_b = \frac{MRT_0^2}{\Delta \bar{H}_{vap}}$$

where M is the solvent molar mass,

T$_0$ its boiling point,

$\Delta \bar{H}_{vap}$ its molar heat of vaporization,

and   R the gas constant.

The derivation of this equation from thermodynamic relationships is sketched in the introduction to this chapter and can be found in physical chemistry texts, for example, reference 1.

REFERENCES

1. G. W. Castellan, *Physical Chemistry,* 3d ed., Addison-Wesley Publishing Co.: Reading, Massachusetts (1983).
2. R. C. Weast, Ed., *CRC Handbook of Chemistry and Physics,* 67th ed., CRC Press: Boca Raton, Florida (1987).

9.23

At the Water's Edge: Surface Spreading and Surface Tension

When a drop of mineral oil is placed on the surface of water, it forms a bead. When a drop of olive oil is placed on the water's surface, it spreads out (Procedure A). A razor blade rests on the surface of water until several drops of liquid detergent are added (Procedure B).

MATERIALS FOR PROCEDURE A

overhead projector

ca. 25 mL distilled water

pinch of lycopodium powder, or other finely divided powder (e.g., finely ground paprika)

1 drop mineral oil

1 drop olive oil (Do not substitute another vegetable oil for olive oil.)

petri dish

MATERIALS FOR PROCEDURE B

overhead projector (optional)

ca. 600 mL tap water

several drops dishwashing detergent

600-mL beaker

double-edged razor blade, or large sewing needle

PROCEDURE A

Preparation and Presentation

Set a petri dish on the stage of the overhead projector. Pour enough distilled water into the dish to cover the bottom. Sprinkle lycopodium powder onto the water. The powder spreads out on the surface. Place a drop of mineral oil on the water. The oil beads up and forms a lens, which shows as a bright spot in the projected image. Place a drop of olive oil on the water. It spreads out, driving the powder to the edge of the dish.　**301**

PROCEDURE B [1, 2]

Preparation and Presentation

Fill the 600-mL beaker nearly full with tap water. Set the beaker on the overhead projector. Carefully pick up the double-edged razor blade by its dull edges. Hold the razor blade horizontally about 5 mm above the surface of the water and drop it onto the water. The razor blade will rest on the surface of the water. Add several drops of dish-washing detergent to the water. After several seconds, the razor blade will sink.

DISCUSSION

The behavior of liquids and a solid on the surface of water is presented in this demonstration. In Procedure A, two different oils are placed on the surface of water; one of them forms beads and the other spreads into a thin film. In Procedure B, a solid denser than water is placed on the surface of water, and it does not sink. After detergent is added to the water, the solid sinks.

The behavior of oils on the surface of water has been studied for some time. Benjamin Franklin studied the spreading of oil [3, 4]. The difference in the behaviors of mineral oil and olive oil can be attributed to differences in their chemical structures. Mineral oil is a homogeneous mixture of hydrocarbons of moderate viscosity. The molecules of mineral oil are essentially nonpolar and, therefore, not attracted to water molecules. Because mineral oil is hydrophobic, small amounts of it tend to minimize the surface contact by forming beads on the surface of water.

Olive oil is a mixture of triglycerides of fatty acids. The structure of a typical molecule is represented by the formula below.

$$
\begin{array}{c}
\qquad\qquad\quad O \\
\qquad\qquad\quad \parallel \\
CH_2-O-C-R_a \\
\mid \qquad\qquad O \\
\qquad\qquad\quad \parallel \\
CH\ -O-C-R_b \\
\mid \qquad\qquad O \\
\qquad\qquad\quad \parallel \\
CH_2-O-C-R_c
\end{array}
$$

As this formula shows, there are polar portions in the molecule, specifically the ester groups. These polar portions can interact attractively with water molecules. Drops of olive oil spread into thin layers on water in order to maximize these interactions.

In the structural formula of a triglyceride molecule, R—(C=O)— represents a fragment of a fatty acid. The molecules of olive oil contain portions of several fatty acids. The four major constituent acids and typical proportions are listed in Table 1.

Table 1. Composition of Olive Oil [5]

Acid	Formula	Wt. %
Oleic	$CH_3(CH_2)_7CH{=}CH(CH_2)_7COOH$	84.4
Palmitic	$CH_3(CH_2)_{14}COOH$	6.9
Linoleic	$CH_3(CH_2)_4CH{=}CHCH_2CH{=}CH(CH_2)_7COOH$	4.6
Stearic	$CH_3(CH_2)_{16}COOH$	2.3

Table 2. Interfacial Tensions and Spreading Coefficients for Several Liquids on Water at 20°C (in J/m^2) [6–8]

Property	Liquids			
	n-Hexane	n-Octane	n-Tetradecane	Benzene
γ_B	1.854×10^{-2}	2.169×10^{-2}	2.654×10^{-2}	2.89×10^{-2}
γ_A	7.275×10^{-2}	7.275×10^{-2}	7.275×10^{-2}	7.275×10^{-2}
$\gamma_{B/A}$	5.080×10^{-2}	5.168×10^{-2}	5.332×10^{-2}	3.50×10^{-2}
$S_{B/A}$	3.41×10^{-3}	-6.2×10^{-4}	-7.11×10^{-3}	8.85×10^{-3}
$\gamma_{B(A)}$	—	—	—	2.88×10^{-2}
$\gamma_{A(B)}$	—	—	—	6.22×10^{-2}
$\gamma_{B(A)/A(B)}$	—	—	—	3.50×10^{-2}
$S_{B(A)/A(B)}$	—	—	—	-1.6×10^{-3}

Whether a drop of liquid (such as an oil) will spread out or bead up on a surface (such as water's) is determined by the relative tensions of the three interfaces involved. These tensions are the water-air interfacial tension, γ_A, the oil-air interfacial tension, γ_B, and the oil-water interfacial tension, $\gamma_{B/A}$. These interfacial tensions can be combined into a spreading coefficient for B on A (oil on water):

$$S_{B/A} = \gamma_A - \gamma_{B/A} - \gamma_B$$

When this spreading coefficient is positive, the oil will spread on water, and when it is negative, the oil will bead up. Interfacial tensions and spreading coefficients for several water-immiscible liquids are given in Table 2. The spreading co-efficients show that a drop of n-hexane will spread on the surface of water, whereas n-octane and n-tetradecane will bead up. The case of benzene is an interesting one, because it is slightly soluble in water, and water is slightly soluble in benzene. When a drop of benzene is placed on the surface of water, it will spread out. However, soon saturated solutions of water in benzene and benzene in water will form. The saturated solutions have different interfacial tensions from those of the pure liquids. The interfacial tension between benzene-saturated water and air, $\gamma_{A(B)}$, falls to 6.22×10^{-2} J/m^2, and that between the water-saturated benzene and air, $\gamma_{B(A)}$, falls only slightly to 2.88×10^{-2} J/m^2. Then, the spreading coefficient of the saturated solutions, $S_{B(A)/A(B)}$ is negative, the water-saturated benzene pulls back into a bead.

If the conditions for spreading are met, the surfactant spreads until a monolayer film is formed. This explains why small amounts of, say, olive oil can cover a very large area—a fact considered incredible in Franklin's day before the molecular nature of liquids was understood. Practical applications of surface spreading include the use of hexadecanol ($C_{16}H_{33}OH$) to retard evaporation in reservoirs. The water cannot diffuse through the oil layer and the vapor pressure of oil is very low, so the vapor barrier remains in place. The addition of a small amount of vegetable oil to cooking pasta helps to prevent its boiling over. Here a reasonably thick film forms and prevents the frothing that leads to boiling over. (Furthermore, when the cooked pasta is lifted out of the water, the oil coats the pasta, preventing it from sticking together.)

In Procedure B, a razor blade or sewing needle is laid on the surface of water. Because both of these objects are denser than water, it is surprising that they do not sink. The reason they do not sink involves intermolecular forces in water. The molecules are strongly attracted to each other, and for an object to sink in water, it must separate water molecules from each other. Although the amount of energy required to do this is quite small on the macroscopic scale of razor blades and needles, by gently placing the razor blade on the water's surface, it is possible to avoid providing the required energy. When a detergent is added to water, its molecules come between the water molecules, reducing the intermolecular forces in the water. As a result, the amount of energy required to separate the molecules even farther is decreased, and the gravitational force on the razor blade is sufficient to cause it to sink.

A series of demonstrations dealing with interfacial phenomena has been developed by J. P. Wightman, Department of Chemistry, Virginia Tech, Blacksburg, Virginia 24061. These demonstrations are designed to enhance advanced undergraduate and graduate courses in surface science and technology. A classification and summary of some interesting interfacial phenomena has appeared in the literature [9].

REFERENCES

1. D. Herbert and H. Ruchlis, *Mr. Wizard's 400 Experiments in Science,* Book-Lab: North Bergen, New Jersey (1983).
2. L. R. Summerlin, C. L. Borgford, and J. B. Ealy, *Chemical Demonstrations: A Sourcebook for Teachers,* Vol. 2, American Chemical Society: Washington, D.C. (1987).
3. B. Jirgensons and M. E. Straumanis, *A Short Textbook of Colloid Chemistry,* John Wiley and Sons: New York (1954).
4. B. Franklin, *Phil. Trans. Roy. Soc.,* June 2d (1774); reprinted in part in M. J. Jaycock and G. D. Parfitt, *Chemistry of Interfaces,* John Wiley and Sons: New York (1981).
5. R. C. Weast, Ed., *CRC Handbook of Chemistry and Physics,* 66th ed., CRC Press: Boca Raton, Florida (1985).
6. G. W. Castellan, *Physical Chemistry,* 3d ed., Addison-Wesley Publishing Co.: Reading, Massachusetts (1983).
7. J. Timmermans, *Physico-Chemical Constants of Pure Organic Compounds,* Elsevier Publishing Co.: New York (1950).
8. A. W. Adamson, *Physical Chemistry of Surfaces,* 4th ed., John Wiley and Sons: New York (1982).
9. A. E. Anwander, R. P. J. S. Grant, and T. M. Letcher, *J. Chem. Educ.* 65:608 (1988).

9.24

Will a Tissue Hold Water?
Interfacial Tension

Water is poured into a facial tissue and the tissue breaks. Water is poured into a second tissue treated with water repellant, and this tissue holds the water [1].

MATERIALS

surfactant water repellant spray (e.g., 3M Scotchgard or Amway Dri-Fab)

100 mL water

2 facial tissues

2 250-mL beakers

2 rubber bands to fit around beakers

PROCEDURE

Preparation

Drape a facial tissue over the mouth of a 250-mL beaker and secure it with a rubber band. The tissue should sag into the beaker. Prepare another of these tissue-covered beakers.

Presentation

Pour 50 mL of water into one of the tissues. The tissue will break and dump the water into the beaker. Spray the second tissue with the water repellant and allow it to dry. Pour 50 mL of water into the treated tissue. The water will stay in the tissue for at least several minutes, provided the tissue is coated thoroughly.

HAZARDS

The spray may contain hydrocarbons and be flammable.

DISCUSSION

The spaces between the fibers in textiles act as capillaries into which a liquid that wets the surface of the fibers will penetrate. The phenomenon of wetting is closely related to the spreading of liquids as described in Demonstration 9.23 [*2, 3*]. The factors that determine whether a liquid wets a fiber are the fiber-liquid interfacial tension and the liquid-vapor interfacial tension. To avoid wetting, a large fiber-liquid and a small liquid-vapor interfacial tension are needed. The pure water liquid-vapor interfacial tension is large, which promotes wetting. Therefore, to prevent wetting of a textile, the fibers of the textile must have an even larger textile-water interfacial tension. Such a large textile-water interfacial tension can be produced by attaching a hydrophobic coating to the fibers of the textile. Aerosol water-repellant treatments do this by applying a silicone coating to the fibers of the textile. Silicones are polymers of alkylated silica. The structure of a simple silicone is

$$
\begin{array}{ccccc}
CH_3 & & CH_3 & & CH_3 \\
| & & | & & | \\
-Si & -O- & Si & -O- & Si- \\
| & & | & & | \\
CH_3 & & CH_3 & & CH_3 \\
\end{array}
$$

The outer surface of the silicone molecules is nonpolar and hydrophobic.

Cotton cloth contains a natural coating of oils and waxes on the fibers that makes it inherently water repellant. However, washing with detergents removes this coating, and the fabric loses its water repellancy. Spraying the fabric with a water repellant regenerates the hydrophobic surface of the fibers. The fabric is water repellant, not water proof; the capillaries remain open, and the textile can "breathe." Water will flow through the fabric if sufficient pressure is applied.

REFERENCES

1. R. H. Hanson, *J. Chem. Educ.* 53:577 (1976).
2. M. J. Jaycock and G. D. Parfitt, *Chemistry of Interfaces,* John Wiley and Sons: New York (1981).
3. A. W. Adamson, *Physical Chemistry of Surfaces,* 4th ed., John Wiley and Sons: New York (1982).

9.25

The Shape of Drops:
Surface and Gravitational Work

The shape of a drop of mercury is determined by the interplay of surface and gravitational work on the drop.

MATERIALS

0.5 mL mercury

petri dish, with a diameter of 10 cm and a depth of 2 cm

overhead projector

single-edged razor blade, or index card

PROCEDURE

Preparation and Presentation

Place the petri dish on the overhead projector. Pour about 0.5 mL of mercury into the dish. With the razor blade or index card, divide the mercury into as many drops as possible. These tiny drops are nearly spherical. Then push the drops together. The tiny drops will combine into one large flattened (sessile) drop. To see that the drop is sessile, observers must look directly at the drop in the dish, rather than at the projected image.

HAZARDS

Mercury is extremely toxic and should be handled with care to avoid prolonged or repeated exposure to the liquid or vapor. Continued exposure to the vapor can result in severe nervous disturbance, insomnia, and depression. Continued skin contact also can cause these effects, as well as dermatitis and kidney damage. Mercury should be handled only in well-ventilated areas. Mercury spills should be cleaned up immediately by using a capillary attached to a trap and an aspirator (see Figure 1). Small amounts of mercury in inaccessible places should be treated with zinc dust to form a nonvolatile amalgam.

DISPOSAL

Carefully return all of the mercury to a stoppered container.

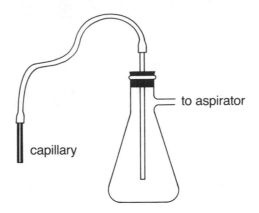

Figure 1. Device for recovering spilled mercury.

DISCUSSION

Tiny drops of mercury on glass are almost spherical, because the air-mercury and glass-mercury interfacial tensions are very large. The relationship between the size of the spheres and the interfacial tensions may be appreciated by considering the various contributions to the Gibbs function, G. The gravitational contribution to G for a collection of identical drops is Mgr, where M is the total mass of the mercury, g the acceleration of gravity, and r is the radius of the drops. However, a collection of n drops of radius r have a surface energy contribution of $n4\pi r^2\gamma$, where γ is the air-mercury interfacial tension. The variables n, M, and r are related by $M = 4n\pi r^3\rho/3$, where ρ is the density of mercury. Eliminating n and minimizing the sum of the surface and gravitational contributions to G with respect to r yields $r = (3\gamma/\rho g)^{1/2}$ as the optimal radius of the drops. Naturally, the result does not depend on M, because increasing M increases the number of drops, not their size. Using $\gamma = 0.48$ J/m^2, $\rho = 1.35 \times 10^4$ kg/m^3, and $g = 9.8$ m/sec^2 yields $r = 3.3 \times 10^{-3}$ m. This model suggests that a collection of drops with a radius of 3.3 mm should form, and that they should not aggregate on contact. This is not what happens, however, indicating a flaw in the preceding model. In fact, the drops are so large that they do not maintain spherical form, which means that the surface energy contribution to G is no longer $n4\pi r^2\gamma$. Instead, the mercury forms a single sessile drop with a contact angle, θ, of about 128 degrees [1, 2]. The contact angle is the angle between the surface of the glass and the tangent to the mercury drop where it meets the glass (see Figure 2). A large sessile drop has the property that its height does not change, but its diameter increases, as mercury is added. The limiting height of the large drop is given by

$$1 - \cos\theta = \rho g h^2/2\gamma$$

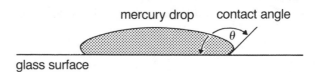

Figure 2. Contact angle of mercury drop on glass surface.

This equation comes from a force balance [1]. For mercury on glass at 20°C, this yields h = 3.56×10^{-3} m, which is the height of the sessile drop formed in this demonstration.

A more familiar example of sessile drop formation, controlled by the same principles, is the formation of beads of water on a newly waxed car when it rains.

REFERENCES

1. L. I. Osipow, *Surface Chemistry. Theory and Industrial Applications,* ACS Monograph Series, Reinhold Publishing Co.: New York (1962).
2. *International Critical Tables,* 1st ed., McGraw-Hill: New York (1928).

9.26

The Ice Bomb: Expansion of Water as It Freezes

Water inside a cast-iron globe is frozen in a dry ice bath. When the water freezes, the iron globe explodes.

MATERIALS

1 liter ice water

500 cm^3 dry-ice chunks

500 mL acetone, CH_3COCH_3

1-liter beaker

cast-iron ice bomb with plug†

safety shield (See Procedure section for description.)

unbreakable pan, metal or plastic, 20–25 cm in diameter, ca. 5 cm deep (e.g., a 3-pound coffee can cut to a depth of 5 cm)

stirring rod

PROCEDURE

Preparation

Fill the ice bomb with ice water by immersing it in the 1-liter beaker filled with ice water. Screw the plug loosely into the immersed opening of the bomb.

Prepare a safety shield with which to cover the bomb during the presentation. This safety shield can be a wooden box without a bottom, about 30 cm on a side and 15 cm deep. The cover for the box can be composed of several wooden slats simply laid in place across the top. The slats absorb some of the force of the explosion, yet are tossed about in a dramatic fashion. A less spectacular, but serviceable, alternative is a heavy towel.

Fill the unbreakable pan to a depth of about 5 cm with small chunks of dry ice. Make a slush by stirring 500 mL of acetone into the dry ice.

Presentation

Before presenting this demonstration, see the Hazards section. Remove the bomb from the ice bath and tighten its plug. Quickly place the bomb in the center of the pan of dry-ice slush, cover the bomb with the slatted box or a heavy towel, and stand

---

†Cast-iron ice bombs are available from American Scientific Products, 1210 Waukegan Road, McGaw Park, Illinois 60085.

back. In less than a minute, the bomb will rupture into several large pieces of shrapnel, throwing the wooden slats upward to a height of about 50 cm. Display the cast-iron fragments, which are about 8 mm thick.

HAZARDS

Do not use a glass vessel to hold the dry-ice slush. The rupturing bomb can throw glass fragments as far as 7 m [1].

Caution! If the bomb does not detonate within a minute, DO NOT immediately uncover it to see what has happened. Allow the entire assembly (dry ice bath, bomb, and cover) to warm to room temperature before touching it.

Acetone is flammable and should be kept away from open flames.

DISPOSAL

Discard the shards of the ice bomb in a solid-waste receptacle. The box can be used in repeat performances of this demonstration.

DISCUSSION

At 0°C the density of water is 1.00 g/cm³ and that of ice is 0.9168 g/cm³. Therefore, water expands as it freezes. For 50.0 g of water at 0°C, the volume increase is from 50.0 cm³ to 54.5 cm³. When the water is sealed inside the ice bomb, its volume is constrained. The bomb exerts pressure on the water to maintain its volume. An idea of the magnitude of the pressure required to keep the water from expanding can be gained from the compressibility of water. The isothermal compressibility of a substance, as defined in the introduction to this chapter, is

$$\beta = -\frac{1}{V}\left(\frac{\partial V}{\partial p}\right)_T \approx \frac{1}{V_1}\frac{(V_1 - V_2)}{(p_2 - p_1)}$$

In this equation, V_1 is the volume of a liquid under pressure, p_1, and V_2 is its volume at p_2 at a fixed temperature, T. For water at 0°C, $\beta = 4.7 \times 10^{-5}$ atm⁻¹ [2]. An estimate of the pressure change produced by the freezing water is

$$p_2 - p_1 = \frac{1}{V_1}\frac{(V_1 - V_2)}{\beta} = \frac{1}{54.5 \text{ cm}^3} \cdot \frac{(54.5 - 50.0) \text{ cm}^3}{4.7 \times 10^{-5} \text{ atm}^{-1}} = 1700 \text{ atm}$$

It is not surprising that the cast iron cannot withstand this immense pressure!

The expansion of water on freezing is responsible for the breakdown of rocks, for the cracking of pavement, and for the bursting of water pipes in cold climates. If water were to contract on freezing, as do most other liquids, this would not occur. However, then ice would sink and settle to the bottom of lakes and the oceans, and this would have a drastic effect on climate and fish.

The dynamics of the freezing of lakes is rather complex [3]. At the interface between ice and water, the temperature is nearly 0°C. The temperature drops linearly through the ice sheet and rapidly and nonlinearly through the thermal boundary layer (a thin layer of still air). The rate of freezing of ice is determined by the value of the transport coefficient for heat in each part of the system, by the enthalpy of freezing, and by the temperature gradient. Because freezing is an exothermic process, as liquid water freezes to ice, the rate of freezing slows. The ease with which chunks of ice of considerable thickness can be produced in a refrigerator might lead one to suspect that ice sheets on lakes grow to be very thick. However, this does not happen, because the thickness of thick sheets of ice is proportional to the square root of the freezing time, not to the freezing time directly.

Water is a very common but very unusual substance. Its properties are described in detail in the introduction to this chapter.

REFERENCES

1. L. B. Leopold and K. S. Davis, *Water,* Time-Life Books: New York (1966).
2. J. A. Dean, Ed., *Lange's Handbook of Chemistry,* 13th ed., McGraw-Hill: New York (1985).
3. C. F. Bohren, *Clouds in a Glass of Beer,* John Wiley and Sons: New York (1987).

9.27

Flow of Liquids Through Pipes:
Liquid Viscosities

The relative viscosities of a series of liquids are shown by the time of flow through a matched set of capillary tubes.

MATERIALS

video camera and monitor (optional)

5 viscometers, each filled with 10 mL of one of these liquids: *n*-hexane, water, *n*-butanol, olive oil, and glycerol (See Procedure section for materials and assembly instructions.)

ring stand and clamp

timer capable of measuring seconds

PROCEDURE

Preparation

Construct five viscometers like the one shown in the figure. For each one a 10-cm length of capillary tubing with an inside diameter of 0.7 mm is joined to about 50 cm of tubing with an inside diameter of 15 mm to form a loop. Place 10 mL of a different one

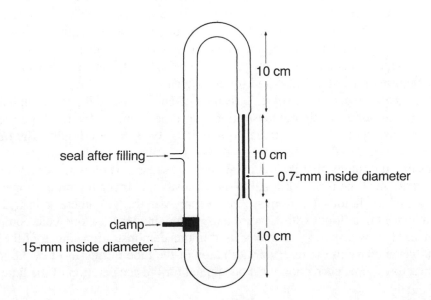

of the following liquids in each tube: *n*-hexane, water, *n*-butanol, olive oil, and glycerol. Seal the side arm on each viscometer.

Tilt each viscometer so all of the liquid collects at one end.

Presentation

If the demonstration is to be presented in a large room, display a viscometer (any one but that containing glycerol, because this liquid flows so much more slowly than the others) on the video monitor. Turn the viscometer so the liquid begins to run through the capillary tubing, and start the timer. Clamp the viscometer to the ring stand. Time how long it takes for all of the liquid to flow through the capillary.

Repeat the procedure for the other viscometers, ending with the one containing glycerol.

HAZARDS

Hexane and butanol are flammable. Care must be taken when sealing these tubes to prevent their ignition. This can be accomplished by freezing these liquids in liquid nitrogen and sealing the tubes while the liquid is frozen.

DISPOSAL

The sealed viscometers can be stored and reused in repeat performances of this demonstration. Or the side arms can be snapped open, the liquids drained into a waste solvent container, and the viscometer discarded in a solid-waste receptacle.

DISCUSSION

Viscosity is the property of fluids responsible for their resistance to flow. A high viscosity is characteristic of a liquid that flows slowly (e.g., molasses), and a relatively low viscosity characterizes liquids that flow freely (e.g., water). The measurement of viscosity is made by a number of methods, including passing the liquid through a narrow tube or dropping a ball through the liquid. The measurements are usually made by measuring the time required for a flow process to be completed and comparing this with the time required for a standard liquid (often water).

For a measurement that involves passing a given volume of liquid through a tube, such as the one used in this demonstration, the time required for the liquid to flow through the viscometer is proportional to the viscosity, η, of the liquid. The time required for the liquid to flow through the viscometer is also inversely proportional to the force pushing it through, that is, gravity acting on its mass. Therefore, for a given volume of liquid, the time is proportional to the viscosity and inversely proportional to the density, ρ, of the liquid. Thus, $\eta = A\rho t$, where A is the viscometer constant. This equation applies to a liquid whose entire mass is moving, that is, one held completely in the tube. However, in the viscometer used in this demonstration, much of the liquid is virtually stationary in the reservoir at the top of the tube during most of the time of flow. Therefore, some correction must be applied for the acceleration of the liquid as it flows from the reservoir into the tube. This correction is $-B\rho/t$, where B is the correction factor. Then, the expression for the viscosity becomes

$$\eta = A\rho t - B\rho/t$$

The correction factor, B, for the viscometer used in this demonstration has a value of approximately $V/8\pi L$, where V is the volume of the liquid that flows through the capillary, and L is the length of the capillary [1, 2]. Using water at 20°C in a viscometer for this demonstration, the flow time is 30.5 sec. Based on a literature value of 1.01×10^{-3} Pa·sec for the viscosity of water [3],

$$A = \frac{\eta}{\rho t} + \frac{B}{t^2} = 3.73 \times 10^{-8} \text{ m}^2 \text{ sec}^{-2}$$

for this viscometer. Expected flow times for the other liquids used in this demonstration can be calculated and are shown in the table. Experimentally measured flow times obtained in this demonstration are given for comparison. The olive oil will be in no hurry in this apparatus; and the flow time for glycerol is so long that it shows the unsuitability of the capillary-flow technique for determining viscosities above about 0.1 Pa·sec. For these the falling-ball method is preferable. In this method the liquid is placed in a tube and a ball is dropped into it. The viscosity is determined from the length of time required for the ball to sink through a calibrated distance on the tube.

Viscosities, Densities, and Flow Times of Liquids at 20°C

	Viscosity (Pa sec)	Density (kg/m³)	Calculated flow time (sec)	Measured flow time (sec)
n-Hexane	3.26×10^{-4}	6.6×10^2	19	20
Water	1.01×10^{-3}	1.00×10^3	30.5	30.5
n-Butanol	2.95×10^{-3}	8.1×10^2	99	90
Olive oil	8.4×10^{-2}	9.2×10^2	2,400	2,200
Glycerol	0.49	1.26×10^3	32,000	—

The expression above relating viscosity to flow time is only approximate, and a more precise one can be derived from a more accurate theoretical treatment of viscosity. The precise treatment requires that the flow of the liquid be well controlled and not turbulent. A well-controlled flow is characterized by a smooth velocity gradient from the liquid in the center of the capillary to the liquid at the outer edge. When such a smooth gradient exists, the liquid is said to be in laminar flow. The Reynolds number, R, is used to estimate when liquids are undergoing laminar flow and when they become turbulent. The Reynolds number, R, is calculated by $du\rho/\eta$, where d is the inside diameter of the tube, and u is the velocity of the liquid. When the Reynolds number is below 2000, the liquid is in the laminar flow region. Turbulence begins with Reynolds numbers in the range of 3000–10,000. The liquid with the highest Reynolds number in this demonstration is n-hexane, with a value of about 1000.

The viscosity of a liquid is an important design parameter in numerous practical applications. The pipe dimensions and size and number of pumping stations on the Alaskan pipeline are determined to a large extent by the viscosity of crude oil. The oil is heated to reduce the viscosity. The viscosity of blood affects the throughput in artificial heart–lung machines. When honey is packaged in a plastic dispenser bottle, its high viscosity requires an increase in pressure, by squeezing, to force the honey through the small nozzle.

REFERENCES

1. R. B. Bird, W. E. Stewart, and W. N. Lightfoot, *Transport Phenomena,* John Wiley and Sons: New York (1960).
2. F. Daniels, R. A. Alberty, J. W. Williams, C. D. Cornwell, P. Bender, and J. E. Harriman, *Experimental Physical Chemistry,* 7th ed., McGraw-Hill: New York (1970).
3. R. C. Weast, Ed., *CRC Handbook of Chemistry and Physics,* 66th ed., CRC Press: Boca Raton, Florida (1985).

9.28

Molecules in Slow Motion:
Diffusion in Liquids

A cylinder contains a colorless liquid and a blue solid on the bottom. Over a period of days, the liquid at the bottom of the cylinder turns blue and color slowly advances upward through the liquid (Procedure A). Another cylinder contains a dark brown liquid on the bottom and a colorless liquid above it, with a red-brown region where the two meet. Over a period of hours, the red-brown region gradually expands upward into the colorless liquid (Procedure B). A third cylinder shows a similar process, but the bottom liquid is violet (Procedure C).

MATERIALS FOR PROCEDURE A

10 g copper sulfate pentahydrate, $CuSO_4 \cdot 5H_2O$

200 mL distilled water

250-mL graduated cylinder, with glass, rubber, or cork stopper

MATERIALS FOR PROCEDURE B

1 mL bromine, Br_2

15 mL distilled water

gloves, plastic or rubber

25-mL graduated cylinder, with glass, rubber, or cork stopper

MATERIALS FOR PROCEDURE C

0.03 g iodine, I_2

3 mL dichloromethane, CH_2Cl_2

15 mL 0.1M potassium iodine, KI (To prepare 1 liter of solution, dissolve 17 g of KI in 600 mL of distilled water and dilute the resulting solution to 1.0 liter.)

gloves, plastic or rubber

25-mL graduated cylinder, with glass, rubber, or cork stopper

PROCEDURE A

Preparation

Arrange 10 g of large crystals of $CuSO_4 \cdot 5H_2O$ in a layer on the bottom of the 250-mL graduated cylinder. Slowly pour 200 mL of distilled water into the cylinder, taking care not to agitate the crystals. Stopper the cylinder.

Presentation

Display the cylinder where it will be undisturbed for several weeks or months. Over this time, the liquid at the bottom of the cylinder will become blue, and a blue front will gradually move up through the cylinder.

PROCEDURE B

Preparation

Wearing gloves, put 1 mL of bromine in the 25-mL graduated cylinder. Carefully pour 15 mL of distilled water onto the bromine in the cylinder. Stopper the cylinder.

Presentation

Display the cylinder where it will be undisturbed and free from rapid temperature fluctuations for several hours. The aqueous layer will turn red-brown at the interface between the two liquids, and the red-brown front will advance upward through the solution at a rate of about 0.5 cm per hour.

PROCEDURE C

Preparation

Wearing gloves, dissolve 0.03 g of iodine in 3 mL of dichloromethane. Put this iodine solution in the 25-mL graduated cylinder. Carefully pour 15 mL of 0.1M KI solution onto the iodine solution in the cylinder. Stopper the cylinder.

Presentation

Display the cylinder where it will be undisturbed and free from rapid temperature fluctuations for several hours. The aqueous layer will turn orange-brown at the interface between the two liquids, and the brown front will advance upward through the solution at a rate of about 0.5 cm per hour.

HAZARDS

Wear gloves while handling bromine and iodine containers. Bromine and iodine are very strong oxidizing agents, and skin contact with liquid bromine or solid iodine can result in severe burns. Because bromine and iodine vaporize readily at room temperature to yield toxic fumes, adequate ventilation must be provided.

Copper compounds are harmful if taken internally. Dust from copper compounds can irritate mucous membranes.

DISPOSAL

The bromine and iodine should be reduced to bromide or iodide ions before disposal. To do so, carefully, add 1 g of sodium thiosulfate to the mixture containing bromine or iodine and stir it until the dark color disappears. Separate any dichloromethane from the mixture by decanting the aqueous layer, leaving the dichloromethane layer in the cylinder. The dichloromethane should be discarded in a waste receptacle for chlorine-containing organic waste. The aqueous phase should be flushed down the drain with water. The copper sulfate solution should be flushed down the drain with water.

DISCUSSION

In Procedure A, blue, solid copper sulfate at the bottom of a cylinder of water slowly dissolves in the water, and the dissolved copper sulfate diffuses into the water, gradually increasing the intensity of the blue color of the solution. In Procedure B, a similar dissolving and diffusing process occurs, but the dissolving substance is liquid bromine, and the solution is red-brown rather than blue. Procedure C employs a violet solution of iodine in dichloromethane on the bottom of the cylinder and a colorless aqueous solution of potassium iodide above it. Here, the iodine reacts with iodide ions in the aqueous layer forming triiodide ions, which gradually diffuse into the aqueous layer.

Demonstrating diffusion in liquids is similar to watching grass grow. Whereas the diffusion coefficient of a typical species in air is about $10^{-5} m^2/sec$ [1], in water diffusion coefficients are around $10^{-9} m^2/sec$. Diffusion is 10,000 times slower in water than in air! This demonstration requires a long time for the effects of diffusion to become apparent. It is best to display the demonstration where it will be undisturbed and can be observed over a span of several hours. Alternatively, it can be set up several hours in advance and observed after a noticeable change has taken place. The effect can be displayed to a large audience via closed-circuit television. An overhead projector laid on its side can also be used to project an enlarged image onto a screen.

The copper sulfate system is most suitable for long-term display. Because diffusion time is proportional to the square of the distance travelled, doubling the dimension of the apparatus quadruples the time required to achieve the same degree of equilibration of the concentration at the larger scale. Thus, a large apparatus suited for corridor display will show concentration changes over a period of months.

The iodine system develops a typical sigmoidal concentration-gradient profile, which covers a range of about 1 cm after 2 hours. If the system is placed on a stable bench top, such as might be used for an analytical balance, no obvious convection effects, such as whorls of color, will be visible. The orange-brown color is due to the I_3^- ions, which, because the aqueous solution contains concentrated I^-, bind 99% of the iodine entering the aqueous layer. The equilibrium constant (molar basis) for $I_3^- \rightleftarrows I_2 + I^-$ is 0.0014 at 25°C [2, 3].

The reactions of bromine with water are not significant [4], and the red-brown color observed in Procedure B is due to $Br_2(aq)$. The concentration gradient in the bromine system develops at about the same rate as in the iodine system, and the two look quite similar.

The diffusion process that occurs in this demonstration can be treated theoretically in an approximate fashion by adopting a one-dimensional model [5–7]. In this model, the diffusing species moves along a line. Position along the line is represented by x, time by t, and the concentration of the diffusing species by c. Initially, at $t = 0$, $c = c_0$ when $x \leq 0$ and $c = 0$ when $x > 0$. The variation of concentration with time is given by Fick's second law:

$$\frac{\partial c}{\partial t} = D\left(\frac{\partial^2 c}{\partial t^2}\right)$$

where D is the diffusion constant.

The solution to this differential equation for the stated initial conditions is an expression relating the concentration of the diffusing species as a function of diffused distance, x, and time, t:

$$c(x,t) = \frac{c_0}{2} \, erfc\left(\frac{x}{2\sqrt{Dt}}\right)$$

(The function erfc(y) is a definite integral whose values are tabulated in standard mathematical tables.) More complicated and more accurate models of the diffusion process are possible, but the solution for c(x,t) is nearly the same when the value of x is small. The familiar equation for the mean squared displacement is obtained from the above solution.

$$\overline{x^2} = 2Dt$$

The following demonstrations, which appear in Volume 2 of this series, illustrate the much more rapid diffusion in gases:

Demonstration 5.14. Flow of Gases Through a Porous Cup. Samples of gas are allowed to flow through a porous cup. The relative rates of flow cause liquid to be forced from a flask of air or air to be drawn into the flask. The demonstration illustrates qualitatively that the rate of diffusion is a monotone decreasing function of the molecular weight.

Demonstration 5.15. Ratio of Diffusion Coefficients: The Ammonium Chloride Ring. Ammonia vapor is introduced at one end of an air-filled tube, and hydrogen chloride gas at the other. After about 20 minutes, a ring of solid ammonium chloride forms inside the tube, nearer the end at which the hydrogen chloride was introduced.

Demonstration 5.16. Molecular Collisions: The Diffusion of Bromine Vapor. Bromine vapor is allowed to fill two containers, one containing air and the other evacuated. The bromine fills the first container only gradually, but it fills the second very rapidly, illustrating the effect of molecular collisions on the rate of gas diffusion.

Demonstration 5.17. Graham's Law of Diffusion. A volume of gas contained in a buret tube flows into the atmosphere through a porous glass frit, while air flows in the opposite direction. From the initial volume of the gas and the final volume of air in the tube, the ratio of the flow rates of the two gases can be determined.

Demonstration 5.18. Graham's Law of Effusion. The time required for a fixed volume of a number of gases to effuse into a vacuum is measured. These times are proportional to the molecular masses of the gases.

REFERENCES

1. *International Critical Tables,* Vol. 5, 1st ed., McGraw-Hill: New York (1928).
2. G. Jones and B. B. Kaplan, *J. Am. Chem. Soc.* 50:1845 (1928).
3. F. Daniels, R. A. Alberty, J. W. Williams, C. D. Cornwell, P. Bender, and J. E. Harriman, *Experimental Physical Chemistry,* 7th ed., McGraw-Hill: New York (1970).
4. T. Moeller, *Inorganic Chemistry—an Advanced Textbook,* John Wiley and Sons: New York (1952).
5. J. Crank, *The Mathematics of Diffusion,* 2d ed., Clarendon Press: Oxford (1975).
6. R. B. Bird, W. E. Stewart, and E. N. Lightfoot, *Transport Phenomena,* John Wiley and Sons: New York (1960).
7. H. S. Carslaw and J. C. Jaeger, *Conduction of Heat in Solids,* 2d ed., Clarendon Press: Oxford (1959).

9.29

Facilitated Transport of Carbon Dioxide Through a Soap Film

A soap bubble grows and changes color when it is immersed in a box filled with carbon dioxide gas. When the bubble is removed, it shrinks. A soap bubble floats on carbon dioxide gas and grows as it floats. This is described in Demonstration 6.29 in Volume 2 of this series. The growth of the bubble results from diffusion of carbon dioxide gas into the interior of the bubble, where the concentration is low. This process occurs reasonably quickly, because the soap film is thin and because the solubility of carbon dioxide in water is great compared with the other major components of air, oxygen, and nitrogen.

9.30

Equilibration of Liquid Density
via Diffusion

A colorless liquid in a large U-tube has a different level in each arm of the tube.

MATERIALS

75 mL distilled water

95 mL methanol, CH_3OH

U-tube, with each arm ca. 1 m long and an inside diameter of 1 cm (The openings of the arms can be tapered to minimize evaporation.)

clamp and stand for U-tube

PROCEDURE

Preparation

Attach the U-tube to the stand with a clamp. Pour 75 mL of distilled water into one arm of the U-tube. Pour 95 mL of methanol into the other arm. The level of the liquid in the arm into which the methanol was poured will be higher than the level of the liquid in the other arm. Bubbles will appear at the junction of the water and methanol, but these will soon rise to the surface and escape.

Presentation

Display the U-tube where it will be undisturbed. The tube appears to contain a single liquid, because the interface between water and methanol is invisible.

HAZARDS

Methanol is poisonous when ingested or inhaled, as well as being flammable.

DISPOSAL

The waste liquids should be flushed down the drain with water.

DISCUSSION

This demonstration shows both the differing densities of methanol and of water and the slowness of their interdiffusion. Because the average density of the methanol column is less than that of the water column, its volume is larger so that the same mass is found on each side of the U-tube. If equal volumes of methanol and water were placed on either side of the U-tube, the levels in the tube would shift so that equal masses would be found on both sides.

As time passes, the column of methanol will be diluted with water, becoming denser. This causes the column to become shorter. On the other hand, the column of water becomes diluted with methanol, and its density decreases. Therefore, the column of water becomes longer. Ultimately, the heights of the columns in the two arms of the U-tube will become the same. The progress of this equalization of heights is very slow, requiring more than several weeks to occur. This shows the slow rate of diffusion in liquids. The mutual diffusion constant, D, of the water-methanol system is about 10^{-9} m^2/sec [1]. The value of D depends on the composition of that part of the system in which the diffusion is occurring, and therefore varies throughout the system in this demonstration. The dynamics of the production of a homogeneous solution is followed by observing the difference in height as a function of time. The height difference is caused by the equilibration of hydrostatic pressure in the U-shaped column of liquid, in which the density increases from the methanol-rich side to the water-rich side. Mixing of the column has the effect of eliminating the density gradient, and thereby causing the height difference to disappear. Mixing takes place by diffusion, a slow process in liquids.

If the apparatus is displayed with no explanation of its contents, the height difference of the liquid in the two arms seems paradoxical. The U-tube appears to contain a single homogeneous liquid. The interface between two liquids is usually visible because of refractive effects, but the interface between water and methanol is invisible because their indices of refraction are very similar, so the refractive effects are only slight. Because liquids generally flow quite freely in a tube of such a large diameter, viewers may be surprised that the heights of the columns in the two arms are not the same. (See Demonstration 9.27 on liquid viscosity.)

A concentrated solution of sucrose in water could be used in place of methanol in this demonstration. For this system, the mutual diffusion constant is about 10^{-10} m^2/sec, so the process is even slower [2]. However, the interface between these liquids is more easily seen.

A measure of the time scale of the diffusion process is afforded by the equation

$$\overline{x^2} = 2Dt$$

where x^2 is the root-mean-squared distance travelled by a diffusing species from time 0 to t [3, 4]. With $D = 1 \times 10^{-9}$ m^2/sec and a root-mean-squared distance of 0.64 m^2, the time is 3.2×10^8 sec, or 10 years. It is clear that the equilibration process for this system is a stick-in-the-mud. A full-blown numerical solution of the working equations (including variation of D with concentration) yielding column heights as a function of time is possible, but it is not needed to interpret the demonstration.

REFERENCES

1. *International Critical Tables*, 1st ed., McGraw-Hill: New York (1928).
2. A. C. English and M. Doyle, *J. Am. Chem. Soc.* 72:3261 (1950).
3. J. Crank, *The Mathematics of Diffusion*, 2d ed., Clarendon Press: Oxford (1975).
4. R. B. Bird, W. E. Stewart, and E. N. Lightfoot, *Transport Phenomena*, John Wiley and Sons: New York (1960).

9.31

Electrical Conductivity of Liquids

Several liquids are used with a conductivity cell in series with an incandescent lamp. With distilled water, the lamp does not light. With each of two solutions a dim glow is observed. When a mixture of the two solutions is placed in the cell, the lamp glows brightly.

MATERIALS

250 mL distilled water

250 mL 0.1M acetic acid, $HC_2H_3O_2$ (To prepare 1 liter of solution, pour 6 mL of glacial [17.5M] $HC_2H_3O_2$ in 600 mL of distilled water and dilute the resulting solution to 1.0 liter.)

250 mL 0.1M aqueous ammonia, NH_3 (To prepare 1 liter of solution, pour 7 mL of concentrated [15M] NH_3 into 600 mL of distilled water and dilute the resulting solution to 1.0 liter.)

5 600-mL beakers

110-volt conductivity tester consisting of two graphite electrodes ca. 15 cm long, 6 mm in diameter, separated by 2.5 cm, in series with a switch to line voltage and a 150-watt unfrosted incandescent light bulb †

wash bottle filled with distilled water

stirring rod

PROCEDURE

Preparation

Pour 250 mL of distilled water into one of the 600-mL beakers, 250 mL of 0.1M acetic acid into a second, and 250 mL of 0.1M aqueous ammonia into a third.

Presentation

Turn on the conductivity tester and immerse its electrodes in the beaker of distilled water. The lamp will not light.

Immerse the electrodes of the tester in the beaker of acetic acid. The lamp will glow faintly. Turn off the tester, hold the electrodes over an empty 600-mL beaker, and rinse the electrodes with distilled water from the wash bottle.

---

† A commercial model of this type of conductivity tester is available from Sargent-Welch Scientific Company, 7300 N. Linder Avenue, Skokie, Illinois 60077.

Turn on the conductivity tester and immerse its electrodes in the beaker of aqueous ammonia. The lamp will glow faintly. Turn off the tester, hold the electrodes over the rinse beaker, and rinse the electrodes with distilled water.

Pour 125 mL of the acetic acid into the remaining 600-mL beaker. Add 125 mL of the aqueous ammonia solution. Stir the mixture. Turn on the conductivity tester and immerse its electrodes in the mixture. The lamp will shine brightly.

HAZARDS

Do not touch the electrodes while power is on; this can result in severe electrical shock.

Concentrated aqueous ammonia can irritate the skin, and its vapors are harmful to the eyes and mucous membranes.

Glacial acetic acid can irritate the skin, and its vapors are irritating to the eyes and respiratory system.

DISPOSAL

The waste solutions should be flushed down the drain with water.

DISCUSSION

In this demonstration the electrical conductivities of several liquids are compared qualitatively with a light-bulb conductivity tester. The more brightly the light bulb glows when the tester electrodes are immersed in the liquid, the more electrically conductive the solution. The conductivity of pure water is compared with that of 0.1M solutions of acetic acid and aqueous ammonia and with that of a mixture of the two solutions. Each of the solutions is more conductive than pure water, and the mixture of the acetic acid with aqueous ammonia is more conductive than either of these solutions by itself.

Liquids conduct electricity by a number of mechanisms. Mercury conducts because of the relatively free motion of electrons (metallic conduction). Most other pure liquids conduct through the motion of ions, usually present as impurities (water contains a small definite concentration of hydronium and hydroxyl ions because of the dissociation equilibrium). Solutions of electrolytes conduct by the motion of solute ions.

The conductivity of a liquid can be determined easily by measuring the resistance to current flow of a square or cylindrical sample of liquid between electrodes that are connected to a voltage source operating at radio frequencies [1–3]. The specific conductivity, σ, of an idealized sample of thickness, L, and area, A, is obtained from the observed resistance via L/RA. The value of σ varies over a wide range for liquids. For mercury, it is 1.04×10^6 ohm^{-1}m^{-1} at 20°C [4]. For "nonconducting" liquids, the specific conductivities are 1×10^{-8} ohm^{-1}m^{-1} or less. Depending on the method of purification, water has a specific conductivity of about 2×10^{-8} ohm^{-1}m^{-1}. Specific conductivities on the order of 1×10^{-18} ohm^{-1}m^{-1} are common for ultrapure nonconducting liquids. Although the conductivity of water is low, electrocution by electrical appliances falling into bathtubs is far from rare.

When strong electrolytes are dissolved in water, the conductivity is nearly proportional to the concentration (except for a small effect due to ion-ion interactions). Weak electrolytes exhibit a strong concentration dependence of conductivity, because the extent of dissociation into ions depends on the electrolyte concentration.

In the procedures of this demonstration, a weak glow is observed in the light bulb in 0.1M acetic acid or aqueous ammonia; no light is observed when water is used as the conducting medium; normal light output is seen for the ammonium acetate solution.

Using equivalent ionic conductivities from the literature [4] and the equilibrium constant of 1.8×10^{-5} (molar basis, 25°C) for both the acetic acid and the aqueous ammonia systems to calculate the concentrations of ionic species, one finds that the contribution of the solution between the electrodes to the series resistance of the circuit is about 1000 ohms for the acetic acid solution, 1500 ohms for the ammonia solution, and 0.1 ohms for the 0.05M ammonium acetate solution made by mixing the other two solutions. The results are as expected. The inner resistance of the light bulb is about 60 ohms, which limits the current when the ammonium acetate is used between the electrodes. With the weak electrolytes the circuit resistance is almost two orders of magnitude larger (1000 ohms versus 60 ohms), and the current is down by a similar amount. Because the power, P, drawn by the circuit is proportional to the square of the current ($P = I^2R$), the power drops by four orders of magnitude, and the 150-watt bulb operates at only a tenth of a watt—it barely glows. In order to see a glow in a light bulb where water is the conducting liquid, the internal resistance of the bulb must be comparable to the resistance of the water between the electrodes (about 5×10^5). Neon glow bulbs, readily available from electronic supply houses, have appropriate specifications ($\frac{1}{25}$ watt, 3×10^5 ohms internal resistance, for example). Such a bulb (they are very small) replacing the 150-watt bulb shows light with water in the circuit.

REFERENCES

1. F. Daniels, R. A. Alberty, J. W. Williams, C. D. Cornwell, P. Bender, and J. E. Harriman, *Experimental Physical Chemistry,* 7th ed., McGraw-Hill: New York (1970).
2. N. E. Hill, W. E. Vaughan, A. H. Price, and M. Davies, *Dielectric Properties and Molecular Behaviour,* Van Nostrand–Reinhold: London (1969).
3. R. A. Alberty and F. Daniels, *Physical Chemistry,* 5th ed., John Wiley and Sons: New York (1980).
4. R. C. Weast, Ed., *CRC Handbook of Chemistry and Physics,* 63d ed., CRC Press: Boca Raton, Florida (1982).

9.32

Moving Liquids with Electricity: Dielectric Properties of Liquids

A stream of water is deflected in the electric field of a charged rubber rod and a charged glass rod.

MATERIALS

overhead projector

50 mL water

mirror, ca. 15 cm square

50-mL buret, with stopcock and stand

250-mL beaker

hard rubber rod

cat's fur

glass rod

silk cloth

hard rubber comb (optional)

PROCEDURE

Preparation

Place the overhead projector on its side and attach the mirror to its lens head to deflect its beam onto the screen. Mount the buret in the stand and position it in the field of the projector so the image of its tip is in focus on the screen. Fill the buret with water. Place the beaker below the buret to catch the water when the buret's stopcock is opened.

Presentation

Open the stopcock of the buret. Rub the rubber rod with the cat's fur and hold the rod about 3 cm from the stream of water. The stream deflects toward the rod. Repeat the presentation with a glass rod rubbed with a silk cloth. The deflection is again toward the rod. A rubber comb run through someone's hair can be used as well.

DISCUSSION

In this demonstration, a stream of water is deflected by an electrically charged rod. The rod is either glass or rubber. The glass rod is charged by rubbing it with a piece of silk, and the rubber rod is charged by rubbing it with cat's fur.

When a liquid composed of polar molecules is placed in an electric field (generated by a separation of charges, for example) the molecules tend to align, producing a macroscopic separation of charge that opposes the charges generating the electric field. Even liquids composed of molecules without permanent dipole moments undergo shifts of electron and nuclear positions, which produce a macroscopic charge separation, although the separation is smaller than that in polar molecules. These displacements typically occur in a time scale of picoseconds or less. A rough gauge of the magnitude of the charge separation is the dielectric constant of the liquid [1]. (The dielectric constant of a liquid is the ratio of the capacitance of a condenser filled with the liquid to the capacitance of the condenser when it is empty.) Although the water molecule has only a moderate dipole moment, the hydrogen-bonded structure of liquid water allows a large charge separation and a concomitantly large dielectric constant.

In a uniform electric field, the force tending to move the positive charges which develop is exactly balanced by oppositely directed force, which attempts to move the negative charges. Therefore the liquid remains in position. In a nonuniform electric field these forces are unequal, and the liquid moves toward the region of higher field strength. This effect, dielectrophoresis, has been used in separation methods [2]. Dielectrophoresis is related to the polarity of the molecules which compose the liquid via the dielectric constant. Onsager's equation relates the dielectric constant, the molecular dipole moment, and the polarizability, which is determined by the ease of distorting the nuclear- and electron-charge distributions [1]. If the dipole moment, μ, is 0, the relative dielectric constant is about 2. For substances with typical values for permanent dipole moments, the relative dielectric constant is larger by about a factor of 3 or more. The relative dielectric constant of water is unusually large (78 at 25°C), a fact that has stimulated numerous attempts to compute the relative dielectric constant of water via statistical mechanics (with mixed success). Thus, dielectrophoresis occurs in liquids of both polar and nonpolar molecules, although the effect is enhanced in polar liquids. A fairly prevalent misconception is that a stream of oil cannot be deflected in the manner of this demonstration because the molecules of oil are nonpolar. This claim focusses on the wrong parameter, the molecular dipole moment instead of the dielectric constant of the liquid. In fact, an easily noticeable deflection of a stream of mineral oil can be produced using a narrow glass rod (8 mm in diameter) rubbed with silk. By using a narrow glass rod, the electric field can be concentrated and the nonuniformity enhanced. The dielectrophoretic effect is fairly feeble relative to the force on bodies carrying a net charge. However, measurements with electrometers show that sufficient charge densities are found on the rods to produce deflections in the stream of liquid. The effect is larger when the rod is placed closer to the liquid stream, because the electric field gradient across the sample increases. Greater charging of the rod and smaller rod diameters enhance the effect. Thus, a highly charged rod with a narrow diameter held close to the liquid stream can produce a deflection in a liquid with a small dielectric constant.

When cat fur and the rubber rod are placed in contact, a negative charge develops

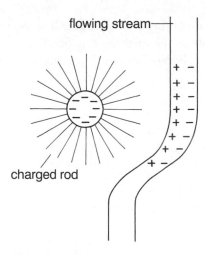

Deflection of a stream of liquid in a nonuniform electric field.

on the rubber rod. Rubbing enhances the effect, which is called triboelectricity. The charged rod is a source of a nonuniform electric field. The field intensity is proportional to $1/r$, where r is the distance from the axis of the rod. The distance between the lines shown in the figure is a representation of this dependence. The stream of water is polarized in the field as shown in the figure. The attractive electric force (the product of the charge multiplied by the electric field) exceeds the repulsive force, and the column is deflected toward the rod.

With the glass rod rubbed with silk, a positive charge is produced on the rod. The polarization of the water molecules in the stream is reversed, but the imbalance of forces again creates a net attraction of the rod for the liquid [3, 4].

The charging of neutral matter by contact electrification and rubbing is a component of the process of electrophotography—a major commercial process [5]. The physics of the charging process is incompletely understood. A list of the "triboelectric series" showing the direction of charge transfer when two materials are rubbed together has been published [5].

If the liquid to be deflected is electrically conductive—that is, contains mobile ions—a charge separation as shown in the figure can occur by ion displacement, and the stream will be deflected toward the rod in the *nonuniform* electric field as with the dielectrophoretic mechanism. A discussion of this possibility appears in the literature [6], where it is claimed that induced (net) charge is the primary cause of the deflection. Partial support for this point of view is that drops from the falling stream collected in a metal cup connected to an electroscope carry a charge opposite in sign to that of the rod. We would explain this observation not in terms of acquisition of a net charge, but by noting that the charged rod causes a vertical as well as horizontal movement of ions, resulting in the bottom part of the stream (which breaks off and is collected) being charged oppositely to the top of the stream and to the rod. The (separation of charge) effect is readily observable with water and is even larger with salt water. Our experiments did not indicate that the deflection of salt water was significantly greater than that obtained with pure water under the same experimental conditions. In addition, we collected drops from a deflected stream of n-hexane (commercial grade) and observed no separation of the electroscope leaves, indicating no detectable charge separation (or net charge acquisition). n-Hexane has an immeasurably low conductivity and cannot be deflected by a mechanism involving ionic charges. It is, however, an unfavorable liquid for observing dielectrophoresis. Nevertheless, deflection of the stream can be observed, suggesting that the dielectrophoretic mechanism is predominant in the demonstration as performed.

REFERENCES

1. N. E. Hill, W. E. Vaughan, A. H. Price, and M. Davis, *Dielectric Properties and Molecular Behaviour,* Van Nostrand–Reinhold: London (1969).
2. H. A. Pohl, *Sci. Am.* 203 (Dec.): 106 (1960).
3. A. D. Moore, *Sci. Am.* 226 (Mar.): 46 (1972).
4. H. A. Pohl, *Dielectrophoresis: the Behavior of Neutral Matter in Nonuniform Electric Fields,* Cambridge University Press: New York (1978).
5. D. M. Burland and L. B. Schein, *Phys. Today* 39: 46 (1986).
6. I. D. Brindle and R. H. Tomlinson, *J. Chem. Educ.* 52: 382 (1975).

9.33

The Tubeless Siphon

Liquid from one beaker is siphoned into another without the use of tubing.

MATERIALS

1 liter distilled water

several drops 2% thymol in ethanol (optional) (To prepare 100 mL of solution, dissolve 1.6 g thymol in 100 mL of ethanol.)

25 g poly(ethylene oxide), with molar mass of 7×10^6 g/mol †

200 mL 95% ethanol

600-mL beaker

2 2-liter beakers

$\frac{1}{4}$-horsepower paddle stirrer

watch glass to cover beaker

PROCEDURE

Preparation

Pour 1 liter of distilled water into a 2-liter beaker. Mount the paddle stirrer over the beaker so the paddle is near the bottom of the beaker. Turn on the stirrer and adjust the speed to create a deep vortex. Several drops of a 2% solution of thymol in alcohol can be added to the water as a preservative antibacterial agent. Combine 25 g of poly(ethylene oxide) having an average molar mass of about 7×10^6 g/mol with 200 mL of 95% ethanol in a 600-mL beaker. Swirl the beaker to make a slurry and quickly pour the slurry into the vortex. Reduce the speed of the stirrer to about 15 revolutions per minute, and raise the paddle until it is just below the surface of the mixture. Continue stirring for 12 to 24 hours. Remove the stirrer, cover the mixture with the watch glass, and allow the mixture to rest for another 12 to 24 hours. Some undissolved polymer may rise to the surface of the liquid. If this happens, scoop out the undissolved polymer, stir the mixture for several more hours, and then allow it to rest for several more hours.

† Poly(ethylene oxide) with a molar mass of 7×10^6 g/mol is available from the Aldrich Chemical Company, 940 W. Saint Paul Avenue, Milwaukee, Wisconsin 53233.

Presentation

Place the 2-liter beaker containing the poly(ethylene oxide) solution at the edge of a table. Tip the beaker and start to pour the solution into the second 2-liter beaker. Lower the second beaker to the floor and return the first to the upright position. The solution will continue to flow from the upper beaker into the lower one.

HAZARDS

Ethanol is flammable and should be kept away from open flames.

The toxicological properties of poly(ethylene oxide) are unknown. Therefore, care should be taken to avoid direct contact with the material and inhalation of the dust.

DISPOSAL

The poly(ethylene oxide) solution should be flushed down the drain with water, or it can be stored in a sealed jar. It will keep indefinitely if it contains thymol as a preservative.

DISCUSSION

With a Newtonian liquid such as water, the siphoning action stops once the surface of the liquid in the first beaker drops below the lip. With the poly(ethylene oxide) solution, the siphoning continues until the first beaker is nearly emptied. A general discussion of the dynamics of polymeric liquids, including a number of striking experimental results, can be found in reference 1. The tubeless siphon is described and secondary references are provided. A photograph of the tubeless siphon can be found in reference 2.

REFERENCES

1. R. B. Bird, R. C. Armstrong, and O. H. Hassager, *Dynamics of Polymeric Liquids. Fluid Mechanics*, Vol. 1, 1st ed., John Wiley and Sons: New York (1977).
2. A. A. Collyer, *Phys. Educ.* 8:111 (1973).

9.34

Rod Climbing
by a Polymer Solution

When a normal liquid is stirred with a rod, a vortex forms in the liquid. When an appropriately chosen polymer solution is stirred in the same fashion, the solution climbs the rod.

MATERIALS

1 liter glycerol, $CHOH(CH_2OH)_2$

5 g poly(acrylamide), $+CH_2CH(CONH_2)+_n$, with average molar mass ca. 5×10^4 g/mol †

2 1-liter beakers

hot plate

electric paddle stirrer

thermometer, $-10°C$ to $+110°C$

hand-cranked paddle stirrer, with a shaft 1 cm in diameter and a paddle 8 cm in diameter at the end of the shaft

PROCEDURE [1]

Preparation

Prepare the poly(acrylamide) solution as follows: Place a 1-liter beaker on a hot plate and mount the electric paddle stirrer over the beaker. Pour 500 mL of glycerol into the beaker and heat it to 35°C. Stir the glycerol fast enough to form a small vortex. Add 5 g of poly(acrylamide) to the heated, stirred glycerol and continue the stirring for several hours, until all of the polymer has dissolved. Stir the mixture slowly to prevent mechanical degradation of the polymer. After the stirring is completed, allow the mixture to rest overnight before use.

Pour 500 mL of glycerol into the other 1-liter beaker.

Presentation

Insert the hand-cranked paddle stirrer in the glycerol and, using the handle, turn the paddle at a sufficient speed to generate a vortex in the liquid.

† Poly(acrylamide) is available from the Aldrich Chemical Company, 940 W. Saint Paul Avenue, Milwaukee, Wisconsin 53233.

Insert the hand-cranked paddle stirrer in the poly(acrylamide) solution and turn it with the handle until the solution climbs up the rod. The rate of stirring required for rod climbing is much lower than that required to form a vortex in the glycerol.

HAZARDS

The toxicological properties of poly(acrylamide) are unknown. Therefore, care should be taken to avoid direct contact with the material.

DISPOSAL

The polymeric liquid can be stored in a sealed glass jar for use in repeated performances of this demonstration. Otherwise, the solution should be diluted by about a factor of 4 by adding water, slowly and with stirring. The resulting solution should be flushed down the drain with water.

DISCUSSION

In this demonstration, the polymer solution shows unusual behavior for a liquid. When stress is applied to the polymer solution by the paddle, the liquid moves in a direction perpendicular to the stress, that is, upward. Such behavior indicates that the liquid is non-Newtonian—its properties cannot be explained in terms of the equations used to describe flow in simple liquids.

Movement perpendicular to an applied stress is evidence of extended structure in a liquid. Polymers of a higher molecular weight show larger effects (a weight average molar mass to 5×10^6 g/mol is suitable). The dynamics of non-Newtonian liquids have been analyzed [2]. The analysis employs a model system with axial annular flow, and the results of the analysis indicate how rod-climbing occurs. The origin of the effect involves the values of the normal components of the stress tensor. These are zero for normal (Newtonian) liquids but nonzero for most polymer solutions. The analysis is of practical significance in the design of mixing tanks used, for example, in the manufacture of paints.

Rod-climbing behavior is commonly observed by bakers. When dough is mixed in an electric mixer, it often climbs up the beaters and collects around their shafts. Dough is a polymeric liquid mixture. The solvent is water and the polymer is a mixture of starches from the flour. Starches are polymers of simple sugars, and their molar masses can be very large.

REFERENCES

1. A. S. Lodge, *Elastic Liquids*, Academic Press: New York (1964).
2. R. B. Bird, R. C. Armstrong, and O. H. Hassager, *Dynamics of Polymeric Liquids. Fluid Mechanics*, Vol. 1, 2d ed., John Wiley and Sons: New York (1987).

9.35

Snappy Liquid:
Elastic Properties of a Soap Solution

As a liquid is poured from a jar into a beaker, the stream of flowing liquid is cut with scissors. The stream of liquid above the cut recoils back into the jar.

MATERIALS

100 g lauric acid, $CH_3(CH_2)_{10}CO_2H$

1 liter 0.75M sodium hydroxide, NaOH (To prepare 1 liter of solution, dissolve 30 g of NaOH in 600 mL of distilled water, and dilute the resulting solution to 1 liter.)

85 g potassium aluminum sulfate dodecahydrate, $KAl(SO_4)_2 \cdot 12H_2O$

500 mL distilled water

ca. 20 g phosphorous pentoxide, P_4O_{10}

1 liter purified toluene

2-liter beakers

stirring hot plate, with stir bar

thermometer, $-10°C$ to $+110°C$

Büchner funnel, with filter paper

filter flask

drying oven

desiccator

2-liter wide-mouth glass jar, with cover

1-liter distillation apparatus (See procedure to determine if required.)

scissors

PROCEDURE [1–4]

Preparation

Prepare an amount of aluminum hydroxydilaurate (aluminum soap) sufficient to make 1 liter of elastic liquid as follows: Combine 100 g of lauric acid and 1 liter of 0.75M NaOH solution in a 2-liter beaker. On the stirring hot plate, slowly heat the mixture to 50°C and stir it until a clear, pale yellow solution of sodium laurate forms. Dissolve 85 g of $KAl(SO_4)_2 \cdot 12H_2O$ in 500 mL of distilled water. Slowly add the $KAl(SO_4)_2$ solution to the stirred sodium laurate solution. A white precipitate of aluminum soap will form. Filter the precipitate. Place the precipitate in a drying oven at 65°C for 2–3 days. Complete the drying by sealing the aluminum soap in a desiccator over about 20 g of P_4O_{10} for 1 week.

Prepare the elastic liquid as follows: Place 40 g of dry aluminum hydroxydilaurate in a 2-liter glass jar. Add 1 liter of purified toluene. (The toluene should be colorless. If it is not it should be distilled from $CaCl_2$ before using.) Seal the jar and shake it continuously until a gel forms and the settling rate of particles is slow (this will take about 5 minutes of shaking). Then, shake the gel periodically until it becomes homogeneous. This may take several days.

Presentation

Begin to pour the solution from the jar into a 2-liter beaker, so a stream of liquid about 3–4 cm in diameter results. With scissors, cut the stream of flowing liquid about 10 cm from the mouth of the jar. The liquid above the cut will spring back into the jar. (Some practice may be needed to determine the optimum stream diameter for the most dynamic recoil.) Photographs of this demonstration appear in reference 2.

HAZARDS

Toluene is toxic and has been identified as mildly carcinogenic. It should be used only with adequate ventilation.

Phosphorus pentoxide is a strong irritant and is corrosive to the skin, eyes, and mucous membranes. It reacts violently with water, producing phosphoric acid, which is also irritating to the skin and eyes.

Solid sodium hydroxide is caustic and can cause burns to the skin, eyes, and mucous membranes. Dust from solid sodium hydroxide is irritating to the eyes and respiratory system.

DISPOSAL

Return the soap solution to the jar and seal the jar. The solution can be kept for use in repeated performances of this demonstration. Otherwise it should be discarded in a container for waste organic solvents.

DISCUSSION

The recoiling of a stream of pouring liquid back into its container against the force of gravity indicates that the liquid still in the container must exert some pull on the pouring stream. This pull is evidence of three-dimensional structure in the liquid. Generally, the forces between molecules in a liquid are weak, relative to those in solids. However, the liquid in this demonstration has intermolecular forces stronger than those of a typical liquid. Yet, it is nonrigid and flows like a typical liquid, so the intermolecular forces are not as strong as those in a solid. This liquid is intermediate between a typical liquid and a solid.

The flow properties of aluminum dilaurate–toluene gels have been investigated as a function of aluminum dilaurate (soap) concentration and shear rate [4]. At concentrations over about 1% by weight, aluminum soaps dispersed in hydrocarbons form geis with three-dimensional structure [1]. The elasticity of the gel is a result of this network structure. The structure of the soap has been investigated by infrared spectroscopy [3]. The conclusion is that the network is held together by covalent bonds, probably involving Al—O linkages.

REFERENCES

1. A. S. Lodge, *Elastic Liquids,* Academic Press: New York (1964).
2. R. B. Bird, R. C. Armstrong, and O. H. Hassager, *Dynamics of Polymeric Liquids. Fluid Mechanics,* Vol. 1, 1st ed., John Wiley and Sons: New York (1977).
3. W. W. Harple, S. E. Wiberley, and W. H. Bauer, *Anal. Chem.* 24:635 (1952).
4. N. Weber and W. H. Bauer, *J. Phys. Chem.* 60:270 (1956).

9.36

Fog:
An Aerosol
of Condensed Water Vapor

When chunks of dry ice are dropped into a pan of warm water, fog (a colloidal suspension of liquid water in air) immediately fills and overflows the pan. (This is described in Procedure C in Demonstration 6.2 in Volume 2 of this series.) When water is boiled in a flask whose opening is fitted with a copper coil, fog emanates from the coil. When the coil is heated by the flame of a Bunsen burner, the fog disappears. The water vapor emanating from the heated coil is superheated steam, which can char paper. This procedure illustrates the difference between fog and steam. (This is described in Demonstration 5.25 in Volume 2 of this series.)

9.37

Clean Air
with the Cottrell Precipitator

A smoke of ammonium chloride generated in a glass tube is precipitated in an electrostatic field.

MATERIALS

50 mL concentrated (12M) hydrochloric acid, HCl

50 mL concentrated (15M) ammonia, NH_3

2 solid rubber stoppers to fit 4-cm diameter glass tube

drill or borer to make holes in stoppers for glass tubing with outside diameter of 6 mm

ice pick or awl

2.5 m 14-gauge uninsulated copper wire

Pyrex tube, with diameter of 4 cm and length of 0.5 m

4 right-angle glass bends, with outside diameter of 6 mm and the length of each arm ca. 5 cm

1 m electrical tape

stand, at least 1 m tall, with clamp

high-voltage transformer, 115-volt primary, ca. 10,000-volt and 2-milliampere secondary (e.g., neon-sign transformer)

2 1.5-m leads with clips to connect transformer to wire electrodes

2 right-angle glass bends, with outside diameter of 7 mm and the length of one arm ca. 5 cm, length of other arm ca. 15 cm

2 2-holed rubber stoppers to fit 125-mL Erlenmeyer flasks

2 125-mL Erlenmeyer flasks

2 m rubber tubing to fit right-angle bends

glass Y-adapter to fit rubber tubing

water aspirator, with ca. 1 m of vacuum tubing

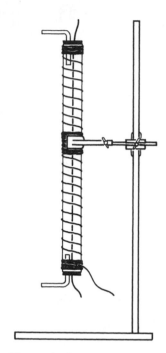

Figure 1. Cottrell precipitator.

PROCEDURE [1]

Preparation

Assemble the Cottrell precipitator as indicated in Figure 1. Wrap the Pyrex tube with 14-gauge copper wire so the loops of the coil are 2–3 cm apart. Fasten the wire coil in place by wrapping the ends of the tube with electrical tape; a length of this wire should also extend beyond the bottom of the tube. Also wrap a section of the center of the tube with electrical tape and fasten the clamp to the ring stand over this electrical tape to insulate the stand from the wire. Drill a hole near the circumference of each stopper to hold the 6-mm glass bends. With an ice pick or similar tool, pierce a hole all the way through the center of each stopper. Insert a 0.7-m length of 14-gauge copper wire through the center of one stopper, then through the 0.5-m Pyrex tube, then through the center of the other stopper; a length of wire will extend from the bottom stopper. Fasten the stoppers in the openings of the Pyrex tube and tighten the wire so it runs up the center of the tube. Insert one of the 6-mm glass bends in the hole near the edge of each of the stoppers.

The transformer has two connecting terminals. Use the leads with clips on each end to attach one of the copper wires to one terminal and the other copper wire to the second terminal. Turn on the transformer momentarily to see whether there is any sparking between the copper wires or the leads. If sparks do form, adjust the positions of the wires and the leads with the transformer turned off, moving them apart until no sparking occurs when the transformer is on.

Insert two glass bends, one of the smaller and one of the larger, through each stopper for the Erlenmeyer flasks, as shown in Figure 2. Adjust the glass bends so the 6-mm bend extends just beyond the base of the stopper and the 7-mm bend extends to the bottom of the flask when the stopper is seated in the mouth of the flask. Pour 50 mL of

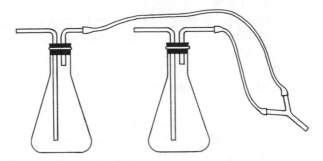

Figure 2. Smoke-generating flasks.

12M HCl into one of the 125-mL Erlenmeyer flasks and 50 mL of 15M NH₃ into the other. Seal each flask with a rubber stopper assembly. Attach rubber tubing to each of the 6-mm right-angle bends, and join these lengths of rubber tubing to the Y-adapter. With another piece of rubber tubing, attach the Y-adapter to the glass bend at the bottom of the precipitator. Attach the glass bend at the top of the precipitator to a water aspirator.

Presentation

Adjust the water aspirator to create a gentle bubbling in the flasks. White smoke will appear where the glass bend enters at the bottom of the Pyrex tube, and it will ascend to fill the tube. Turn on the transformer. The smoke will quickly precipitate to the bottom and sides of the apparatus. Turn off the transformer, and the tube will again fill with smoke. Turn on the transformer to precipitate the new smoke. Turn off the aspirator and the transformer.

HAZARDS

A painful shock can result if both terminals of the high-voltage transformer are touched simultaneously. The shock can be harmful if the transformer is capable of delivering more than several milliamperes of current.

Hydrochloric acid can cause severe burns. The vapors are extremely irritating to the skin, eyes, and respiratory system.

Concentrated aqueous ammonia can irritate the skin, and its vapors are harmful to the eyes and mucous membranes.

DISPOSAL

Rinse the apparatus with water to dissolve the ammonium chloride. Flush the rinse and other waste solutions down the drain with water.

DISCUSSION

In this demonstration, smoke is generated by the reaction of gaseous hydrogen chloride with gaseous ammonia to form a suspension of tiny particles of ammonium chloride in air. This smoke is drawn into a tube wrapped with a wire and having another wire running through it. When these two wires are connected to the terminals of a high-voltage transformer, the smoke particles are drawn to the walls of the tube and to the wire inside the tube, clearing the air of the suspended particles.

The transformer used in this demonstration produces a strong oscillating electric field, which causes some of the air molecules inside the glass tube to become ionized. These ions attach to the smoke particles, giving them a charge. Some of the particles are positively charged, some negatively charged. The oppositely charged particles are attracted to each other, and they coalesce into larger particles that are drawn by gravity to the bottom of the precipitator. Some charged particles are attracted to the electrodes, where they adhere [2]. The apparatus used in this demonstration differs somewhat from commercial precipitators in that it uses alternating voltages on the electrodes. This causes solid to precipitate at both electrodes. Commercial precipitators use electrodes that are constantly polarized the same way, so the solid precipitates at only one electrode, making collection of the solid more convenient.

The smoke generated for this demonstration is produced by mixing the hydrogen chloride vapors emitted from hydrochloric acid with ammonia gas from aqueous ammonia. The combination forms tiny particles of ammonium chloride, in the form of smoke:

$$HCl(g) + NH_3(g) \longrightarrow NH_4Cl(s)$$

The life story of Frederick Cottrell is interesting. Both short and detailed biographies have been written [3, 4]. In addition to developing a number of useful commercial processes, Cottrell founded the Research Corporation, which has provided a great measure of support to American science.

REFERENCES

1. F. T. Weisbruch, *Lecture Demonstration Experiments for High School Chemistry,* St. Louis Education Publishers: St. Louis, Missouri (1951).
2. B. Jirgensons and M. E. Straumanis, *A Short Textbook of Colloid Chemistry,* John Wiley and Sons: New York (1954).
3. A. B. Costa, *J. Am. Chem. Soc.* 62:135 (1985).
4. F. Cameron, *Cottrell, Samaritan of Science,* Doubleday: Garden City, New York (1952).

9.38

A Collection of Foams: Some Suds for Drinking and Some for Washing

Egg whites are whipped to a foam (Procedure A). Two flasks containing dishwashing detergent, one with hard water, the other with soft water, are shaken; more suds form in the flask of soft water (Procedure B). Two flasks of beer, one also containing simethicone, are shaken, and less foam appears in the flask containing simethicone (Procedure C).

MATERIALS FOR PROCEDURE A

fresh egg

glass mixing bowl, ca. 1-liter volume

egg beater, electric or manual, or whisk

MATERIALS FOR PROCEDURE B

200 mL hard water (hard tap water, or water to which several crystals of $CaCl_2$ or $MgCl_2$ have been added)

200 mL distilled water

4–10 mL dishwashing detergent (Do not use detergent for automatic dishwashers; it will not form suds.)

2 1-liter flasks

2 stoppers for flasks

MATERIALS FOR PROCEDURE C

12-ounce can of beer

40 mL liquid antacid containing simethicone

2 1-liter flasks

2 stoppers for flasks

PROCEDURE A

Preparation and Presentation

Separate the white of a fresh egg from its yolk and place the white in a glass bowl. Using beaters or a whisk, whip the white until a stiff foam forms.

PROCEDURE B

Preparation and Presentation

Pour 200 mL of hard water into one of the 1-liter flasks and 200 mL of distilled water into the other. Add 2–5 mL of dishwashing detergent to each flask. Stopper the flasks and shake each one vigorously. Copious soap bubbles form in the flask of distilled water, very few in the hard water.

PROCEDURE C

Preparation and Presentation

Pour half a can of beer into each of the 1-liter flasks, pouring down the side to minimize the formation of a head. Add 40 mL of a commercial antacid containing simethicone to one of the flasks. Stopper and shake both flasks. Foaming in the flask containing simethicone will be less extensive than in the other flask.

DISPOSAL

The waste solutions should be flushed down the drain with water.

DISCUSSION

The procedures in this demonstration show the formation of a foam by incorporation of air or carbon dioxide into a liquid phase. Pure liquids do not form foams; a third material, the foaming agent, is needed [1, 2]. Egg whites contain proteins as foaming agents and produce persistent foams (meringues). Beer contains proteins and carbohydrates which promote foaming. Strong foaming is also caused by natural soaps (sodium carboxylates of fatty acids with more than 12 carbon atoms, e.g., sodium laurate, $CH_3(CH_2)_{10}CO_2Na$) or synthetic detergents (sodium alkyl sulfonates, e.g., sodium lauryl sulfonate, $CH_3(CH_2)_{10}CH_2SO_3Na$).

The stability of foams is a subject of great commercial interest. Beaten egg whites produce a very stable foam—lemon meringue pie keeps for several days without losing visual appeal. Soap suds keep for the minutes needed during cleaning. The ions in hard water, mainly calcium and magnesium, and oils in dirt combine with soap or detergents, reducing their ability to stabilize the suds. Beer suds are somewhat fragile; the amount of foam depends on the manner in which the beer is poured, the shape of the glass, and how clean the glass is. A poorly cleaned glass, used previously for milk, for example, will suppress the formation of a head of foam. Factors affecting foam stability have been investigated for many years [1]. The procedures in this demonstration provide a range of foam stabilities.

Some pictures of the change in beer foam with time appear in the literature [3]. Under the influence of gravity, liquid drains from the head of foam on a glass of beer. As this happens, the shape of the carbon dioxide bubbles composing the foam changes from roughly spherical to polyhedral. Because the pressure inside a bubble drops as its size increases, carbon dioxide diffuses through bubble walls from smaller bubbles into the larger ones. The surface molecules in the bubble films also move, and this affects interfacial tension. These factors are incorporated into mathematical models of foam evolution.

Foam is not universally desirable. Its elimination in lubricating oils, sugar beet juice, and upset stomachs is a matter of commercial concern. Defoaming agents are numerous and varied, but the choice of defoaming agent for a particular system is still largely an empirical process. Esters with high molecular weight are added to latex paints to avoid foaming and the pits this causes in the painted surface. Procedures B and C show that alkaline earth ions destroy soap foams. Silicones (e.g., methyl polysiloxane, $—((CH_3)_2SiO)_n—$) have wide use. One has only to watch television commercials to hear claims for the effectiveness of simethicone, a polysiloxane formulation, in antiflatulent preparations (*caveat emptor*).

REFERENCES

1. J. J. Birkerman, J. M. Perri, R. B. Booth, and C. C. Currie, *Foams: Theory and Industrial Applications,* Reinhold Publishing Co.: New York (1953).
2. A. E. Alexander and P. Johnson, *Colloid Science,* Vol. 2, Clarendon Press: Oxford (1949).
3. J. H. Aubert, A. M. Kraynik, and P. R. Rand, *Sci. Am.* 254 (May):74 (1986).

9.39

Solid Foams

Two viscous liquids are mixed and, after a short time, they foam, increasing in bulk 20–30 times and hardening to a solid (Procedure A). An expanded polystyrene coffee cup is placed in a clear colorless liquid and rapidly "disappears" (Procedure B). Polystyrene in the form of packing "peanuts" is dissolved in a clear colorless liquid, and that liquid is used to apply a coating to the inside of a beaker, from which the solidified coating can be unmolded (Procedure C).

MATERIALS FOR PROCEDURE A

See Materials for Demonstration 3.2 in Volume 1 of this series.

MATERIALS FOR PROCEDURE B

50 mL acetone, CH_3COCH_3, or acetone-containing nail polish remover

2 identical, disposable, foam coffee cups

petri or crystallizing dish, with diameter larger than that of cup

paper towel

glass stirring rod

MATERIALS FOR PROCEDURE C

55 mL ethyl acetate, $CH_3CH_2OCOCH_3$

ca. 2 liters foam packing "peanuts"

2 250-mL beakers

glass stirring rod

watch glass to cover beaker

spatula or table knife

PROCEDURE A

See Demonstration 3.2 in Volume 1 of this series for the preparation of polyurethane foam.

PROCEDURE B

Preparation

Test one of the cups to be sure that it disintegrates in acetone. Hold the foam cup over the petri dish and pour 10 mL of acetone or nail polish remover into the cup. In a few seconds the liquid should leak from the cup into the dish. If it doesn't, the cups are unsuitable for this demonstration. Empty any acetone from the petri dish onto the paper towel and allow the acetone to evaporate in a well-ventilated area.

Presentation

Hold the remaining foam cup over the petri dish and pour 40 mL of acetone or nail polish remover into the cup. In a few seconds the liquid will leak from the cup into the dish. Set the cup in the dish. The cup will gradually sink into the dish as the foam disintegrates in the acetone. After the cup has disappeared, use a stirring rod to lift the residue of the cup from the dish; it is limp and translucent.

PROCEDURE C[†]

Preparation

Test the packing peanuts to see if they are suitable for this demonstration. Pour several milliliters of ethyl acetate into one of the beakers and add two or three of the packing peanuts to the beaker. If the peanuts immediately collapse and dissolve, they are suitable for the demonstration. Empty any ethyl acetate from the petri dish onto the paper towel and allow the ethyl acetate to evaporate in a well-ventilated area.

Presentation

Pour 50 mL of ethyl acetate into a 250-mL beaker. Add foam packing peanuts to the beaker and stir the mixture. The peanuts will disappear as they dissolve in the ethyl acetate. Continue to add peanuts until the liquid in the beaker has become quite syrupy, with the consistency of honey. This will require a volume of about 2 liters of peanuts.

Swirl the beaker to coat its inside surface with the syrupy liquid, and then pour the liquid into a second 250-mL beaker. Cover the second beaker with the watch glass and set the first beaker on its side on the bench for several minutes to allow most of the ethyl acetate to evaporate. Pour the liquid from the second beaker back into the first, and swirl the first beaker to coat it again. Pour the liquid back into the second beaker. Allow the first beaker to rest on its side again. The interior of the first beaker can be coated with several more layers, as time permits.

After all of the ethyl acetate has evaporated from the coated first beaker, use a spatula or knife to loosen the coating from the rim of the beaker. Pull the loosened coating from the beaker.

† We wish to thank Jeanne Dyer of Vivian Gaither High School, Tampa, Florida, for providing us with this procedure.

HAZARDS

Acetone and ethyl acetate are highly flammable; they should be kept away from sparks or open flames and should be used only in well-ventilated areas.

DISPOSAL

The liquid residues from Procedures B and C should be placed on a paper towel, and the liquid in the residue allowed to evaporate in a well-ventilated area away from sparks or flames. The hard, dried residue should be discarded in a solid-waste receptacle.

DISCUSSION

This demonstration shows the formation and destruction of solid foams, which are suspensions of small pockets of gas in a solid. Polyurethane foam is produced by forming the polymer in the presence of a boiling liquid (Procedure A). Polystyrene foam is destroyed by softening the solid and allowing the trapped gas to escape from it (Procedure B). Polystyrene foam is also destroyed by dissolving the solid, which also allows the gas to escape. Furthermore, the solid can be recovered without the trapped air by allowing the solvent to evaporate (Procedure C).

Foams can be formed in a solid matrix as well as a liquid one (see Demonstration 9.38 for liquid-matrix foams). Bread is an example of a solid foam. Polyurethane foam produced in Procedure A is another, and foamed polystyrene (e.g., Styrofoam) is a third.

In Procedure A a solid-matrix foam is produced by polymerization under controlled conditions in the presence of a blowing agent (low-boiling fluorocarbon). A polyurethane foam is prepared by reacting a polyol with a polyfunctional isocyanate. The polyfunctional character of the reactants results in a high degree of cross-linking in the product, producing a rigid foam. Because the matrix of the foam turns from liquid to solid as the material polymerizes, an enduring foam is produced.

The foam coffee cup in Procedure B is another example of a foam of indefinite stability. It is stable because the liquid phase has been removed, leaving only the foaming agent, polystyrene. When a liquid capable of wetting the polystyrene is added to the cup, the foam becomes unstable and collapses. As Procedure B shows, the polystyrene in the cup does not dissolve in the acetone. Cups made of polystyrene that has not been blown into a foam can hold acetone. (Such cups are transparent and brittle and show colorful patterns when placed between crossed polarizing filters [see Demonstration 9.51].) Some foamed cups are coated with a thin film impervious to acetone. Therefore, the cups must be tested for suitability in this demonstration before they are used.

9.40

Oil-Water Emulsions

Two clear, colorless solutions are mixed in each of two graduated cylinders, and two layers form in each. In one, the lower layer becomes turbid, indicating emulsion formation. In the other, the top layer becomes turbid.

MATERIALS

40 mL distilled water

30 mL toluene, $C_6H_5CH_3$

10 mL absolute ethanol, C_2H_5OH

2 50-mL graduated cylinders

25-mL graduated cylinder

10-mL graduated cylinder

2 50-mL beakers

stirring rod

PROCEDURE [1]

Preparation and Presentation

Pour 20 mL of distilled water into each of the two 50-mL graduated cylinders. With the 25-mL graduated cylinder measure 18 mL of toluene into one of the 50-mL beakers and 12 mL into the other beaker. Pour 2 mL of ethanol into the beaker containing 18 mL of toluene and 8 mL into the beaker containing 12 mL of toluene. Stir the contents of each beaker. Pour the contents of one beaker into one of the 50-mL graduated cylinders and the contents of the other beaker into the other cylinder. Each cylinder contains two layers of liquids with water on the bottom in each. In the cylinder containing the 40-60 ethanol-toluene mixture, the bottom layer becomes turbid. In the cylinder containing the 10-90 ethanol-toluene mixture, the upper layer becomes turbid. The turbidity is evidence of emulsion formation, which proceeds rapidly and is complete in 5–10 minutes.

HAZARDS

Ethanol and toluene are flammable and should be kept away from open flames.

Toluene is toxic and has been identified as carcinogenic. It should be used only with adequate ventilation.

DISPOSAL

The mixtures of toluene and ethanol should be discarded in a receptacle for combustible organic solvents.

DISCUSSION

Emulsions are dispersions of one liquid in another, most commonly oil and water [2]. Almost invariably a third component, the emulsifying agent, is added to attain reasonable stability. As this demonstration illustrates, it is sometimes possible to "invert" an oil-in-water emulsion to a water-in-oil emulsion. The ethanol-toluene-water system used here converts from an oil-in-water emulsion to a water-in-oil emulsion when the amount of ethanol in the mixture is increased. In this system, emulsification is spontaneous and does not require the use of an emulsifying agent. Cream is an example of an emulsion of oil (butterfat) in water. Butter is a water-in-oil emulsion. To produce butter, cream is strongly agitated, and during this process, the emulsion inverts. In the process of homogenizing milk, the size of the oil particles is reduced mechanically to improve the stability of the emulsion. Mayonnaise and sauces are emulsions of the oil-in-water type. In mayonnaise, egg yolk is used as the emulsifying agent. In gravies, flour is the emulsifying agent for an oil-in-water emulsion; meat drippings cooked with flour and stirred with water yield gravy. Natural and synthetic soaps emulsify oils in water, allowing them to remove oil and grease from the object being cleaned.

An article in reference 3 describes the principles underlying the stability of sauce béarnaise, a dispersion of butter and egg yolk in vinegar and white wine with salt, shallots, and tarragon. The article's authors, although they followed a standard recipe, produced an unacceptable heterogeneous product. They then applied their theory of emulsion formation, which held that the addition of vinegar with vigorous stirring would produce a stable homogeneous product, and with this they were successful. Because emulsifying agents act at the interface, the advent of the food processor has aided amateurs in the preparation of sauces (foodstuff emulsions): its rapid mixing action creates a large interfacial area, thus helping the emulsifying agent to work.

REFERENCES

1. D. Sievers, *Chem 13 News* 158:4 (1985).
2. A. E. Alexander and P. Johnson, *Colloid Science*, Vol. 2, Clarendon Press: Oxford (1949).
3. C. M. Perram, C. Nicolau, and J. W. Perram, *Nature* 270:572 (1977).

9.41

Color of the Sunset:
The Tyndall Effect

When a beam of white light passes through a sol, the beam appears blue when viewed from the side and orange-red when viewed on end.

MATERIALS FOR PROCEDURE A

water to fill aquarium

for each gallon of water:

either

20 mL saturated sodium thiosulfate, $Na_2S_2O_3$ (To prepare ca. 100 mL of stock solution, combine 50 mL of distilled water and 90 g of $Na_2S_2O_3 \cdot 5H_2O$. Warm the mixture until it is liquid and allow it to cool; decant the liquid, which is the saturated solution.)

5 mL 6M hydrochloric acid, HCl (To prepare 1 liter of stock solution, pour 500 mL of concentrated [12M] HCl into 300 mL of distilled water, and dilute the resulting solution to 1.0 liter.)

or

50 mL milk

small aquarium, 1–5 gallons

slide projector

mirror, ca. the size of a smaller side of the aquarium

stirring rod

MATERIALS FOR PROCEDURE B

overhead projector, with screen

400 mL 10% sodium thiosulfate, $Na_2S_2O_3$ (To prepare the solution, dissolve 68 g of $Na_2S_2O_3 \cdot 5H_2O$ in 375 mL of distilled water.)

10 mL 1M hydrochloric acid, HCl, in dropper bottle (To prepare 1 liter of stock solution, pour 83 mL of concentrated [12M] HCl into 600 mL of distilled water, and dilute the resulting solution to 1.0 liter.)

single-edged razor blade

poster board, slightly larger than the stage of the overhead projector

600-mL beaker

glass stirring rod

MATERIALS FOR PROCEDURE C

1 liter 1M sodium chloride, NaCl (To prepare 1 liter of solution, dissolve 58 g of
    NaCl in 600 mL of distilled water and dilute the resulting solution to 1.0 liter.)

enough unflavored gelatin mix to prepare 1 liter (140 g; 5 oz.)

2 1-liter transparent containers with flat sides

slide projector

PROCEDURE A [1]

Preparation

Arrange the aquarium, slide projector, and mirror as shown in the figure. The
beam of the projector should pass through the aquarium side to side, so the audience
can see the length of the beam as it passes through. The mirror should be placed at the
side opposite from the projector and adjusted so the reflection of the beam shines toward the audience. Add water to the aquarium until it is nearly full.

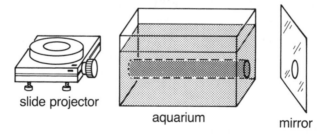

slide projector   aquarium   mirror

Presentation

Turn on the projector. The beam will be virtually invisible where it passes through
the aquarium and bright white where it emerges and shines toward the viewers.

If the effect of the gradual formation of a sol is desired, follow the steps described
in the next paragraph of Procedure A. If the effect of the immediate formation of a sol is
desired, follow the steps described in the last paragraph of Procedure A. Both effects
cannot be shown with a single aquarium.

To produce a sol gradually, for each gallon of water in the aquarium, add 20 mL of
saturated sodium thiosulfate solution and 5 mL of 6M hydrochloric acid. Stir the contents of the aquarium. The mixture will become turbid in 1–2 minutes. The turbidity
will gradually increase for about 10 minutes. As the turbidity increases, the beam will
become visible where it passes through the mixture. The beam passing through the solution appears blue, and where it emerges it is orange. Furthermore, the blue beam will
gradually widen.

To produce a sol quickly, for each gallon of water in the aquarium, add 25 mL of milk. Stir the contents of the aquarium. The beam will become visible where it passes through the mixture. The beam passing through the solution is blue, and where it emerges it is orange-red. Add another 25 mL of milk. The beam will still appear blue, but it will be wider than before.

PROCEDURE B [2]

Preparation

With the razor blade, cut a round hole slightly smaller than the diameter of a 600-mL beaker in the center of the poster board. Place the poster board on the stage of the overhead projector and center the 600-mL beaker over the hole.

Presentation

Pour 400 mL of 10% sodium thiosulfate solution into the beaker. Focus the overhead projector to produce a bright spot on the screen. Note that, when viewed from the side, the solution in the beaker is clear. Add 2–3 drops of 1M HCl to the solution in the beaker and stir the mixture. The solution in the beaker will become slightly turbid, and it appears brighter when viewed from the side of the beaker. At the same time, the projected spot on the screen will become less bright. Add several more drops of 1M HCl to the beaker and stir the mixture. The mixture will become more turbid, the beaker will appear brighter when viewed from the side, and the projected spot will become dimmer. Continue to add drops of 1M HCl to the beaker while stirring the mixture. Eventually, the projected spot will become too dim to see, and the mixture in the beaker will appear very bright at the bottom, and dim at the top.

PROCEDURE C [3]

Preparation

Fill one of the containers with 1M sodium chloride solution. Prepare 1 liter of gelatin according to the directions on the package and pour it into the second container before it sets. Allow the gelatin to set.

Presentation

Arrange the containers side by side and shine the beam of the slide projector through both containers simultaneously. Where the beam passes through the gelatin it is visible, but not where it passes through the NaCl solution.

HAZARDS

Hydrochloric acid can cause severe burns. The vapors are extremely irritating to the skin, eyes, and respiratory system.

DISPOSAL

The waste mixtures should be flushed down the drain with water.

DISCUSSION

It has been known for a long time that colloidal suspensions scatter light strongly, producing the Tyndall effect, named after the British physicist John Tyndall (1820–1893), who extensively studied this phenomenon. This phenomenon is shown here for sols, dispersions of a solid in a liquid phase, and for gels, dispersions of a liquid in a solid phase. The theory of light scattering from these suspensions is well developed [4]. Shorter wavelengths are scattered more, and thus, the scattered light is rich in the blue region of the visible spectrum. The transmitted light is correspondingly poor in this region, so it is relatively rich in the red. According to Rayleigh's limiting law for a system dilute in particles that are small relative to the wavelength of the light, the intensity of the scattered light is inversely proportional to the fourth power of its wavelength, that is, it is proportional to λ^{-4}. Because the wavelength of orange light is a factor of 2 longer than that of blue light, the scattering of blue is an order of magnitude greater than that of orange. For coarser particles, the dependence is weaker. Thus, blue light scatters more than orange light, because the wavelength of blue light is shorter than that of orange light. The color of a colloid need not result only from light scattering; gold sols containing relatively large particles absorb as well as scatter light, and are red instead of blue [5].

In Procedure A, the beam gradually widens (or appears wider after the second addition of milk) because the amount of scattering increases. At first the light is scattered only once, and the beam appears narrow. As the number of suspended particles in the colloid increases, the scattering increases. Some of the light is scattered more than once, and the beam appears wider.

Lasers should not be used, because even a small helium-neon laser has a power output 100 times higher than the threshold of eye damage.

Nature provides its own sunset demonstration daily [6]. When viewed at an angle to the light source (the sun), the atmosphere appears blue because it scatters the shorter (blue) wavelengths of light preferentially. When viewed directly toward the light source, the atmosphere appears orange because the longer wavelengths are not scattered as much. This is most obvious at sunset or sunrise because the direct light of the sun passes through the greatest thickness of atmosphere at these times. It is inadvisable to attempt to look directly at the light of the sun at any time other than at sunrise or sunset, because of potential damage to the retina of the eye.

Scattering is also responsible for the color of clouds [7]. When a thick cloud is viewed upward toward the sun, it appears dark, because most of the light heading toward the observer is scattered and some is absorbed. On the other hand, when the same cloud is viewed from above, for example, from an aircraft or a mountaintop, reflected light from the cloud makes it appear white.

REFERENCES

1. W. G. Lamb, *Sci. and Children* (Jan.):101 (1984).
2. R. H. Goldsmith, *J. Chem. Educ.* 65:623 (1988).
3. D. D. Ebbing, *General Chemistry,* Houghton Mifflin Co.: Boston (1984).
4. H. R. Kruyt, Ed., *Colloid Science,* Vol. 1, *Irreversible Systems,* Elsevier: Amsterdam (1952).
5. B. Jirgensons and M. E. Straumanis, *A Short Textbook of Colloid Chemistry,* John Wiley and Sons: New York (1954).
6. J. Trefil, *Meditations at Sunset,* Charles Scribner's Sons: New York (1987).
7. C. F. Bohren, *Clouds in a Glass of Beer,* John Wiley and Sons: New York (1987).

9.42

An Ancient Colloid: India Ink

An example of a sol that has been in commercial use for nearly five millennia is India ink, a dispersion of carbon in oil or water.

MATERIALS

ca. 1 g carbon black †

ca. 8 g boiled linseed or tung oil

mortar and pestle

artist's paint brush

sheets of white paper

Bunsen burner and porcelain evaporating dish (optional)

PROCEDURE

Preparation and Presentation

Place 8 g of linseed oil in the mortar and grind the carbon black into the oil, either by rubbing the stick of carbon black in the oil or by grinding the powder into the oil with the pestle. Use the ink with a brush and draw pictures with it. If desired, you can make the carbon black by adjusting a Bunsen burner so the gas burns incompletely on a porcelain surface.

HAZARDS

Carbon black is a flammable solid; its airborne dust should be kept away from open flames.

DISPOSAL

Wash up with soap and water. India ink, when dry, is extremely difficult to remove from clothing.

† Carbon black is available in stick form from many art supply stores or in powder form from chemical supply houses.

DISCUSSION

Chinese, or "India," ink has been made by a process that has changed little since 2700 B.C. A recipe for ink was given by Vitruvius, a Roman engineer, at the time of the birth of Christ. Inks were initially made with lampblack, the soot from the smoke produced by the incomplete combustion of oils or resins. The historical procedure used in China produces a product that is competitive with modern technology. The lampblack is mixed with a drying oil, such as linseed oil or tung oil, which forms an oxidative film that sets the ink. The lampblack can also be mixed with water. Oil-based ink is commonly used in printing (block printing), whereas the water-based ink is used with pens [1]. Around 1864 carbon black began to replace lampblack as the ink pigment. Carbon black is formed by the deposition of carbon on a surface (such as metal or porcelain) during the incomplete combustion of hydrocarbons. Natural gas is the current fuel of choice. Lampblack particles are about 100 nm in size; carbon black contains particles in the range 20–30 nm [2]. In addition lampblack is flocculent and amorphous, but carbon black is semicrystalline (as revealed by X-ray diffraction patterns). The ink is made by mixing and grinding the pigment in boiled drying oils with other ingredients to control the flow and other properties of the ink. The formulation depends on the desired application. The small particle sizes of carbon black allow the formation of a very smooth, rapid-drying, free-flowing ink, with high covering power. The large surface area of the collection of particles and the particle sizes are typical of colloidal sols.

REFERENCES

1. J. Watrous, *The Craft of Old Master Drawings,* University of Wisconsin Press: Madison (1957).
2. C. L. Mantell, *Industrial Carbon. Its Elemental, Absorptive and Manufactured Forms,* 2d ed., Van Nostrand: New York (1946).

9.43

Canned Heat: Alcohol Gels

Two liquids are poured into a container, and after a short time, the container is inverted without spilling its contents.

MATERIALS

150 mL 95% ethanol, C_2H_5OH

30 mL saturated calcium acetate, $Ca(C_2H_3O_2)_2$ (To prepare about 100 mL of solution, place 40 g of $Ca(C_2H_3O_2)_2$ in a stoppered glass bottle with 100 mL of water and shake occasionally over a period of a few days. If all of the solid dissolves, add more. The saturated solution should be decanted from the solid which settles to the bottom of the bottle.)

metal can, with capacity of ca. 250 mL, or 250-mL beaker

matches (optional)

PROCEDURE

Preparation and Presentation

Pour 150 mL of 95% ethanol and 30 mL of saturated calcium acetate simultaneously into a clean, dry 250-mL beaker or metal can. Do not stir the mixture. Wait 10 seconds and invert the beaker. The gel will remain in place.

If the gel is in a metal can, it can be ignited with a match. (If the gel is ignited in a glass container, the container may shatter from the heat.) Darken the room so the faint blue flame can be seen.

HAZARDS

Ethanol is flammable, and the flame is difficult to see. Care should be taken to avoid burns.

DISPOSAL

Allow the ethanol to evaporate, or burn it off. Then, dissolve the calcium acetate in water and flush the solution down the drain with water.

DISCUSSION

The addition of ethanol to a saturated solution of calcium acetate in water reduces the solubility of the salt. The calcium acetate precipitates rapidly, forming a network of solid throughout the liquid. This network traps the liquid within it and forms a firm gel almost as soon as the liquids are mixed. The product is easy to make and relatively inexpensive. Similar preparations are available commercially as "canned heat" for portable cookstoves.

9.44

"Slime": Gelation of Poly(vinyl alcohol) with Borax

Two clear, colorless liquids are mixed and almost immediately form a gel. The gel can be formed into a ball, but if left unhandled, it flattens and runs.

MATERIALS

50 mL distilled water

2 g poly(vinyl alcohol), $+CH_2CH(OH)+_n$ (The polymer should be >99% hydrolyzed with an average molar mass of at least 10^5 g/mol.)†

5 mL of a 4% (by weight) aqueous solution of sodium tetraborate, $Na_2B_4O_7$ (To prepare 100 mL of solution, dissolve 7.6 g of $Na_2B_4O_7 \cdot 10H_2O$ [borax] in 100 mL of distilled water.)

250-mL beaker

stirring hot plate, with stir bar

thermometer, $-10°C$ to $+110°C$

100-mL disposable plastic cup, or beaker

flat-bladed wooden stirrer

PROCEDURE [1]

Preparation

Pour 50 mL of distilled water into the 250-mL beaker, place a stir bar in the beaker, and set the beaker on the stirring hot plate. Warm the water to a temperature not to exceed 90°C. While stirring the warm water, slowly sprinkle 2 g of poly(vinyl alcohol) on the surface of the water. This procedure prevents the formation of a sticky mass of polymer that is difficult to dissolve.

† Such polymers are available from the Eastman Kodak Company, Laboratory and Research Products Division, Rochester, New York 14650, and from the Aldrich Chemical Company, 940 W. Saint Paul Avenue, Milwaukee, Wisconsin 53233.

Presentation

Pour the poly(vinyl alcohol) solution and 5 mL of sodium borate solution together into the plastic cup and stir the mixture vigorously with the wooden stirrer. Gel formation begins almost immediately, and a material, "slime," with interesting physical properties results. Knead the gel into an elastic ball. Hold the ball in the palm of your hand and tip your hand. The ball will stretch out into a long column. Attempt to stretch the column abruptly; it will break.

DISPOSAL

The plastic cup and wooden stirrer used to prepare the gel should be discarded in a solid-waste receptacle.

The polymer solution may develop mold, in which case it should be discarded by mixing it with plenty of water and flushing the mixture down the drain. The gel can be discarded in a solid-waste receptacle, or by mixing it with plenty of water and flushing the mixture down the drain with more water. The gel is difficult to remove from carpeting. However, it comes out of clothing with laundering.

DISCUSSION

Poly(vinyl alcohol) is a polymer with a structure composed of repeating vinyl alcohol units:

$$\left[-\underset{\underset{\text{H}}{|}}{\overset{\overset{\text{H}}{|}}{\text{C}}} - \underset{\underset{\text{OH}}{|}}{\overset{\overset{\text{H}}{|}}{\text{C}}} - \right]_n$$

The average value of n for the poly(vinyl alcohol) used in this demonstration is at least 2300.

Sodium borate hydrolyzes in water to form a boric acid–borate ion buffer with a pH around 9.

$$B_4O_7{}^{2-}(aq) + H_2O(l) \leftrightarrows HB_4O_7{}^-(aq) + OH^-(aq)$$

In this demonstration, the borate ions react with the hydroxyl groups of the polymer to form cross links possibly with the elimination of water. These cross links probably involve hydrogen bonds that break and reform under flow. Such relatively weak cross-linking is necessary to produce the viscoelastic gel. The capability of poly(vinyl alcohol) to engage in extensive hydrogen bonding is suggested by its high solubility in water.

REFERENCE

1. E. Z. Casassa, A. M. Sarquis, and C. H. Van Dyke, *J. Chem. Ed.* 63:57 (1986).

9.45

Shake It and Move It: Thixotropy and Dilatancy

Steel balls fall more quickly through catsup after it has been stirred (Procedure A). A mixture of cornstarch and water flows freely when poured, but becomes stiff when stirred (Procedure B). A mixture of a white powder and water forms a gel when it is allowed to rest, but the gel liquefies when the mixture is shaken (Procedure C).

MATERIALS FOR PROCEDURE A

500 mL catsup

600-mL beaker

stand with ring small enough to support beaker

mirror, ca. 20 cm in diameter

10 (or more) small steel balls, ca. 6 mm in diameter (e.g., bicycle ball bearings)

stopwatch

stirring rod

MATERIALS FOR PROCEDURE B

125 mL cornstarch

50 mL water

2 600-mL beakers

stirring rod

spoon

MATERIALS FOR PROCEDURE C

150 mL distilled water

23 g fumed silica (e.g., CAB-O-SIL) †

500-mL Erlenmeyer flask

rubber stopper for flask

---

†CAB-O-SIL is manufactured by and available from the Cabot Corporation, 125 High Street, Boston, Massachusetts 02110.

PROCEDURE A [1, 2]

Preparation

Pour 500 mL of catsup into the 600-mL beaker. Set the beaker on the ring stand and place a mirror under the beaker so the bottom of the beaker can be seen. Allow the catsup to rest for at least 5 minutes.

Presentation

From about 3 cm above the surface of the catsup, drop one of the steel balls into the beaker of catsup and time how long it takes for the ball to reach the bottom of the beaker. Repeat this with four more balls, and compute the average of the times. Do not drop more than one ball at any one point at the surface of the catsup, because each ball leaves a track through the catsup that another ball may follow. A significant variation in the times is to be expected.

Stir the catsup for 1 minute. Drop another ball and time its travel. It will sink much more quickly than the previous balls. Time the falls of several more balls dropped at 1-minute intervals. The time required for the ball to sink to the bottom increases with each ball.

PROCEDURE B

Preparation and Presentation

Place 125 mL (about half a cup) of cornstarch in a 600-mL beaker. While stirring the cornstarch, slowly add about 50 mL of water. Stir the mixture with the stirring rod until it is homogeneous. The mixture will be very stiff and difficult to stir. Pour the mixture into the other beaker. It pours freely, although slowly. Stir the mixture again, and again it will be difficult to stir. Pour some of the mixture into the palm of your hand. Strike the mixture in your hand with the spoon. It will not splatter. Invite members of the audience to stir and pour the mixture, and contrast its resistance to stirring with its ability to flow freely.

PROCEDURE C

Preparation and Presentation

Pour 150 mL of distilled water into a 500-mL Erlenmeyer flask. Sprinkle about 5 grams of fumed silica onto the water, stopper the flask, and shake it until the silica is dispersed in the water. Repeat this process until all 23 g of the fumed silica have been added to the water. After the last addition, allow the flask to rest for several minutes, and the liquid in the flask will form a gel. After the liquid has gelled, tip the flask to show that its contents do not flow. Shake the flask vigorously, and the contents will become a viscous liquid. Allow the flask to rest for several minutes, and its contents will again solidify.

HAZARDS

Fumed silica can be irritating to the respiratory system. Care should be taken to avoid breathing the dust from fumed silica.

DISPOSAL

The balls can be removed from the beaker by using a bar magnet to draw them up the side of the beaker and out of the catsup. The catsup can be stirred and poured into a wide-mouthed jar for storage and use in repeated performances of this demonstration. Otherwise it can be diluted with water and flushed down the drain.

The cornstarch mixture should be diluted with water and flushed down the drain.

DISCUSSION

Thixotropy is the ability of certain substances to liquefy when agitated and to return to a gel form when at rest. The term *thixotropy* is derived from the Greek words *thixis,* meaning "the act of handling," and *trope,* meaning "change." Thixotropic substances are colloidal gels when solid and sols when liquefied. Examples of thixotropic substances include catsup, some hand creams, certain paints and printer's inks, and suspensions of clay in water. The reversibility and essentially isothermal nature of the gel-sol-gel transformation distinguish thixotropic materials from those that liquefy upon heating—for example, gelatin.

Thixotropic systems are quite diverse. Therefore, it is unlikely that a single descriptive theory can include them all. However, in general, the phenomenon is found only in colloidal suspensions. Particle shape seems to be important, and thixotropy is enhanced when the shapes of the particles are nonuniform. Particle shapes in thixotropic clays include disks, plates, and rod-like forms. In gels of high polymers, the polymer particles tend to be branched. Other important factors include the pH of the medium and the nature and concentration of electrolytes. For a particular system, the time required for gel formation can be modified by adjusting the electrolyte concentration. For example, the gelling time in one system can be adjusted by varying the proportion of water mixed with a fixed ratio of ammonium molybdate and thorium nitrate [3].

Various mechanisms can cause thixotropic behavior. For a gel system, agitation disrupts the three-dimensional structure that binds the system into a gel. Agitation might also introduce order into the system. In a system containing long polymeric molecules, these molecules can be disordered in the gel. When the gel is agitated, the molecules can align in the direction of flow, reducing the resistance to flow.

Some substances possess a property which is nearly the opposite of thixotropy. This property is called dilatancy. A dilatant substance is one that develops increasing resistance to flow as the rate of shear increases. A household example of a dilatant material is a thick dispersion of cornstarch in water, as displayed in Procedure B. This appears to be a free-flowing liquid when poured, but when it is stirred, it becomes very firm. Another familiar example of dilatancy is the phenomenon of wet sand appearing to dry and become firm when it is walked on.

A very versatile commercial product that can perform multiple functions simultaneously in colloidal formulations is CAB-O-SIL. This is a submicroscopic, fire-dried fumed silica. The product is generated in the gas phase according to the reaction

$$SiCl_4(g) + 2\ H_2O(g) \xrightarrow{1100°C} SiO_2(s) + 4\ HCl(g)$$

The particles produced have diameters on the order of 20 nm and are sintered together in branched chains with enormous surface areas of typically 200 m^2/g. The surface of the particles binds water and allows hydrogen bonds to form between individual particles. Because the index of refraction of CAB-O-SIL is close to that of many organic liquids, dispersions are usually transparent or translucent. The morphology of the product leads to its unusual properties. A few typical uses are described below.

Because fumed silica is inert, up to 2% by weight is allowed in foods. CAB-O-SIL is used as an emulsifier in salad dressings. The great efficiency in this application allows incorporation of more water in the product, resulting in "light" dressings. CAB-O-SIL can be added to catsup to make it thixotropic. It serves as an anticaking agent in cocoa, nondairy creamers, malted milk powder, baking soda, and so on. In fact, vinegar can be changed to a powder by adding 33% CAB-O-SIL. The powder can be added, for example, to dry sweet-sour mix; the acid is released, and the CAB-O-SIL amount drops to the 2% level when the mix is used. CAB-O-SIL acts as a thickener in heat-resistant margarine.

Lubricating oil is a nonpolar hydrocarbon liquid. The viscosity is increased by orders of magnitude with the addition of CAB-O-SIL. This allows the grease formed to be used at elevated temperature without chemical degradation and loss of viscosity (see Demonstration 9.27 on viscosity). The change of viscosity with temperature in bearing lubricants can be stabilized by adding a few percent of CAB-O-SIL, allowing higher speeds, temperatures, and pressures without oozing out. Paints can be stabilized in the same manner, permitting applications at room temperature and drying out at higher temperatures without runs. Also thicker films can be applied. Combined control of thixotropy and viscosity can be achieved.

Gelling of liquid cleaners by adding CAB-O-SIL allows the product to remain where it is applied so chemical action can take place where intended. Gels of wood preservative can be formed, so thick, nonsagging layers can be applied to the underground parts of wood piles. The preservative is released from the gel to the pile over a period of years. Plant food tends to cake under humid conditions; CAB-O-SIL prevents this by nesting between the fertilizer particles, thus keeping them from direct contact with each other.

REFERENCES

1. J. Walker, *Sci. Am.* 239 (Nov.): 186 (1978).
2. A. A. Collyer, *Phys. Educ.* 9:313 (1974).
3. D. Sievers, *Chem 13 News* 158:4 (1985).

9.46

Staying Dry: Phase Transitions of a Poly(acrylamide) Gel†

A clear colorless liquid is poured from one beaker into another containing a small amount of white powder. As the mixture is poured back and forth between the beakers, it thickens, and within a minute the mixture becomes too thick to pour. Another white powder is sprinkled on the thickened mixture, and when the mixture is stirred, it again becomes fluid.

MATERIALS

either

> 3 g superabsorbent poly(sodium acrylate) polymer‡

or

> 3 large "ultra-absorbent" disposable diapers (e.g., Ultra Pampers)
>
> 5-gallon plastic bag
>
> kitchen sieve

800 mL distilled water

ca. 1 mL food coloring (optional)

10 g sodium chloride, NaCl

2 1000-mL tall form beakers

glass stirring rod

PROCEDURE

Preparation

If superabsorbent polymer is available, no further preparation is required. If the water-absorbent material is to be extracted from the diapers, follow the instructions in this paragraph. Cut the diapers open, and without shredding the cotton-like filling, remove it and place it in the 5-gallon garbage bag. Note that the filling feels gritty; this

† We wish to thank several contributors to this demonstration. The procedure was developed by Professors C. M. Lang and D. L. Showalter of the University of Wisconsin–Stevens Point. Valerie Wilcox of the Boston Science Museum suggested the use of material found in ultra-absorbent diapers.

‡ A suitable polymer is J-550, available from The AbTech Company, P.O. Box 632, Muscatine, Iowa 52761.

grittiness results from the particles of water absorbent in the filling. When it is inside the plastic bag, shred the filling into small pieces. Close the bag and shake it vigorously for a minute to loosen the water-absorbent particles from the filling. Without opening the bag, manipulate the filling by rubbing it against itself. Shake the bag again. Open the bag and check the filling. If it still feels gritty, repeat the manipulation and shaking steps several times. When the filling no longer feels gritty, gently shake the bag to settle the water absorbent to the bottom of the bag. Open the bag and remove the filling. The particles of water absorbent can be separated from most of the remaining small pieces of filling by passing them through the sieve.

Presentation

Pour 800 mL of distilled water into one beaker, and, if desired, add about 1 mL of food coloring. In the other beaker place 3 g of the superabsorbent polymer powder or the water absorbent extracted from the diapers. Rapidly pour the (colored) water onto the superabsorbent polymer and quickly pour the mixture back and forth from one beaker to the other. Soon the mixture will become a gelatinous mass that is difficult to pour. To "unlock" the bound water, sprinkle 5–10 g of sodium chloride onto the surface of the gel, and stir it into the gel. As the gel begins to incorporate the salt, water is released until the entire beaker contains a rather viscous fluid.

HAZARDS

Superabsorbent polymer is very slippery when wet, and spills should be cleaned up at once. The toxicological properties of the polymer are unknown. Therefore, caution should be exercised to guard against inhalation of the powder's dust.

DISPOSAL

The final fluid should be flushed down the drain with water.

DISCUSSION

The superabsorbent polymer is a formulation of poly(acrylamide-co-sodium polyacrylate). The large reversible volume changes in chemical gels of this type with variations in solvent composition, temperature, and additives have been studied for many years [1–3].

The rapid, large volume changes in ionized poly(acrylamide) gels make this system a candidate for a "mechanochemical" switch, that is, a device in which large changes in dimensions are triggered by small changes in chemical composition [2]. Consumer uses of superabsorbent polymers occur in agriculture, industry, and personal-care products. In seed coating and hydro-mulching, water is drawn to and retained by the polymer which contacts the plant and helps maintain a uniform moisture level for germination and growth. Water can be removed from diesel and aviation fuels by contact with superabsorbent polymer. Aqueous wastes can be contained by binding to a relatively small amount of polymer. (Fumed silica is another candidate for this application; see Demonstration 9.45 for a discussion of its properties.) Some disposable diapers and feminine napkins contain superabsorbent polymer [4]. (Other materials used in disposable diapers are discussed in reference 5.) Novelty items in the shapes of fish, insects, and other things are also fabricated from this material; they swell to many times their original volume when they are immersed in water [6].

To prepare the poly(acrylamide) gel, the vinyl acrylamide monomers are polymerized by initiation with a mixture of tetramethylethylenediamine $((CH_3)_2N(CH_2)_2-N(CH_3)_2)$ and ammonium peroxydisulfate $((NH_4)_2S_2O_8)$. The tetramethylethylenediamine acquires an unpaired electron that attacks the carbon-carbon double bond of the acrylamide monomer, leaving a free radical as a product that continues the chain reaction. The result is poly(acrylamide) (see figure). If the solution were to contain only

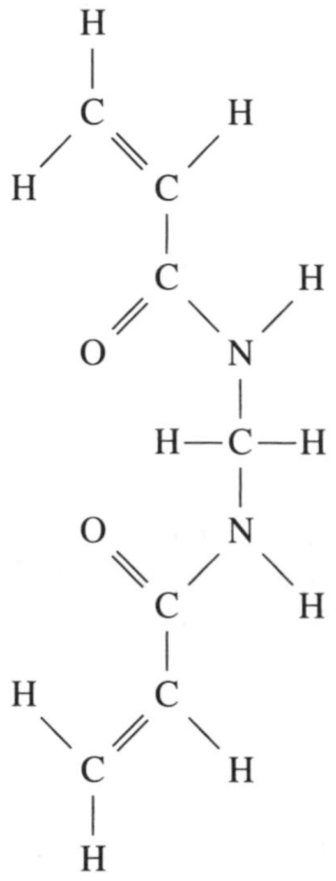

bis(acrylamide)

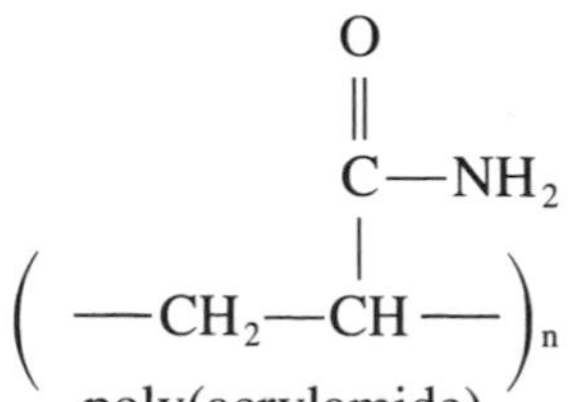

poly(acrylamide)

acrylamide monomers, the chain would not be branched and no gel would form. To generate the crosslinked "chemical" gel, bis(acrylamide) (see figure) is added to the formulation. This allows permanent links to be formed between separate chains. The polymerization takes about 30 minutes, at which time the gel is soaked in water to wash out unreacted initiator and monomer. An important final step is the soaking of the polymer in base (0.01M NaOH, for example) for a period of days to weeks. During this time, some (25% for 60 days of soaking) of the amide groups hydrolyze to carboxylate (sodium salt). The hydrolyzed gels have many times the original volume. They can be caused to collapse reversibly by changing the solvent composition and temperature or by adding electrolyte. Photographs appear in reference 1.

The swelling of the powdered polymer network when water is added can be understood as a balance of three factors [1]. They can be expressed in units of pressure and their sum (termed the osmotic pressure of the gel) is the independent variable that determines the swelling ratio (final volume/initial volume).

The *first* factor is the rubber elasticity of the network. If the gel is swollen, most polymer strands are stretched and there is a force tending to contract the gel and a negative contribution to the osmotic pressure. The *second* factor is the polymer-polymer affinity that derives from the difference between polymer-solvent and polymer-polymer interactions. The interactions with solvent can be either attractive or repulsive. If the interaction is attractive (polymer "likes" solvent), the gel will tend to incorporate solvent and swell. If the polymer prefers to contact polymer, there will be a sharp increase in the magnitude of the negative contribution to the pressure when the last amounts of solvent are excluded. This effect is short-ranged. If large amounts of solvent are incorporated into the gel, the resistance to expansion is small, because the polymer chains cannot contact each other appreciably. This behavior is a contributor to the very dramatic volume changes observed with small changes in conditions: the osmotic pressure–swelling ratio curves are similar to p-V curves for liquid-vapor equilibria where first-order phase transitions are observed. The *third* factor is termed the hydrogen ion (sodium ion) pressure that is associated with ionization of the polymer network. The counterions are forced to remain in the gel because of the electrostatic attraction of the polyelectrolyte network. This is a positive contribution to the osmotic pressure. If electrolyte (sodium chloride) is added, the electrostatic interactions are screened, the ion pressure is reduced, the network contracts, and a suspension of polymer particles in solution results, as in the second part of the demonstration.

REFERENCES

1. T. Tanaka, *Sci. Am.* 244 (Jan.): 124 (1981).
2. Y. Hirose, T. Amiya, Y. Hirokawa, and T. Tanaka, *Macromolecules* 20:1342 (1987).
3. R. Buscall, T. Corner, and J. F. Stageman, *Polymer Colloids,* Elsevier Publishing Co.: London (1985).
4. D. Tanis, *Diapers and Polymers,* presented at CEA Convention, South Bend, Indiana, Oct. 1987.
5. J. Cleary, *J. Chem. Educ.* 63:422 (1986).
6. W. Bleam, Radnor High School, Radnor, Pennsylvania, personal communication to Valerie Wilcox, 1987.

9.47

Growing Colorful Crystals
in Gels†

Over a period of several days, large crystals of various shapes and colors grow in gels in test tubes.

MATERIALS FOR PROCEDURE A

15 mL 1M acetic acid, $HC_2H_3O_2$ (To prepare 1 liter of solution, pour 57 mL of glacial [17.5M] $HC_2H_3O_2$ into 600 mL of distilled water and dilute the resulting solution to 1.0 liter.)

2 mL 1M lead acetate, $Pb(C_2H_3O_2)_2$ (To prepare 1 liter of solution, dissolve 380 g of $Pb(C_2H_3O_2)_2 \cdot 3H_2O$ in 600 mL of distilled water and dilute the resulting solution to 1.0 liter.)

15 mL aqueous sodium silicate, Na_2SiO_3, with a density of ca. 1.06 g/mL (To prepare about 1 liter of solution, dilute 160 mL of commercial 38–40% sodium silicate solution to 1.0 liter with distilled water and stir the mixture. Alternatively, the solution can be prepared by dissolving 244 g of Na_2SiO_3 $\cdot 9H_2O$ [sodium metasilicate] in 600 mL of distilled water and diluting the resulting solution to 1.0 liter.)

2 mL 2M potassium iodide, KI (To prepare 1 liter of solution, dissolve 332 g of KI in 600 mL of distilled water and dilute the resulting solution to 1.0 liter.)

50-mL beaker

gloves, plastic or rubber

stirring rod

test tube, 15 mm × 200 mm

test tube rack

stopper for test tube

MATERIALS FOR PROCEDURE B

See Materials for Procedure A.

†This demonstration was developed by Professor Earle Scott of Ripon College, Ripon, Wisconsin.

MATERIALS FOR PROCEDURE C

15 mL 1M acetic acid, $HC_2H_3O_2$ (For preparation, see Materials for Procedure A.)

2 mL 2M potassium iodide, KI (For preparation, see Materials for Procedure A.)

15 mL aqueous sodium silicate, Na_2SiO_3, with a density of ca. 1.06 g/mL (For preparation, see Materials for Procedure A.)

2 mL 1M mercury(II) nitrate, $Hg(NO_3)_2$ (To prepare 1 liter of solution, dissolve 325 g $Hg(NO_3)_2$ in 600 mL of distilled water and dilute the resulting solution to 1.0 liter.)

50-mL beaker

stirring rod

test tube, 15 mm × 200 mm

test tube rack

gloves, plastic or rubber

stopper for test tube

MATERIALS FOR PROCEDURE D

7.5 mL 1M acetic acid, $HC_2H_3O_2$ (For preparation, see Materials for Procedure A.)

7.5 mL distilled water

15 mL aqueous sodium silicate, Na_2SiO_3, with a density of ca. 1.06 g/mL (For preparation, see Materials for Procedure A.)

5 mL 0.2M mercury(II) chloride, $HgCl_2$ (To prepare 1 liter of solution, dissolve 50 g $HgCl_2$ in 800 mL of distilled water and dilute the resulting solution to 1.0 liter.)

50-mL beaker

stirring rod

test tube, 15 mm × 200 mm

test tube rack

gloves, plastic or rubber

stopper for test tube

MATERIALS FOR PROCEDURE E

15 mL 1.5M tartaric acid, $H_2C_4H_4O_6$ (To prepare 1 liter of solution, dissolve 225 g of $H_2C_4H_4O_6$ in 600 mL of distilled water and dilute the resulting solution to 1.0 liter.)

15 mL aqueous sodium silicate, Na_2SiO_3, with a density of ca. 1.06 g/mL (For preparation, see Materials for Procedure A.)

2 mL of *one* of the following:

> 1M copper sulfate, $CuSO_4$ (To prepare 1 liter of solution, dissolve 250 g of $CuSO_4 \cdot 5H_2O$ in 600 mL of distilled water and dilute the resulting solution to 1.0 liter.)

> 1M potassium chloride, KCl (To prepare 1 liter of solution, dissolve 75 g of KCl in 600 mL of distilled water and dilute the resulting solution to 1.0 liter.)

> 1M potassium nitrate, KNO_3 (To prepare 1 liter of solution, dissolve 100 g of KNO_3 in 600 mL of distilled water and dilute the resulting solution to 1.0 liter.)

50-mL beaker

stirring rod

test tube, 15 mm $\times$ 200 mm, with stopper

test tube rack

PROCEDURE A [1, 2]

Preparation

Pour 15 mL of 1M acetic acid into the 50-mL beaker. Wearing gloves, add 2 mL of 1M lead acetate and stir the mixture. While stirring the mixture vigorously, slowly pour 15 mL of the sodium silicate solution into the beaker. The mixture must be stirred continuously while the silicate is added, because if any region of the mixture becomes basic, the lead will precipitate. Pour the mixture into the test tube and place the tube in a rack until a gel forms (5–30 minutes).

Presentation

Pour 2 mL of 2M potassium iodide solution onto the top of the gel and stopper the tube. Place the tube where it can be observed undisturbed over a period of several days. Compact yellow crystals of lead iodide will form, growing from the top of the gel toward the bottom.

PROCEDURE B

Preparation

Pour 15 mL of 1M acetic acid into the 50-mL beaker. Add 2 mL of 2M potassium iodide and stir the mixture. While stirring the mixture vigorously, slowly pour 15 mL of the sodium silicate solution into the beaker. Pour the mixture into the test tube and place the tube in a rack until a gel forms (5–30 minutes).

Presentation

Wearing gloves, pour 2mL of 1M lead acetate solution onto the top of the gel and stopper the tube. Place the tube where it can be observed undisturbed over a period of several days. Yellow fern-like crystals of lead iodide will form in the gel, starting at the top and spreading toward the bottom.

PROCEDURE C

Preparation

Pour 15 mL of 1M acetic acid into the 50-mL beaker. Add 2 mL of 2M potassium iodide and stir the mixture. While stirring the mixture vigorously, slowly pour 15 mL of the sodium silicate solution into the beaker. Pour the mixture into the test tube and place the tube in a rack until a gel forms (5–30 minutes).

Presentation

Wearing gloves, pour 2 mL of 1M mercury(II) nitrate solution onto the top of the gel and stopper the tube. Place the tube where it can be observed undisturbed over a period of several days. Bright orange crystals of mercury(II) iodide will form in the gel, starting at the top and spreading toward the bottom.

PROCEDURE D

Preparation

Pour 7.5 mL of 1M acetic acid and 7.5 mL of distilled water into the 50-mL beaker. While stirring the mixture vigorously, slowly pour 15 mL of the sodium silicate solution into the beaker. Pour the mixture into the test tube and place the tube in a rack until a gel forms (2–10 minutes).

Presentation

Wearing gloves, pour 5 mL of 0.2M mercury(II) chloride solution onto the top of the gel and stopper the tube. Place the tube where it can be observed undisturbed over a period of several days. Crystals of various forms of mercury(II) oxychlorides will form in the gel, starting at the top and spreading toward the bottom, with gradual conversion between forms.

PROCEDURE E

Preparation

Pour 15 mL of 1.5M tartaric acid into the 50-mL beaker. While stirring the mixture vigorously, slowly pour 15 mL of the sodium silicate solution into the beaker. Pour the mixture into the test tube, stopper the tube, and place it in a rack until a gel forms (up to several days).

Presentation

Pour 2 mL of 1M copper sulfate, 1M potassium chloride, or 1M potassium nitrate onto the top of the gel and stopper the tube. Place the tube where it can be observed undisturbed over a period of several days. With copper sulfate, bright blue crystals of various copper(II) tartrates will form in the gel. With potassium chloride or nitrate, large clear crystals of potassium tartrate will form. The crystals will start growing at the top of the gel and spread downward.

HAZARDS

Mercury and all of its compounds are poisonous, and toxic effects can result from inhalation, ingestion, or skin contact. Chronic effects can result from exposure to small concentrations over an extended period of time. The dust from salts of mercury is quite poisonous and can irritate the skin and eyes.

Lead acetate is harmful if taken internally. The dust from lead salts should not be inhaled. The effects of exposure to small concentrations can be cumulative, causing loss of appetite and anemia.

Glacial acetic acid can irritate the skin, and its vapors are harmful to the eyes and respiratory system.

Sodium silicate solutions can be quite basic. Therefore, prolonged exposure of the skin to these solutions should be avoided.

DISPOSAL

The stoppered tubes can be stored indefinitely. The crystals will gradually change over many months, eventually becoming static. Those tubes that contain lead or mercury should be discarded in receptacles for solid, heavy metal wastes. The others can be discarded in a standard waste receptacle.

DISCUSSION

The art of growing crystals in gels is old. The original interest was in growing large crystals for other experiments, in probing the spatial distribution of crystal formation, and in the inherent beauty of the phenomenon. In some cases, the crystals do not form randomly distributed throughout the gel. Instead they form in layers or rings (Liesegang rings). A demonstration of layer formation by crystallization in a gel is described in Demonstration 9.48. In the absence of convection, the components of the growing crystals reach them by diffusion. Equations describing the diffusion-controlled growth rates for several assumed geometries appear in the literature [3].

The rates and processes of the formation of silica gel are also understood [4]. The gel used in this demonstration is prepared by acidifying a solution of sodium silicate (water glass). Water glass is prepared by dissolving silica (sand, SiO_2) in hot aqueous sodium hydroxide. This produces a solution of sodium metasilicate (Na_2SiO_3). This solution is quite basic. When the solution is acidified, the metasilicate hydrolyzes, and SiO_2 precipitates out in an amorphous gelatinous form called silica gel. The silica gel is a three-dimensional polymer of SiO_2. There has been a recent study of the hydrolysis reaction, motivated by the advantages of producing a high-quality glass from silica gels [5].

REFERENCES

1. H. K. Henisch, *Crystal Growth in Gels,* Pennsylvania State University Press: University Park, Pennsylvania (1970).
2. S. L. Suib, *J. Chem. Educ.* 62:81 (1985).
3. F. C. Frank, *Proc. Roy. Soc.* 201A:586 (1950).
4. R. K. Iler, *The Colloid Chemistry of Silica and Silicates,* Cornell University Press: Ithaca, New York (1955).
5. I. Artaki, M. Bradley, T. W. Zerda, and J. Jones, *J. Phys. Chem.* 89:4399 (1985).

9.48

Liesegang Rings: Spatial Oscillation in Precipitate Formation

Over a span of several weeks, bands of crystals appear at regular intervals down a gel-filled test tube. The gel is a sodium silicate (Na_2SiO_3) solution, commercial water glass, and the gel is doped with potassium chromate (K_2CrO_4) and silver nitrate ($AgNO_3$) in acetic acid ($HC_2H_3O_2$). The bands are a form of static spatial oscillation in precipitate formation, a result of competition between diffusion and chemical reaction in controlling the reactant concentrations in different regions of the test tube. This is described in Demonstration 7.14 in Volume 2 of this series.

9.49

Colorful Stalagmites:
The Silicate Garden

Crystals of various salts are added to a sodium silicate solution, and colorful columnar membranes grow up from the surface of each crystal, some reaching the surface of the solution.

MATERIALS

450 mL commercial sodium silicate solution (water glass)

750 mL distilled water

crystals of hydrated forms of several of the following (color of membrane produced is indicated in parentheses):

aluminum(III) chloride, $AlCl_3$ (white); cobalt(II) chloride, $CoCl_2$ (dark blue); chromium(III) chloride, $CrCl_3$ (dark green); copper(II) chloride, $CuCl_2$ (light blue-green); iron(III) chloride, $FeCl_3$ (yellow); nickel(II) chloride, $NiCl_2$ (light green); tin(IV) chloride, $SnCl_4$ (white); iron(II) sulfate, $FeSO_4$ (grayish white); nickel(II) sulfate, $NiSO_4$ (green); aluminum(III) nitrate, $Al(NO_3)_3$ (white); cobalt(II) nitrate, $Co(NO_3)_2$ (dark blue); and chromium(III) nitrate, $Cr(NO_3)_3$ (dark green)

1500-mL beaker

watch glass to cover beaker

PROCEDURE

Preparation and Presentation

Dilute 450 mL of commercial sodium silicate solution with 750 mL of distilled water in a 1500 mL beaker. Drop crystals of the salts into the solution so that they are distributed evenly on the bottom of the beaker. Cover the beaker with the watch glass. The formation of colloidal semipermeable membranes of insoluble silicate salts will begin as soon as the salts start to dissolve (almost immediately) and will continue to grow into columnar formations for several days. The color of each column is determined by the identity of the salt crystal from which it grows.

HAZARDS

Aluminum chloride is caustic and can cause burns to the skin on prolonged contact.

Chromium(III) nitrate is a cancer-suspect agent. Contact with the skin and breathing of the dust should be avoided.

Copper compounds are harmful if taken internally. Dust from copper compounds can irritate mucous membranes.

Nickel salts and their solutions irritate the eyes upon contact. Dust from solid nickel compounds is harmful, and the compounds are toxic if ingested. Nickel salts are suspected carcinogens.

Tin(IV) chloride is corrosive and can cause burns to the skin on prolonged contact.

DISPOSAL

The garden should be flushed down the drain with water.

DISCUSSION

This demonstration is well known, and kits can be purchased at toy stores. A description with photographs of ripe gardens appears in reference 1.

Crystals of various salts are dropped into a sodium silicate solution (Na_2SiO_3). As the salts dissolve, the metal ions combine with silicate ions and form membranes of insoluble silicates around the crystals. The inside of the membranes contains lower water concentrations and higher salt concentrations than the outside, so water passes inward by osmosis, ultimately causing breaks in the membrane and formation of more membrane surface as the salt solution contacts fresh sodium silicate. Growth takes place with a coral-like morphology in a generally upward direction. This demonstration illustrates the dynamic formation of solid phases controlled by osmosis and diffusion in a system far from equilibrium. Such conditions are required for oscillations in time and space in chemical reactions. Sodium silicate gel can be used as a medium for producing spatially separated bands of silver chromate crystals—an example of Liesegang rings (see Demonstration 9.48).

REFERENCE

1. G. B. Kauffman and C. A. Ferguson, *J. College Sci. Teach.* 491 (1986).

9.50

Colors and Shapes of Soap Films and Bubbles

The geometry, elasticity, and color patterns of soap films are explored (Procedure A). A bubble is formed over a person's head, and the person can look out from the inside of the bubble (Procedure B).

MATERIALS FOR PROCEDURE A

one of the following solutions:

> commercial soap bubble kit (available at toy stores)

> 500 mL unscented dishwashing detergent dissolved in water to make 4 liters of solution (Stir gently to minimize suds formation.) (Do not use detergent for automatic dishwashers; it will not form bubbles.)

> 1.4 g triethanolamine, $N(CH_2CH_2OH)_3$, and 2 g oleic acid, $CH_3(CH_2)_7$-$CH:CH(CH_2)_7COOH$, mixed with 100 g of 85% glycerol, CH_2OHCH-$OHCH_2OH$, in water (This formulation permits the formation of particularly thin, long-lasting films [1, 2].)

metal wire frames of various geometries (rectangular, tetrahedral, cubical), with edges ca. 7 cm long

container to hold soap solution with a depth sufficient to allow complete immersion of the frames

MATERIALS FOR PROCEDURE B

200 mL liquid dishwashing detergent (Do not use detergent for automatic dishwashers; it will not form bubbles.)

40 mL glycerin

1920 mL water

40 g table sugar

nail, ca. 8 cm long

board, ca. 1 cm $\times$ 3 cm $\times$ 20 cm

Bunsen burner

round plastic washtub, ca. 60 cm in diameter

cardboard box, large enough for a person to fit inside

utility knife

2 hair dryers, with cold air settings

2-liter beaker

600-mL beaker

4 70-cm pieces of 2×4 lumber

plastic tablecloth, with each edge at least 90 cm long

PROCEDURE A

Preparation and Presentation

Dip the rectangular frame into the soap solution and withdraw the frame. A rectangular planar film will form. Blow gently on the film to show its elasticity. Blow harder and force the film to break. Note the formation of spherical bubbles in the air.

Dip the tetrahedral frame into the soap solution and withdraw the frame. Note that the films do not cover the faces of the tetrahedron but rather form six triangular planar films with a common contact point at the center of the tetrahedron.

Do the same thing with the cubical frame and note the complex film structure that forms.

Note the color patterns formed in the soap bubbles and films, and see how the patterns change when the films are distorted (by blowing) or when the films age and become thinner. If the third soap formulation is tried, the films will ultimately appear black [2, 3].

PROCEDURE B [4]

Preparation

Pound a nail about 8 cm long through a board about 1 cm thick. Heat the nail in the flame of a Bunsen burner. Draw a circle about 50 cm in diameter on the bottom of a round plastic washtub. Use the hot nail to trace the circle, cutting a hole in the washtub large enough so that a person's head can fit through the opening. The nail may need to be reheated several times to accomplish this.

Cut a hole of about the same size as that in the washtub in the small side of a cardboard box large enough for a person to fit into. Cut a door in a side perpendicular to that having the hole (see figure). Near the bottom of the box, cut two holes just large enough so the two hair dryers can be mounted in them. Mount the hair dryers and plug them into electrical outlets.

In a 2-liter beaker, dissolve 200 mL of liquid dishwashing detergent and 40 mL of glycerin in 1260 mL of water. In the 600-mL beaker, dissolve 40 g of table sugar in 460 mL of water. Pour the sugar solution into the detergent solution and stir the mixture.

Arrange the 4 pieces of 2×4 lumber in a square on the floor. Drape a plastic tablecloth over the boards to form a large, shallow dish. Pour the detergent solution into this dish.

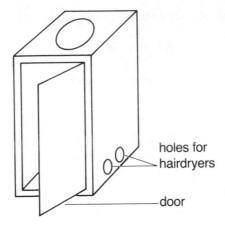

Practice making a soap film over the rim of the washtub by dipping the rim into the detergent solution. This is most easily accomplished by raising one edge of the rim out of the solution and lifting the tub so its opening is held vertically.

Presentation

Have someone enter the box through the door and bend down to keep the head below the hole. Be sure the door is closed. Form a soap film over the rim of the washtub. Set the bottomless tub over the hole in the box. Turn on the hair dryers (set to their cold air settings). As the dryers blow air into the box, a bubble will form at the rim of the washtub. When the bubble is large enough, instruct the person in the box to rise and look out through the bubble from the inside. Turn off the hair dryers. If the door is sealed sufficiently well, the bubble will maintain its size for several minutes after the dryers have been turned off.

HAZARDS

Triethanolamine is a strong base and can irritate the skin and mucous membranes.

DISPOSAL

The wastes should be flushed down the drain with water.

DISCUSSION

The geometry of soap films can be understood by realizing that the films minimize their total area (subject to the constraint that they are attached to the edges of the frame and any other constraints). A useful starting point is Laplace's formula [5]:

$$\Delta P = \gamma\left(\frac{1}{r} + \frac{1}{r'}\right)$$

where ΔP is the pressure difference across the surface,

γ is the interfacial tension,

and   r and r' the radii of curvature of the film in two perpendicular directions.

With the various frames, $\Delta P = 0$, and so $r = r' = \infty$, and the entire film must be composed of planar parts. Because the contribution of the surface to the Gibbs function, G, is γA (A is the area of the film), equilibrium will be reached when A is a minimum causing G to be a minimum.

The film on the rectangular frame has no choice but to form a single plane. The situation is more complicated for the tetrahedral and cubical frames.

The pattern for the tetrahedron forms six equivalent triangular faces. If the length of each edge of the tetrahedron were 1 cm and films covered each of the four faces, the total area would be 1.72 cm^2, whereas the total area of the six triangular films is 1.06 cm^2, a smaller value. Thus, this pattern forms in preference to covering the faces. Similar considerations apply to the cubical geometry where 8 triangular films attempt to contact at the center of the cube (this is sufficient to support the assembly of films, and the other 4 possible films which would produce a completely symmetrical array do not form; in fact, the total area of 8 films is less than that of the area of the six faces of the cube, whereas 12 films would have a larger total area). Often air is trapped and a small cube appears in the center of the film pattern. This helps to support the films, and the size of the inner cube can be reduced by drawing air out with a hypodermic whose needle has been coated with glycerol. Another configuration found has a central square to which the other films are attached.

Free bubbles form spheres in the absence of gravitational effects (see Demonstration 9.25). If the bubble were not spherical, its two radii of curvature would be unequal, and pressure differences would exist inside the bubble. These would disappear when the high-pressure region inside the bubble expands at the expense of the low-pressure region, until the radii become equal and the bubble assumes spherical shape. (Another demonstration illustrating the properties of soap films is Demonstration 6.29 in Volume 2 of this series: Facilitated Transport of Carbon Dioxide Through a Soap Film.)

The color patterns of soap bubbles are caused by interferences among the wavelength components of white light. Depending on the film thickness and the wavelength, the components can interfere constructively and be seen (thickness being an odd multiple of quarter wavelengths) or destructively and be invisible (thickness being an even multiple of quarter wavelengths), with all possibilities between these extremes. If the film thickness is not constant *or* if the angles of incidence vary, different parts of the film will exhibit different colors [2, 6]. The observation and analysis of this phenomenon dates at least back to Newton—a related case is that of reflection from a plano-convex lens placed on a planar surface—the resulting pattern is termed Newton's rings. If the film becomes very thin (about 10 nm thick), as it will if allowed to drain and/or evaporate, nearly complete destructive interference occurs for all wavelengths and the film appears black [2]. It has been claimed that references to black films were written by the Assyrians on clay tablets more than 3000 years ago [3]. (Another colorful phenomenon, but one that is fundamentally different, is birefringence, which is presented in Demonstration 9.51.)

The stability of suspended soap films has been investigated [7]. For an array of films with trapped air inside suspended from two circular frames with a common axis, depending on the difference of pressure inside and outside the films, the film shape can be shifted from spherical to cylindrical or become unstable and break. If the films do not trap air inside, a catenoid surface forms between the frames. Here the film has two radii of curvature with equal magnitude but opposite sign ($r = -r'$).

REFERENCES

1. J. Walker, *Sci. Am.* 257 (Aug.): 104 (1987).
2. K. J. Mysels, K. Shinoda, and S. Frankel, *Soap Films,* Pergamon Press: New York (1959).
3. P-G. de Gennes, *Phys. Today* 40 (July): 7 (1987).
4. S. Sato, *J. Chem. Educ.* 65: 616 (1988).
5. L. L. Landau and E. M. Lifshitz, *Statistical Physics,* Addison-Wesley Publishing Co.: Reading, Massachusetts (1958).
6. A. Sommerfeld, *Optics,* Academic Press: New York (1954).
7. J. Walker, *Sci. Am.* 257 (Sept.): 108 (1987).

9.51

Rotating Rainbows:
A Solution in Polarized Light

The color of light transmitted through a sugar solution depends on the rotational position of two polarizing filters and on the depth of the solution.

MATERIALS

overhead projector

550 mL distilled water

550 mL 50% by weight aqueous sucrose, $C_{12}H_{22}O_{11}$ (To prepare the 50% solution, gradually add 350 g sucrose [table sugar] to 350 mL of distilled water, with stirring. Stir the mixture until all of the sucrose has dissolved.)

poster board, large enough to cover stage of overhead projector

2 polarizing filters, ca. 20 cm × 20 cm†

4 600-mL tall-form beakers

PROCEDURE [1, 2]

Preparation

Cut two circular openings in the poster board, so that the diameter of each opening is slightly smaller than the diameter of a 600-mL tall-form beaker. The openings should be close enough together so both can be covered by one of the polarizing filters.

Presentation

Place the poster board on the stage of the overhead projector. Place one of the polarizing filters on top of the poster board so that it covers both openings. Hold the other filter horizontally over the first and rotate it in the horizontal plane. The projected light will dim or brighten depending on the degree of rotation of the second filter with respect to the first.

Pour distilled water to a depth of about 1 cm in one of the beakers (about 50 mL) and to a depth of about 10 cm in another (about 500 mL). Place both beakers side by side on top of the polarizing filter on the overhead projector. Hold the other filter horizontally above the beakers and rotate it in the horizontal plane. The projected light

---

† Polarizing filters are available from the Edmund Scientific Company, 101 E. Gloucester Pike, Barrington, New Jersey 08077.

passing through the two beakers will dim or brighten to the same degree depending on the degree of rotation of the second filter with respect to the first.

Pour 50-50 sucrose solution to a depth of about 1 cm into the third beaker and to a depth of about 10 cm in the last. Remove the two beakers of water from the overhead projector and replace them with the beakers of sucrose solution on top of the polarizing filter. Hold the other filter horizontally above the beakers and rotate it in the horizontal plane. The projected light passing through the two beakers will have different colors, and the colors will change depending on the degree of rotation of the second filter with respect to the first. Rest the top filter on the two beakers and note both the colors of the light through each beaker and the position of the upper filter. Lift the upper filter, pour about 50 mL of solution from the nearly full beaker into the nearly empty beaker, and position the upper filter as it was previously. The colors of the light through each beaker will change. Repeat the pouring step several times, noting the colors.

DISPOSAL

The sucrose solution can be saved for several weeks for reuse in repeat performances of this demonstration, but mold may eventually grow in it. It should then be flushed down the drain with water.

DISCUSSION

The colors produced by the sugar solution between polarizing filters are a result of the action of both the polarizing filters and the sugar solution on the light that passes through them. Both affect the light, and their combined effects reveal the colors. The effect of both the polarizing filters and the sugar solution on light is interpreted in terms of the wave properties of light.

Visible light is electromagnetic radiation with wavelengths between about 400 nm and 800 nm. Electromagnetic radiation is characterized by electrical and magnetic fields perpendicular to each other. These fields vary sinusoidally (the waves) in both space and time perpendicular to the direction of travel (see figure). If ordinary (white) light were passed through a polarizing filter, as in this demonstration, only radiation with a definite orientation of the electric field (and a definite perpendicular orientation of the magnetic field) would pass through the filter. If a second polarizing filter at an angle, θ, to the first were placed in the path of the light beam, the intensity of the light

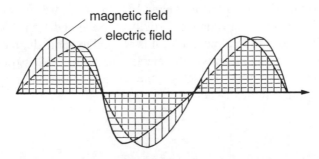

Oscillating electric and magnetic fields of electromagnetic radiation.

passing through the second filter would be $\cos^2\theta$ of that passing through the first, *provided that the plane of the electric field were not rotated by any medium in the intervening space*. This is the case when water is placed between the polarizers. If a sucrose solution were placed between the polarizers, the electric field of the beam would rotate by an amount proportional to the length of the path through the solution, roughly proportional to the concentration of the solution, and only weakly dependent on temperature. For fixed path length, concentration, and temperature, the rotation also depends strongly on the wavelength of the light. (The wavelength is represented by the distance between oscillations in the figure.) The table shows how the specific rotation of a sucrose solution varies with wavelength; $[\alpha]^t_\lambda$ represents the specific rotation at Celsius temperature, t, and wavelength, λ, and it is defined by α/Lc, where α is the rotation in degrees, L is the path length in dm, and c is the concentration in g/mL. The short wavelengths are rotated more than the longer wavelengths. This causes the second polarizer to affect the intensities of various wavelengths differently. Because the color of light is determined by its wavelength, the color of the light transmitted varies with the path length. The dependence of specific rotation on wavelength is termed optical rotatory dispersion.

Specific Rotation of Aqueous Sucrose at 20°C [3]

Wavelength (nm)	Color	Concentration (moles sucrose/liter of solution)		
		1/8	1/16	1/32
		Specific rotation, $[\alpha]^{20}_\lambda$		
656	red	53.18	53.22	53.48
589	orange	66.50	66.71	66.81
535	yellow	82.25	82.76	82.93
508	yellow-green	91.53	91.79	92.59
479	green	104.24	104.67	105.42
447	blue	121.63	122.80	123.80

Colors are revealed when a sugar solution is placed between polarizing filters because light of different colors (wavelengths) is rotated by differing amounts as it passes through the sugar solution. If yellow light is rotated by 90 degrees, it will pass through a polarizer rotated by 90 degrees. If blue light is rotated by 180 degrees, it will not pass through a polarizer rotated by 90 degrees. Yellow light will pass through and blue light will not, causing the emergent light to be colored.

A sugar solution rotates the plane of polarized light because the sugar molecules are unsymmetrical in a special way. Sugar molecules and their mirror images are not superimposable. We can make a comparison with a right hand, whose mirror image is a left hand, which is not identical to a right hand [4]. In this sense, sugar molecules have "handedness." When the electromagnetic field of polarized light interacts with the electric field of the electrons in a molecule, it is rotated to some degree. Because sugar molecules have the same handedness, the electromagnetic field is rotated in the same direction by all of the molecules. Therefore, there is a net rotation of the plane of polarized light, and sugar is "optically active."

REFERENCES

1. G. F. Hambly, *J. Chem. Educ.* 65:623 (1988).
2. D. J. Kolb, *J. Chem. Educ.* 64:805 (1987).
3. *International Critical Tables,* Vol. 2, 1st ed., McGraw-Hill: New York (1928), p. 336.
4. M. Gardner, *The Ambidextrous Universe,* 2d ed., Charles Scribner's Sons: New York (1979).

9.52

Floating and Sinking: Osmosis Through a Copper Hexacyanoferrate(II) Membrane

A blue liquid in a dropper is injected near the bottom of a test tube filled with a yellow liquid. Where the liquids meet, a drop of blue solution surrounded by a red-brown membrane forms. After the dropper is removed, the blue drop enclosed in the membrane slowly rises in the yellow solution, but before it reaches the surface, it stops rising and sinks slowly to the bottom of the tube.

MATERIALS

overhead projector that can be positioned on its side

10.74 g copper sulfate pentahydrate, $CuSO_4 \cdot 5H_2O$

200 mL distilled water

11.83 g potassium hexacyanoferrate(II) trihydrate, $K_4Fe(CN)_6 \cdot 3H_2O$

2 100-mL beakers

2 100-mL volumetric flasks, with stoppers

25-mm × 200-mm test tube

mirror, ca. 20-cm square, or larger

3 stands, each with clamp

water-tight container, with flat parallel sides (See procedure for description.)

ring that can be clamped to stand

dropper, ca. 20 cm long, with bulb

PROCEDURE [1]

Preparation

In a 100-mL beaker, dissolve 10.74 g of $CuSO_4 \cdot 5H_2O$ in 75 mL of distilled water. Transfer the solution to one of the 100-mL volumetric flasks. Rinse the beaker with 10 mL of distilled water and add the rinse to the volumetric flask. Dilute the solution in the flask to the calibration mark with distilled water. Stopper the flask and invert it several times to mix its contents. This solution is 0.430M in $CuSO_4$, and its density at 20°C is 1.0647 g/mL [2].

In the other 100-mL beaker, dissolve 11.83 g of $K_4Fe(CN)_6 \cdot 3H_2O$ in 75 mL of distilled water. Transfer the solution to the remaining 100-mL volumetric flask. Rinse

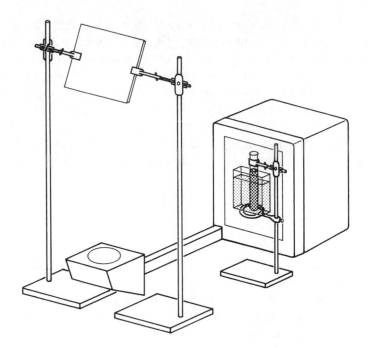

the beaker with 10 mL of distilled water and add the rinse to the volumetric flask. Dilute the solution in the flask to the calibration mark with distilled water. Stopper the flask and invert it several times to mix its contents. This solution is 0.280M in $K_4Fe(CN)_6$, and its density at 20°C is 1.0655 g/mL [2].

Assemble the apparatus as shown in the figure. Set the overhead projector on its side so that its stage is vertical and it is projecting toward the ceiling. Clamp the mirror between two stands and position it over the head of the projector, so that the mirror reflects the projected image onto the screen. Clamp the 25-mm × 200-mm test tube vertically to the third stand and fill it with the yellow $K_4Fe(CN)_6$ solution. Immerse the test tube in a water-filled container with straight parallel sides.† Fasten the ring to the stand to support the container. Position the test tube inside the container about 6 cm from the vertical stage of the overhead projector. Focus the image of the test tube on the screen. The image should be a uniform yellow. (If the test tube is not immersed in the water-filled container, when the image is projected onto the screen it will be bright yellow at the center and dark near the sides of the tube because of the lensing effect of the filled test tube. This nonuniformly bright image makes it difficult to observe the progress of the demonstration.)

Presentation

Fill the dropper with the blue $CuSO_4$ solution.

Insert the dropper into the yellow $K_4Fe(CN)_6$ solution in the test tube and hold its tip 1–2 cm above the bottom of the tube. Gently squeeze the dropper bulb to force a pea-sized drop of $CuSO_4$ solution into the $K_4Fe(CN)_6$ solution. As the $CuSO_4$ solution

---

†A suitable container can be made from ¼-inch acrylic (Plexiglas) sheet. Cut two 4-cm × 18-cm pieces and two 12-cm × 18-cm pieces for the sides and one 4.5-cm × 12-cm piece for the bottom. Glue them together with acrylic cement.

enters the $K_4Fe(CN)_6$ solution, a red-brown membrane of $Cu_2Fe(CN)_6$ will form where the two liquids meet, surrounding the drop of $CuSO_4$. Quickly withdraw the dropper from the solution, leaving the blue drop of $CuSO_4$ in the yellow $K_4Fe(CN)_6$ solution. The blue drop inside the red-brown membrane will begin to rise slowly in the yellow solution. Before the drop reaches the top of the solution, it will stop rising. After several seconds it will begin to sink slowly to the bottom of the test tube.

The procedure described in the preceding paragraph can be repeated several times using the same tube of potassium hexacyanoferrate(II) solution. This solution becomes gradually more dilute and less dense as more drops are injected into it, and eventually the injected drops will no longer rise.

HAZARDS

Copper compounds are harmful if taken internally. Dust from copper compounds can irritate mucous membranes.

Potassium hexacyanoferrate(II) should not be mixed with concentrated acids; this can liberate hydrogen cyanide gas, which is extremely toxic. Potassium hexacyanoferrate(II) itself is of minor toxicity.

DISPOSAL

Pour the potassium hexacyanoferrate(II) solutions into a 1-liter beaker containing 100 mL of water. In a fume hood, add 500 mL of laundry bleach (5% sodium hypochlorite solution) to the beaker. Allow the mixture to rest in the fume hood for at least 8 hours. Flush the mixture down the drain with plenty of water.

The copper(II) solution should be flushed down the drain with plenty of water.

DISCUSSION

In this demonstration, a membrane of copper hexacyanoferrate(II) is formed between a solution of copper sulfate and a solution of potassium hexacyanoferrate. The membrane forms spontaneously when a drop of copper(II) solution is injected into the hexacyanoferrate(II) solution. The equation for the reaction is

$$2\ Cu^{2+}\ (aq) + Fe(CN)_6^{4-}\ (aq) \longrightarrow Cu_2Fe(CN)_6(s)$$

Immediately upon formation, the precipitate is gelatinous and surrounds the drop of copper sulfate solution. Because the density of the copper sulfate solution used here is slightly less than that of the potassium hexacyanoferrate(II) solution, the drop of copper sulfate solution slowly rises in the potassium hexacyanoferrate(II) solution.

However, the drop does not rise all the way to the surface of the hexacyanoferrate(II) solution. Before it reaches the surface, it stops and begins to sink. It sinks all the way to the bottom. After a short while on the bottom of the tube the drop will have shriveled. The drop does not make it to the surface of the yellow solution because the membrane surrounding it is semipermeable. Water can move through the membrane more readily than ions can. Water moves from inside the drop (i.e., from the $CuSO_4$ solution) into the solution surrounding the drop (i.e., into the $K_4Fe(CN)_6$ solution). This indicates that the chemical potential of the water inside the drop is greater than that outside the drop. The chemical potential of the solvent is related to the concentration of the solvent, although the relationship is far from ideal in these concentrated electrolyte solutions. The concentration of water in each solution can be calculated. The mass of 100.0 mL of 0.430M $CuSO_4$ solution is 106.47 g and this solution contains (0.100 liter)(0.430 mol/liter)(159.60 g/mol) or 6.86 g of $CuSO_4$. Therefore, the solution contains 106.47 g − 6.86 g or 99.61 g of water, and the concentration of the water in the solution is (99.61 g)/[(18.01 g/mol)(0.1000 liter)] or 55.3M. The mass of 100.0 mL of 0.280M $K_4Fe(CN)_6$ solution is 106.55 g. It contains 10.31 g of $K_4Fe(CN)_6$. Therefore, it also contains 96.24 g of water, and the concentration of water is 53.4M. Water diffuses through the semipermeable membrane from the solution in which water is more concentrated ($CuSO_4$) to the solution in which it is less concentrated ($K_4Fe(CN)_6$).

As water leaves the copper(II) sulfate solution, the density of the solution increases. Eventually, its density becomes the same as that of the $K_4Fe(CN)_6$ solution, and the drop of solution stops rising. As even more water leaves the drop, the drop becomes denser than the $K_4Fe(CN)_6$ solution, and it sinks.

REFERENCES

1. N. Nicolini and A. Pentella, *J. Chem. Educ.* 65:614 (1988).
2. R. C. Weast, Ed., *CRC Handbook of Chemistry and Physics,* 66th ed., CRC Press: Boca Raton, Florida (1985).

Demonstrations in Volume 1

2   CHEMILUMINESCENCE

3   POLYMERS

Demonstrations in Volume 2

7 OSCILLATING CHEMICAL REACTIONS

400

Demonstrations in Future Volumes

The following demonstration topics will be included in future volumes:

atomic structure
chemical periodicity
chromatography
clock reactions
corridor demonstrations and exhibits
cryogenics
electrochemistry
fluorescence and phosphorescence
kinetics and catalysis
lasers in chemistry
organic chemistry
overhead projector demonstrations
photochemistry
radioactivity
spectroscopy and color